Advances in
PARASITOLOGY

VOLUME 2

Advances in
PARASITOLOGY

Edited by

BEN DAWES

Department of Zoology, King's College,
University of London, England

VOLUME 2

1964

ACADEMIC PRESS

London and New York

ACADEMIC PRESS INC. (LONDON) LTD.
Berkeley Square House
Berkeley Square
London, W.1

U.S. Edition published by
ACADEMIC PRESS INC.
111 Fifth Avenue
New York 3, New York

Library of Congress Catalog Card Number: 62–22124

PRINTED IN THE NETHERLANDS
BY JOH. ENSCHEDÉ EN ZONEN, HAARLEM

CONTRIBUTORS TO VOLUME 2

S. ADLER, *Department of Parasitology, Hebrew University, Hadassah Medical School, Jerusalem, Israel* (p. 35)

CLIFFORD O. BERG, *Department of Entomology and Limnology, Cornell University, Ithaca, New York, U.S.A.* (p. 259)

THOMAS W. M. CAMERON, *Institute of Parasitology, McGill University, Montreal, Canada* (p. 1)

BEN DAWES, *Department of Zoology, King's College, University of London, England* (p. 97)

T. E. GIBSON, *Central Veterinary Laboratory, Weybridge, England* (p. 221)

D. L. HUGHES, *Allen and Hanbury's Ltd., Ware, Hertfordshire, England* (p. 97)

J. D. SMYTH, *Department of Zoology, Australian National University, Canberra, Australia* (p. 169)

PREFACE

In introducing another volume of *Advances in Parasitology* thanks must be expressed to many readers who spoke highly of the inaugural volume. My remark that parasitology has its roots in morphology and taxonomy induced one reviewer to consider these subjects as "shackles" which have isolated parasitologists from the advancing field of biology and to welcome this breakaway from isolation, with mechanisms of infection, physiology, immunology, pathology, fine structure and chemical analysis coming to occupy most interest, giving a picture of developments which are in line with those of modern biology. At a time when a contribution on the subject was in preparation, he also considered that information on the subject of chemotherapy might well have been included! Another reviewer invited potential readers to learn how the "application of the methods of biochemistry, biophysics and experimental biology in general have altered, in a few productive years, the whole structure of parasitology". He also made favourable reference to a "wealth and detail of information" provided, which I regard as an essential if the scope of this series is to extend far enough beyond the inadequate limits of textbooks on the subject to satisfy the needs of modern teaching and research.

It is my hope that an adequate amount of detail will be found in the present volume on the topics of host specificity and the evolution of the helminths, *Leishmania* and leishmaniases, the invasive stages in fascioliasis, the biology of hydatid organisms, anthelmintic treatment of domestic animals and the problem of snail control in trematode diseases. A start has been made on the process of extending the variety of parasitological topics without curtailment of necessary detail.

Professor Cameron deals with his subject in a masterly fashion as an author who is well versed in relevant published information and sufficiently experienced to have formed firm but flexible convictions. Helminthic parasitism is usually a close association between an invertebrate animal and a vertebrate host. However, the early progenitors of the cestodes probably occurred in crustaceans, mites and insects, those of the digenetic trematodes in molluscs and particularly gastropods, long before vertebrates became available as hosts. Some nematodes adopted insects as hosts but many of them became parasitic only when vertebrates appeared on the earth, and many existing forms even awaited the origin of the Tetrapods and attained full development only when herbivors turned from brousing to grazing, probably in Tertiary times, when modern helminths were becoming established.

The early history of vertebrates which serve as hosts of parasites is well documented in the fossil record, although there is little in this record concerning parasites. But the essence of this intimate relationship is that the parasite and the host evolved contemporaneously, so that the palaeontological record does throw some light on the evolutionary history of the parasites, affording a check on the deliberations of systematists who have to deal with the parasitic groups. In the "First Symposium on Host Specificity among Parasites of Vertebrates" (Paul Attinger, Neuchatel, 1957) a distinguished zoologist, Ernst Mayr, considered the evolution of the parasite to be "as closely correlated with that of the host as if it were an organ of the host" and made a plea for sound systematics and reliable host records. Cameron makes the point that caution is required in the interpretation of "specific" and "generic" characters of parasites in classification and phylogeny. Many "species" have been made without reference to breeding characteristics and are represented by stored specimens in museums and the result of subjective judgements bearing on minute degrees of variability. He has chosen, no doubt very wisely, to concentrate his deliberations on higher groupings of both parasites and hosts and on the multiplicity of related parasites in related hosts and also such considerations as palaeoclimate, vertebrate evolution and the significant migrations of the hosts about the face of the earth. His account had to be given in breadth rather than in depth, forming a framework for later detailed consideration and indicating how the study of parasites in relation to their hosts may help in the elucidation of the problems of vertebrate evolution.

Professor Adler has indicated that the problems of the leishmaniases call for the attentions of protozoologists, entomologists, ecologists, pathologists and immunologists and has forecast future intensive research along these lines both in East Africa and South America. He deals with the morphology, immunology, epidemiology and culture of species of *Leishmania* in very practical ways which leave no doubts as to the direction in which research is moving in this field. Morphological differences between such species can rarely be observed in the intracellular stages, although the electron microscope has revealed fresh data concerning the flagellum and the kinetoplast. The identification of the leishmanias must usually be made by the use of antigenic analyses, and in the evolution of the Protozoa antigenic differences represent a step in speciation which is established before novel morphological characters can be recognized under the microscope. In the instance of the leishmanias, antigenic differences are associated with pathological manifestations, forming an instructive example of the epidemiological significance of closely related pathogens which differ only in the slightest degree. It is stressed by Adler that the leishmanias of lizards have been somewhat neglected by protozoologists, although they are valuable and possibly essential to our understanding of related forms which utilize mammals. We can welcome this contribution from one of the few experts on the leishmanias on this side of the Atlantic, and moreover one who has frequently and quite recently crossed this ocean in order to assist in the alleviation of much human suffering caused by these baneful parasites.

Fascioliasis is a subject about which there has long existed an illusion of much knowledge where in truth knowledge has been fragmentary. During the first half of the century the common liver fluke of the sheep, and its life cycle, have been described in textbooks of zoology and biology and also veterinary and medical parasitology in such a manner as to imply that it was necessary only to determine means of eradication. The contribution here submitted indicates that this field of research has surpassed many others in the wealth of unexpected data disclosed by recent research. The reader will be left in no doubt that attempts have been made to allot credit where it is due. The historical review contains much information which has never been presented hitherto in book form and which has in fact been grossly neglected by authors writing on fascioliasis. In consequence of recent researches much more information has become available on the process of excystment of juvenile forms of *Fasciola hepatica*—which depends on the activation of encysted forms—as well as on growth and maturation of the young flukes and their activities in the wall of the intestines, the perivisceral cavities, the hepatic parenchyma and the bile ducts of the definitive hosts. There is a significant demonstration of the tissue feeding activities of the young flukes and of the part played by the local tissue reactions of the host in producing new tissues on which feeding may take place in some locations. Steps have been taken which clear the way for more detailed study using modern techniques in future research, not only in respect of the fluke but also in regard to the pathology of fascioliasis in Man as well as animals. Various methods of attempting to "influence" the invasive stages of *F. hepatica* have been discussed. At the present time there is no available method of preventing the establishment of this fluke in animals placed on natural pastures and no treatment by which the young fluke may be killed before serious damage has accrued in the liver of heavily infected animals. Experimental results also lead us to the conclusion that in the present state of our knowledge and with methods at present available the prospect of immunizing the hosts against this liver fluke are poor. What has been written down, however, may represent a stimulus to the further effort which will be required before fascioliasis can be significantly alleviated or eradicated.

Professor Smyth deals with the biology of hydatid organisms and points out how little is known in a most informative manner. These are cestodes of greater than usual pathogenicity and causal agents of widespread disease in Man and animals which is being studied in progressively increasing detail in various institutions throughout the world. His field covers a very extensive foreign literature, much of it written in Spanish and Russian, and its translation could not have been an easy task. The contribution covers also his own recent researches, which illuminate the topic. Hydatid cysts are of various kinds—unilocular, multivesicular, alveolar and multilocular—and represent mainly two species of *Echinococcus*, namely, *E. granulosus* and *E. multilocularis*, which do not develop in the same intermediate host, although larvae of both species will develop in Man. Ten other species have been reported and named but so little is known about their life cycles that they must be left out of immediate considerations. Smyth shows, however, that speciation is likely to be much more complex than was evident formerly, because strains of these species and possibly "clones" have arisen in consequence of peculiarities of reproduction and development which favour the expression of mutant forms and their selection and establishment in different hosts. This phenomenon is associated with the enormously multiplicative larval stages, a feature of hydatid organisms which is not evinced by most larval cestodes. Host specificity is considered in respect of these two species and their various strains, in definitive and intermediate hosts, and the biochemical basis for host specificity is discussed. Adults and larvae are considered from the viewpoints of the morphologist, cytologist and histochemist, and special sections are devoted to the life cycle, chemical composition and physiology, immunity and *in vitro* culture, so that the reader is provided with a great fund of biological information of inestimable value.

Dr. Gibson, dealing with the anthelmintic treatment of domestic animals, refers to increasing recent tendencies to use as anthelmintics synthetic organic compounds which are supplanting medicaments of vegetal origin in chemotherapy. In his introduction, he considers the questions of the desirable qualities of anthelmintics used by veterinarians and the valuable attributes such as a wide therapeutic index and ease of administration, showing that until recently progress in this field has been slow. He marks the first really significant advance as the introduction of phenothiazine about twenty years ago but indicates also the more rapid progress which has been made during the past four years, and emphasizes the intensity of the quest now being conducted on a global scale for new and better anthelmintic compounds. He then considers the nature and the value of such compounds in respect of diseases such as parasitic gastro-enteritis of cattle and sheep, parasitic bronchitis, fascioliasis, dicrocoeliasis, ascariasis, equine strongylosis and capillariasis in poultry, as well as diseases due to tapeworms and hookworms. In this account important consideration is made of such attributes as particle size and purity, absorptive value and toxicity, and synergism between one drug and other compounds. He deals separately with effects produced in different hosts such as sheep and cattle, and he is not averse from giving us his conclusions in respect of particular hosts or specific anthelmintics, also providing a final summary which indicates future hopes and aims.

In the final contribution, Professor Berg is concerned with the possible value of Sciomyzid larvae as agents of snail control in trematode diseases. However, he has provided first of all a clear picture of the need for new methods for the control of snail populations, discussing the problem in terms of cost of chemical control, pesticide and population dynamics, resistance problems, stable residues and the toxicity of molluscicides to other animals. This is followed by a discussion of biological and integrated control methods, with consideration of the prospects for the biological control of snails, the natural enemies of snails, the complexities of biological control and the risk of foreign introductions, with separate discussion of the possibilities of integrated control programmes that make use of the advantages of chemical and biological methods of control but minimize their disadvantages. In this critical exposition, Professor Berg has been compelled to speak boldly and to risk being misunderstood in some quarters, but his frankness and honesty will no doubt be recognized and valued highly by many readers at this time, when constructive criticism is plainly necessary. His discussions on the salient features of the natural history of the Sciomyzids and recent laboratory and field trials of these snail-eating larval Diptera against the snail hosts of *Schistosoma* and *Fasciola* take us to the limits of existing knowledge in this field of study, at the same time indicating that there may be much more profit than is immediately evident in future researches along these lines.

I am extremely grateful to all these authors who have put aside their personal interests and halted their research activities sufficiently to give us these detailed accounts of progress in the fields of parasitology. To the staff of the Academic Press also I express my thanks for the great care which has been exercised in the production of this book.

BEN DAWES
Professor of Zoology
(Parasitology)

KING'S COLLEGE
UNIVERSITY OF LONDON
STRAND, LONDON, W.C.2

April 1964

CONTENTS

Fascioliasis: the Invasive Stages of *Fasciola hepatica* in Mammalian Hosts

BEN DAWES and D. L. HUGHES

The Biology of the Hydatid Organisms

J. D. SMYTH

Recent Advances in the Anthelmintic Treatment of the Domestic Animals

T. E. GIBSON

Snail Control in Trematode Diseases: the Possible Value of Sciomyzid Larvae Snail-Killing Diptera

CLIFFORD O. BERG

Host Specificity and the Evolution of Helminthic Parasites

THOMAS W. M. CAMERON

Institute of Parasitology,
McGill University, Montreal, Canada

I. Helminths and Evolution

"Geology", wrote Dana, "is a particularly alluring field for premature attempts at the explanation of imperfectly understood data." This is equally the case with Zoology.

The reconstruction of the early history of most invertebrates without exo- or endo-skeletons is very difficult and gets little assistance from palaeontology. In the case of the parasites, however, the situation is a little more promising because the essence of parasitism is the presence of a host and as parasites have obviously evolved coincidentally with their hosts the palaeontological record of the host throws some light on the evolutionary history of the parasite.

To translate the sequence of events into a time scale is equally difficult because "The Palaeontological time-scale, founded on the law of superposition, and independent of any assumption, is not a scale at all but a succession without dimensions. The radiometric scale is a series of charted control points along a succession much of which is already established. Specialists in both fields need constantly to remind themselves that they are not operating exact sciences" (North, 1963).

For the purposes of this contribution the scale proposed by Holmes (1947) has been adopted.

While parasites have been recognized both in man and animals since the beginning of written history, Parasitology as a well defined branch of study is relatively recent and its greatest development has taken place in the last half century. Much of this development has been in the nature of morphology, not only of parasites of man and domestic animals but of wild animals of all kinds. Accompanying the morphology, have been numerous studies on life histories of different kinds and as a result of both of these lines of research, a great many attempts at classification have been undertaken; essentially these have been intended to aid in the identification of particular animals and most are extremely artificial paying little or no regard to phylogeny. However, an enormous number of different kinds of parasites has been described and it is safe to say that virtually no vertebrate, and probably no invertebrate, is without its parasitic fauna. No parasite is universally distributed among all kinds of hosts however, and the evolutionary significance of parasites and parasitism is in consequence of considerable general importance.

Most animals—probably all—have a variety of parasites. Sometimes these consist of closely related forms which obviously must have evolved from a common source. More often they include a number of unrelated kinds. Parasitism has been a most successful type of life and the great variety of kinds bespeaks a very long existence.

Once the principle of evolution by natural selection had been accepted, it became obvious that parasites must have evolved, even if it took some time to dispose of the concept of spontaneous generation. Although Darwin made no reference to parasites beyond designating them as disgusting, his principles applied to them as much as they did to non-parasites. Natural selection is the effect of change in the environment on an existing population already adapted to an existing environment. New varieties—races, species and so on—occur only when there is some change in the environment; then the "fittest" survive to have most offspring. Evolution is merely adaptation to the environment and is a change in the genetic composition of the population. The least specialized form is the most likely to survive, the most specialized the most likely to be eliminated.

As a parasite is an animal which by definition must have a host, it is dependent on the ability of another animal to supply it with food, shelter and the ability to reproduce. All living material—the bioplasm—must have the properties necessary to survive and be adapted to its environment, but equally the environment must also have the properties necessary for the organism to survive. This is the basis of host specificity which, in its essence, is no different from that underlying the survival and ecology of free-living terrestrial, freshwater and marine organisms. This is achieved as the result of the evolution of suitable mechanisms within the parasite to enable its cells to live in a medium which itself is slowly changing as the host itself adapts to a changing external environment.

There are several general *facts* about helminths, an appreciation of which is essential to a discussion of their evolution.

Parasites are, in general, not naturally pathogenic. This is particularly true of the parasitic worms which are unable to multiply within the definitive host and are pathogenic only when a relatively large number have secured admission. This factor operates quite independently of any acquired resistance mechanism. Parasites are adapted to the ecology of the host and if this remains normal the population gaining successful admission also remains normal. If the ecology is altered markedly (as it is under domestication) the parasite population may become abnormal and disease due to excessive numbers may result.

All helminths, with a few obvious exceptions, must spend some time outside of the host in which sexual maturity is attained. This external existence is governed by quite strict ecological requirements of climate, temperature, humidity, oxygen tension and so on, as well as by factors which govern the development in or presence of, an intermediate host or hosts should such play a part in the life cycle.

Most parasites are specialized. They live in a specialized environment of which they have, in the course of evolution, taken full advantage. They have dispensed with unnecessary organs and often with enzyme and endocrine systems which have been replaced by those of the host. In many cases this has resulted in a rather high degree of host specificity, i.e. certain kinds of parasites can live only in certain specific types of animals. This obviously is a product of joint evolution and it follows that in their earlier days these parasites had less strict host requirements. Once adapted to a host however, they evolved with it, becoming more and more specific in their need for host substances. Their progenitors in the early stages of parasitism, must have been able not only to do without specific host substances, but to take advantage of such as they encountered, probably becoming more and more specific by simple mutations—mutations involving losses as well as new biological characters. Morphological characters would also change but the part these would play in adaptation would probably be less significant.

It is necessary to show caution in the interpretation of "specific" and even "generic" characters of parasites in classification and phylogeny. During the past half century, an increasing number of "species" of parasites has been described. These species exist primarily in bottles on the shelves of museums and, in the absence of breeding characteristics, cannot be discussed objectively. A large number is based on variations of a most minute character and quite obviously they must represent a subjective judgement on the part of the worker who has described them. The "genus" in parasitology is also completely subjective and merely represents a grouping of convenience, the degree of which varies with different workers. Classification is on an "either or" basis and only in its higher characters is it of any phylogenetic significance (Chabaud, 1959). Species and even genera are often widely separated on some arbitrary character which has appealed to a specific worker—and which is often quite uncritically accepted by others. It is essential to remember that a parasite *must* have a host and its phylogeny and classification can only be interpreted in terms of the phylogeny and classification of the host. While it often happens that the host's phylogeny and classification are equally subjective, it is obvious that, in general, related hosts tend to have related parasites.

There is a tendency for parasites to be confined to a single species or group of species of hosts. This is the result of evolution and natural selection acting on genetic impurity. New races develop first, then strains, then species, but none must cause the production of a reaction on the part of the host of a violence sufficient to endanger the survival of either party.

Host specificity is one of the fundamental characteristics of parasitism but it is seldom absolute, and to a variable degree it is relative. Even when a single species is apparently limited to a single host, it may be because other hosts have not been exposed to it. It is, however, obvious that host specificity is progressive and with the mutual development and evolution of host and parasite, it becomes more and more strict. As the metabolism of the host changes, so must that of the parasite or the association will be terminated. However, not only must the metabolism of the parasite change but its morphology may also change and, while harmful mutations will be eliminated, harmless as well as useful ones will remain. Consequently a long association will often result in a series of very similar species in essentially the same or a closely similar habitat (Schad, 1963).

Moreover when hosts have evolved from a common ancestor, the parasites of that ancestor will evolve with the hosts, so that the modern descendants of that ancestor will tend to have descendants of its parasites, often in considerable variety. Changes in morphology will always be subsequent to adaptations to new host physiology.

An examination of the evolution of parasites must accordingly disregard minor characters and concentrate on the higher groupings of parasites and of hosts, on multiplicity of related forms in related hosts, on host specificity, and on palaeontological considerations such as palaeoclimatology, vertebrate evolution and migrations. The story has to be told in broad general lines.

II. The Evolution of the Helminths

The parasitic worms form an ecological group rather than a related series. Conventionally, we divide them into the Flatworms and the Roundworms. These two "phyla" are not zoologically related and it is far from clear that a number of the minor groups classified under the term of Platyhelminthes are closely related either. However, there is little doubt that most of the Platyhelminthes—the Trematoda and the Cestoda particularly—are almost certainly monophyletic, having descended from different pre-turbellarian ancestors at different times and in different directions. The cestodes were the earliest and they originated from relatively simple beginnings to develop into a great variety of obviously quite closely related parasites.

A. THE CESTODA

To the zoologist, a cestode is an animal of ribbon-like appearance with a scolex, neck and a series of proglottids each of which contains the sexual organs. There is no digestive canal.

There is no justification whatever for the statement that a cestode is degenerate. This is based on the assumption that it has somehow lost a gut and adopted a fundamentally different type of absorption of food. It is more probable that the proto-cestode never had a gut and was an early turbellarian. Some turbellarians still parasitize Crustacea, molluscs and echinoderms. The genus *Fecampia* when free, has a mouth and a pharynx; it loses them both when it becomes parasitic in a marine crustacean, retaining only an intestine. The free-living *Convoluta*, which also originally had a gut, loses it when it begins to feed on its symbiotic algae.

The scolex contains the nerve ganglia and the connecting commissures which together constitute the brain. There are flame cells scattered throughout the entire worm, but here there is a more or less elaborate plexus. At the base of the scolex is the "neck" from which grow the genital segments serially. The scolex and neck together can be considered as the main body of the tapeworm. *In vitro*, they can often be kept alive for long periods (Webster and Cameron, 1963) but *in vivo*, they serially bud off genital segments, which in all cestodes have basically the same genital plan—sometimes duplicated—of a single ovary and a number of testicles, some yolk glands (although these may be combined with the ovary) and the usual genital canals—sperm duct, vagina, and uterus. The uterus may have no opening to the exterior and become gravid or may have an opening through which eggs can be passed continuously.

Tapeworms are classified mainly on the different kinds of adhesive organs on the scolex (Baer, 1950), but the organs of adhesion are obviously an adaptation to parasitism. The fact that they not only vary in different groups of cestodes but vary quite consistently within host groups, suggests that the modern classification is in effect and in broad general lines, a phylogenetic one. The phylogenetic significance, however, lies in the fact that series of similar types tend to occur in series of related hosts (Yamaguti, 1959).

It is generally assumed that the function of the adhesive organs is to maintain the worm *in situ* in the small intestine. This can be true only when the tapeworm is small or immature. Once full grown, it maintains itself by muscular activity against the peristaltic action of the gut. Most adhesive organs are obviously quite inefficient except to hold a small worm in close contact with the villi, or a young one while it is growing up.

There is a tendency in the larger cestodes for the scolex to become simplified morphologically. In these large forms, its function as a hold-fast organ is obviously greatly reduced. In *Dibothriocephalus*, the bothria are rudimentary. In the large taenias of man and carnivores, the scolex is only used as an attachment organ during early growth and there is a tendency for the hooks to disappear. Thus, in the cysticercus of *T. saginata*, as Leuckart (1866) showed, embryonic hooks start to develop only to be reabsorbed. Lubinsky (1960) has demonstrated the labile characters in the hooks of larval Echinococcus. The variation in the organs of attachment in the different groups strongly suggests that not only are they not part of the original ancestor but that their appearance is relatively recent. This is probably true also of the segments in which the gonads appear, mature, and become detached to spread their eggs abroad.

The scolex-neck with its central nervous system and its complex of canals is the basic animal. The suckers, appendages, hooks and so on, came later—and likewise the segments.

Stunkard (1962) believes that elongation of the body and replication of reproductive organs are correlated with the parasitic habit and the production of enormous numbers of offspring. He is convinced that the cestode is an individual, not a colony, because of the organic unity of the nervous, muscular and excretory systems. Serial repetition of the gonads is a sexual phenomenon within an individual and modern tapeworms quite frequently show this tendency to reduplicate gonads within the same segment (Baer, 1951).

The genital segments of tapeworms are based essentially on one or other of two plans—a closed or an open tubular uterus. The closed uterus is characteristic of tapeworms of terrestrial vertebrates, the open one of aquatic vertebrates. The open uterus has an egg containing a ciliated larva and is obviously the more primitive type. The closed uterus, which does not discharge eggs until the gravid segment containing it disintegrates, has an egg intended to be swallowed, in the great majority of cases, by an arthropod. In the forms in aquatic vertebrates a second intermediary—usually a fish—is required.

The embryo is identical in all cestodes and is known as the onchosphere or hexacanth larva. It develops in the intermediate host into a mobile procercoid which sheds the six larval hooks. The procercoid grows into what is in effect the scolex of the adult tapeworm. It may be a plerocercoid, a cysticercoid, an amphicyst or a cysticercus—or some modification of these; this takes place in an arthropod except in the case of the cysticercus which develops in a mammal.

The plerocercoid has developed organs of attachment to the mucosa—such as suckers and sucking grooves. The cysticercoid has converted the lower middle of the plerocercoid into a capsule consisting of a double fold of cuticle, leaving a solid tail. The amphicyst which is somewhat similar has the scolex invaginated (Spasski, 1951). The cysticercus has obtained its protection by an enlargement of an excretory vesicle into which the scolex is sunk; the scolex develops in an invaginated form. The hydatid cyst is the most complicated of all the larval stages and is a modification of the cysticercus with a great multiplicative potential in which a number of scoleces develops inside a mother cyst; it is probably the most recent of the cestode larvae (Webster and Cameron, 1961).

In all cases, when the young tapeworm is swallowed by the definitive host, the accessory parts are digested away leaving the scolex and neck of the adult.

In a recent review, Stunkard (1962) suggests that the progenitor of the cestodes before they became parasitic had a planula-like larva and that changes in physiology, morphology and life cycles proceeded step by step as evolutionary adaptations to parasitism. He regards the greatest hiatus in knowledge of cestodes as the gap between the original ciliated planula and the onchosphere with its gland cells and hooks—the loss of which and the development of plerocercoids, are correlated with the elaboration of the life history which resulted from the infection of vertebrate hosts. Cestodes, like trematodes, he believes, were originally parasites of invertebrates and the digestive tract was lost before the worms became parasites of vertebrates and strobilization provided the method of maintaining populations after the original hosts were eaten.

Baer (1947) believed that there is no justification whatever for assuming that tapeworms were originally parasites of invertebrates that might today be the intermediate hosts. This is obviously true. However, there is no evidence to conclude that the original hosts of tapeworms could *not* have been invertebrates. As Stunkard (1957) has put in "There are of course no factual data concerning the origin of parasitism by the ancestor of the cestodes". Nevertheless, the balance of probability favours a crustacean origin which may, however, have passed through an equally primitive echinoderm before reaching the early vertebrates.

While it is still true that the origin of tapeworms is speculative, it is a reasonable assumption that there was a free-living turbellarian-like ancestor, probably with a considerably reduced digestive system. It was probably a freshwater form which became parasitic and matured in Crustacea. The hexacanth embryo shows that the young larva was designed to penetrate the gut wall and reach the body cavity in which it developed. The protocestode, with its single segment and no attachment organs, present in infected Crustacea which were ingested by primitive fish-like vertebrates, was able to survive in the intestine and produce eggs; ultimately the young parasite in the crustacean lost the ability to develop gonads and remained a procercoid larva.

The most elaborate and bizarre of the cestodes occur in elasmobranchs and have a life history involving a marine crustacean (Baer, 1957). However, teleosts also became infected—probably much later—and when teleosts became prey for mammals, they became carriers of the still undeveloped tapeworms. The teleost development was not exclusively marine, and from freshwater teleosts, the parasites spread to land tetrapods—reptiles and later, birds and mammals. The life cycle was taken over by land arthropods—particularly mites—and in the cestode eggs the cilia of the embryos were replaced by cuticular envelopes adapted for terrestrial conditions. With the change in intermediate hosts, there came a change in life cycle, the procercoid developing into a cysticercoid type of larva in the arthropod, thereby replacing the second intermediary of the aquatic forms.

Parasitism of the vertebrates probably began in the Palaeozoic—that of Crustacea much earlier. During the Mesozoic, the transition to land animals began and was well under way by the end of the era. Birds easily became infected and as a result of their own sudden species explosion, their tapeworms quickly became group specific and in consequence, the majority of cyclophyllidean tapeworms have avian hosts. The herbivorous and insectivorous mammals also became infected in the late Mesozoic or the very early Tertiary (Baer, 1940, 1950).

The *in vitro* development of certain pseudophyllidean cestodes as described by Smyth (1962), is particularly significant in explaining the transition from an immature form to a sexually mature one. Certain stimuli which can be applied either *in vitro* or *in vivo* are all that are necessary to bring this about.

In *Schistocephalus*, the genitalia are present as well developed primordia in the plerocercoid stage in a fish. When the fish is eaten by a bird, egg-laying commences within 36 h. Given a high temperature (40°C) and certain other physical requirements, it is possible to produce sexual maturity *in vitro*. The physical requirements are correct pH and osmotic pressure, and anaerobiosis as well as the removal of waste products and the presence of compression such as the worm would find in a tubular intestine.

In related *Ligula*, conditions are a little more critical, but maturation can be induced *in vitro*, although it takes longer, few eggs are produced and some host nutrients are probably necessary.

In *Diphyllobothrium dendriticum*, which also matures in birds, the plerocercoid is small, practically undifferentiated and maturation *in vitro* requires a highly specialized solution of nutrients in addition to the necessary physical environment.

In *Caryophyllaeus* and related genera, we have a monozootic pseudophyllidean tapeworm which may well represent the slightly altered original type of parasite. The absence of segments is primitive and the adhesive organs are rudimentary. The eggs are swallowed by an oligochaete worm and the onchosphere becomes a procercoid in the body cavity, to develop into the adult tapeworm when eaten by a teleost. Little change is required in the life cycle to suggest that in the beginning the procercoid became sexually mature in the invertebrate.

To become established a parasite must first get into a host and possess the ability to avoid destruction by digestion or otherwise and to obtain nourishment. It must not materially harm the host and it must develop the ability to remain in the host, either by developing mechanisms (such as suckers or muscles) to counteract peristalsis or by leaving the digestive tract to live elsewhere. It may have to become microaerobic or anaerobic in order to metabolize its food. It also loses certain useless senses and properties.

While all cestodes must absorb their food through their cuticle (although some authorities deny the presence of cuticle in these animals) or ectoderm, there is still no clear evidence as to how that food is assimilated. It has been suggested that the adult tapeworm is really inside out. The processes shown in sections photographed by electron microscopes and which suggest villi, may in reality be artifacts and the spaces between these processes have been filled originally by phospholipids, dissolved out in the technique of preparing the material for sectioning. Consequently quite large molecules of nitrogenous substances may possibly be assimilated through the outer cuticle. Certainly most tapeworms in the host are associated with bile and mucus which may not only account for the host specificity of the adult but provide a diet of macro- instead of micro-molecules.

The high host specificity and the distribution of particular groups of tapeworms in particular host groups, show that the hosts and cestodes have evolved in parallel lines. However, it does not follow that the most primitive types of hosts harbour the most primitive types of cestodes. It simply means that the elasmobranchs, which appeared first, became infected first and because of this, elaborate modifications in morphology had more time to develop (Baer, 1932, 1934).

As long ago as 1869, Krabbe showed that in Iceland certain bird tapeworms were group specific. Fuhrmann (1932) continued this work and showed that genera of tapeworms were characteristic of orders of birds while Joyeux and Baer (1934) extended this concept to other groups of vertebrates.

While host specificity is quite strict in some cases, it is surprisingly low in others. In many of these cases, the parasites depend on the food habits of their hosts rather than on the interior milieu. Thus, tapeworms carried by insects or by mites tend to have a much wider host range than those depending on other kinds of intermediate hosts; quite often, however, this host range is generic rather than specific, indicating that host specificity is becoming more strict. The tapeworm genus *Hymenolepis*, for example, is widespread among insectivorous birds and mammals but its *species* are much more restricted.

The Anoplocephalata is a group of tapeworms carried by mites to herbivorous tetrapods of all kinds. Quite obviously it is a very ancient group—

probably the most ancient—affecting reptiles as well as mammals and birds.

Spasski (1951) in his recent survey of the Anoplocephalata-like tapeworms has suggested that their evolutionary sequence among terrestrial vertebrates is somewhat different from that accepted by most other authors. All regard the group as containing the most primitive of these tapeworms. Early in their evolution they diverged along several lines. One, the Linstowids, developed in primitive mammals and produced the modern Davaineids. Another line gave rise to the Anoplocephaline tapeworms of herbivorous mammals and to the Hymenolepids of birds. A third—probably more modern than these—produced the Taenias of the carnivores, with vertebrates carrying a cysticercus (or its modification) as an intermediate stage; *Mesocestoides* may represent a transition stage between the primitive Anoplocephalines and the Taenias.

Oöchoristica is generally regarded as the most primitive tapeworm of land vertebrates occurring in snakes and lizards in all continents between 55° North and South. There is a single species in turtles although there are no tapeworms in crocodiles.

Very closely related to *Oöchoristica* (and distinguished from it by a rather more advanced type of uterus in the gravid segments) is the genus *Linstowia* (and its very close relations) which occurs in monotremes, Australian and South American marsupials, edentates and a wide variety of other mammals including insectivores, lemurs, cebid monkeys, primitive carnivores and rodents.

So far as Australia is concerned, its species of Anoplocephaline tapeworms are unique: this is true also of South America, which, however, shares with Australia the genus *Linstowia* (although with different sub-genera).

The genus *Bertiella* which also is a rather primitive form, has evolved with the insectivores and primates as well as the Australian marsupials. In South America it occurs in cebid monkeys and in Africa in cercopithic monkeys and in apes (and occasionally in man). Its main distribution, however, is in the Sunda Islands and surrounding territories and its appearance in Australia may be relatively recent.

The genus *Anoplocephala* itself, which also has the primitive tubular uterus, is confined to the Perissodactyla and Paenungulata. The more advanced *Moniezia*, which has a reticulate uterus, occurs in ruminants and camels. The greatest number of tapeworms in this group, however, is found in Holarctic rodents: the uteri are of the reticulate type. The absence of the more advanced kinds in South America is of considerable significance.

It is also significant that birds have few of the more primitive types although they are hosts to a great majority of more advanced types with their very considerable degree of host specificity.

B. THE CESTODARIA

The Cestodaria are usually classified with the Cestoda (Yamaguti, 1959) more for historical reasons than any other because they obviously are not closely related. Their larva—the lycophore—is a decacanth one and the adults are mostly parasitic in the coelom of fishes and freshwater turtles, with one unique representative in the intestine of chimaeroid fish. They have no scolex and the morphology as well as the location of the larger group—the Amphilinids—suggests that they are really neotenic larvae; Janicki (1930) somewhat improbably even suggested that the original final host must have been a Mesozoic reptile! It seems more probable, as Baer (1951) has suggested, that excessive food has produced a precocious development of the gonads; without some such mechanism, the disappearance of an extinct reptile would have meant the disappearance of the parasite.

It is probable that the group arose as an early evolutionary side line from a primitive turbellarian in the Mesozoic or even the Palaeozoic.

C. THE ACANTHOCEPHALA

The Acanthocephala is another small group of elusive affinities and puzzling origin. In spite of the efforts of numerous zoologists to classify them with the roundworms, they are not even remotely related to the Nematoda. In fact, van Cleave (1948) considered that they might even be Platyhelminthes of some sort, with a distant relationship to the Cestoda. If they are so related, the relationship is quite distant. They probably have no living relations. They are exclusively endoparasitic and like the Cestoda, have no digestive tract at any phase of their existence. The sexes are separate. The complex egg-shell contains a hooked embryo (the acanthor) which hatches only after ingestion by the intermediate host—an arthropod. The hooks, like those of the cestode larva, are used only to penetrate the gut of the intermediate host; they are then lost and replaced by the proboscis hooks of the adult. When the intermediate host is eaten, the young adult becomes sexually mature in the intestine. They are absent in elasmobranchs, rare in marine fishes but common in freshwater teleosts. They are equally absent from terrestrial reptiles but occur in Amphibia and freshwater turtles, birds and mammals. They probably originated in freshwater teleosts, after elasmobranchs became marine, from ancestors which cannot now even be suggested.

D. THE TREMATODA

The Trematoda are usually classified in two groups on the basis of their life histories (Yamaguti, 1958).

It is quite doubtful if the smaller of these groups, the Monogenea, have any even relatively close relationship with the Digenea. Most are ectoparasitic on elasmobranchs but as Baer (1951) has pointed out, the distinctive freshwater fishes of Australia are common hosts and these fish are regarded as of marine origin, having immigrated relatively recently from the sea. This, and the relatively straightforward type of life cycle, suggests a very early origin of the stock from a primitive rhabdocoelan turbellarian and their early diversification in marine and freshwater animals.

As freshwater turtles are hosts but marine turtles are not, and as the marine turtles separated from the land about mid-Mesozoic, it would seem that the Monogenea had become quite host specific by that time and that their origin lies in the earliest Mesozoic—very much earlier than the probable appearance of the Digenea, which certainly did *not* evolve from them.

Although the digenetic trematodes also probably arose from an early turbellarian ancestor, they are not closely related to the cestodes. They are less host specific in their adult stages, although as van Cleave and Mueller (1934) pointed out "as a general rule, phylogenetically related fishes may be expected to carry similar parasitic populations. However other factors, such as food habits and habitat preferences frequently prove more important than the taxonomic relationships of the host in determining what kind of parasites occur". Stunkard (1937) agrees that while parallel evolution of hosts and parasites exist, "its uncritical application can result in grossly erroneous conclusions".

If observations are confined to sexually mature helminths, phylogeny is liable to be misinterpreted and it is important to take life cycles and larval morphology into consideration. Adult trematodes are parasites of vertebrates and host specificity is far from strict. This is not true for the "larval" stages which occur in molluscs; they are usually very host specific.

The first larval stage is a ciliated planula-like miracidium, cilia being present in this larva even if the egg does not hatch until it is eaten by a terrestrial snail. In the snail the larva becomes a sporocyst and the germinal cells contained in it multiply and as the result of what is probably polyembryony, produce cercariae. The cercariae are immature adults with some quite specialized larval organs developed to enable them ultimately to become mature adults in a vertebrate. These specialized larval characters are very varied in different groups and suggest a relatively recent evolution; the adult characters on the other hand are much more constant.

It seems quite obvious that the digenetic trematodes were originally parasites of molluscs in which they became sexually mature, only subsequently deferring maturity until they left the mollusc.

Three main methods of development evolved. The immature adult, i.e. the cercaria, after losing its tail and some larval characters, became encysted in a mucous coat either freely in water, on vegetation, or on fish scales. There it remained until the necessary stimuli to develop further were provided by the vertebrate host swallowing it. In the second case, the cercaria actually penetrated the substance of a fish and there became encysted in a mucous coat secreted by itself; its larval organs were lost during the penetration process but it commenced the development of its adult organs. Its subsequent development was in a carnivorous fish-eating vertebrate. The great variety of obviously related species in both birds and mammals, suggests that this is a comparatively old method of development. In the third kind, the cercaria itself penetrated the skin of the final host without encysting and became sexually mature in the bloodstream.

Stunkard (1923) believed that the blood flukes are a very old group and that the ancestors of those of warm-blooded animals are derived from their reptilian ancestors. As birds and mammals originated from entirely different Mesozoic reptilian groups, it would seem that the common ancestors were at least very early Mesozoic, possibly even Palaeozoic. The blood flukes of reptiles and fishes are still hermaphroditic; those of warm-blooded forms are physiologically hermaphroditic even if physically dioecious. Vogel's (1941, 1942) experiments with mixed species of blood flukes (and there is no doubt here that the species were "good" ones) shows that the male of one species and the female of another will pair and produce eggs of the female type; the eggs, however, are produced parthenogenetically. It seems highly probable that bisexualism is a secondary phenomenon, a belief that is reinforced by the fact that although all young from a single miracidium are of the same sex, neither sex will develop to maturity in the absence of the other. In unusual hosts, an infection with only males produces a proportion of imperfect hermaphrodites.

E. THE NEMATODA

The phylum Nematoda contains a very large number of free-living species—forms which live in almost every conceivable habitat—terrestrial, aquatic and marine—and is comparable in numbers to the arthropods. All are relatively simple small animals with a pronounced basic similarity. All possess a digestive tube which, typically, has its anterior portion in the form of a triradiate, muscular sucking pharynx. By a rhythmical contraction of the radial fibres a negative pressure is produced against a food particle, a wave of contraction passes down the pharynx, and the particle enters the intestine through a valvular apparatus at the posterior end. This radial arrangement is further reinforced by a posterior muscular bulb in front of the valve, which by contracting its fibres, increases the negative pressure. This is one of the most characteristic features of the phylum but it is generally modified in the parasitic species by the absence of the posterior bulb, except in some archaic forms living in the large bowel and in the free-living non-parasitic stages of species which feed on comparatively gross material. Its presence in these forms is of no phylogenetic significance. It is merely the most efficient mechanism for ingesting particulate food and almost certainly was present in the ancestors of all parasitic nematodes.

It is quite impossible to say how parasitism began—it certainly occurred a number of times, giving rise to several distinct and unrelated groups of parasitic nematodes—a fact completely obscured by most of our modern systems of classification (Baer, 1946; Cameron, 1956). Undoubtedly, in most cases it was (as it still is) by ingestion; in some cases, however, the well-known thigmotaxis of so many small forms appears to have accounted for an entry through the skin.

1. The Capillarids—a group which includes the trichina worm, the whip-worms and such aberrant forms as *Trichosomoides* of the bladder of rodents—have an isolated evolutionary line. It is probably a very old line, the genus *Capillaria* being widely distributed in all vertebrate groups from fishes to mammals. In mammals, not only does the genus occur with considerable frequency, but it has developed into whipworms, much larger forms with the head end embedded in the mucosa of the large intestine, and into the much smaller trichina worms which have become larviparous. In the genus *Trichosomoides* of rodents, the male has become parasitic in the uterus of the female, which in turn lives in the urinary bladder of the host.

2. The Oxyurids represent a second line of parasites. The genus *Oxyuris* itself occurs in Perissodactyles and Paenungulates and is unfortunately quite atypical. It does not possess the characteristic "oxyurid" double bulb oesophagus and its fourth-stage larva is, as Le Roux showed in 1924, a blood-sucking form, although the adult is not a tissue feeder. The other Oxyurids are probably the oldest of the parasitic nematodes but appear to have only a distant relationship although a similar habitat to *Oxyuris* itself. They occur in lung-fish, teleosts, Amphibia, reptiles (especially turtles and lizards) and in certain restricted groups of mammals—especially rodents, rabbits and the insectivore-primate group. They also occur in cockroaches, beetles and millepedes (Basir, 1956) and these may well represent the stem from which the vertebrates acquired their species. The genus *Enterobius* and its near relations occur in both New and Old World primates, testifying to the common origin of both groups.

Typical Oxyurids are absent from birds being replaced by the genus *Subulura* which probably has an indirect life cycle. This group has also passed into mammals with several related species in monkeys and lower primates.

The rather unusual *Cruzia* is a South American genus with representatives in amphibians, reptiles, opossums and edentates.

3. There is a strong probability that the ascarid worms were of marine origin and that the Anisakidae is the parent family from which the Ascaridae arose. They have no obvious relationship with the Oxyurids and the original progenitor was probably a free-living marine nematode. The Anisakids contain a large variety of forms parasitizing such varied vertebrates as elasmobranchs, birds, and sea mammals of different kinds. These worms, so far as is known, require an intermediate host, necessary to trans-

port the egg or larva from the sea bottom to the alimentary tract of the definitive host. The Ascaridae retained part of this life cycle on parasitizing terrestrial animals but show a tendency to dispense with this phylogenetic reminiscence—as Fülleborn (1921) called it—and have a direct life cycle. *Ascaris* and its close relations still have this peculiar and apparently unnecessary migration through the body of the definitive host before attaining maturity, and this quite definitely points to the presence of an intermediate host in the distant past.

Some of the related forms—such as *Toxascaris* in carnivores and *Ascaridia* in birds—have almost lost this migration instinct while others, like *Toxocara* in carnivores, have so modified it as to enable *in utero* infection of their young to take place (Webster, 1958).

4. The genus *Strongyloides* differs from *Parastrongyloides* in the complete absence of parasitic males, and it has been found in amphibians, reptiles, birds and a large variety of placental mammals. Mackerras (1959) found it also in marsupials in Australia in which, unlike those elsewhere, a second free-living generation could occur although infective larvae developed simultaneously from either generation and from these, parasitic females alone developed. However, in monkeys—both Old and New World—Premvati (1959) showed that not only a single free-living generation existed but that this was one of two kinds producing (a) free males and females which gave rise exclusively to infective larvae, or (b) free males and infective larvae; which course was followed depended on the culture medium. In other words, about half of the offspring of the parthenogenetic parasitic females were genetically females—either to become mature free forms producing more parasitic females or to do this directly. The other half of the offspring were destined to become males, irrespective of the medium.

The related genus *Parastrongyloides* was first found in insectivores (shrews and moles) in England. Its life cycle showed an alternation of generations, parasitic males and females. The parasitic forms had the type of cylindrical oesophagus found in intestinal nematodes, the free-living forms—adults and feeding larvae—had the typical double bulb of free-living nematodes. Other species have subsequently been found in Australian marsupials (Mackerras, 1959); in these, a second generation of free-living adults occurred but from either generation infective larvae and parasitic adults of both sexes could develop.

In both genera, the free-living adults have a very short life whereas the parasitic adults can live for some considerable time. It is interesting to note also that in Australian marsupials, both genera possess more than one free-living generation. This may be the result of long isolation and it suggests an important genetical difference in the constitution of the free-living forms.

In spite of the superficial resemblance of the free-living larvae of these species, there is no evidence to suggest that they are closely related to the largest group of parasitic nematodes—the bursate nematodes.

5. The bursate nematodes are characterized by the possession by all of a basically similar caudal bursa in the male. The bursa is supported on each side by three groups of rays—all of which carry nerves ending in sensory papillae. These are conventionally called ventral, lateral and dorsal rays. There are two ventral, three lateral, and one dorsal which has a lateral accessory ray which also carries a sensory nerve. In most cases the dorsal rays on each side partially fuse, but the terminal portions *each* still carry sensory papillae originating, respectively, from the left and right sensory nerve. The dorsal rays are subject to considerable variation and may carry one or more extra digitations which, however, are *not* enervated; this is most obvious in the bursate nematodes of the more archaic host groups. This arrangement of rays is quite analogous with the five digits of the tetrapod vertebrate limb and while it is subject to considerable minor variations, its pattern is a fundamental character of the group.

All bursate nematodes tend to have a basically similar life cycle involving an egg, four larval stages, and an adult. Two of the larval stages are spent outside of the host, the third enters the host in the food, the fourth stage and the adults are parasitic. Some feed on mucus and other secretions, some on blood, and some on tissue. The tissue feeders have developed a cup-like mouth capsule often of considerable complexity. Irrespective of the complexity of the adult mouth, the fourth-stage larvae all have simple terminal mouth capsules containing three small lacerating teeth. This is also seen in the adult stages in some forms, such as *Amidostomum* of birds and *Ollulanus* of mammals, suggesting that this probably was the original type of mouth capsule in the ancestor of those with the present complicated mouth capsules.

Complexity of mouth capsule tends to accompany complexity of bursal rays and seems to be a function of evolutionary age because such species occur in the more archaic groups of vertebrate hosts.

The simplest but not necessarily the most ancient bursate nematodes are the Trichostrongyles, species without a real mouth capsule. A few occur in amphibians, reptiles and birds. They are absent from the Perissodactyles (except for a modern infection of horses) and Paenungulates. They are present in Australian marsupials (Johnston and Mawson, 1940), however, but they reach their greatest diversification in the Artiodactyles—especially the ruminants.

The genus *Trichostrongylus* itself, because of its morphology and wide host distribution, would seem to be the most primitive genus and its many related species have become more or less modified according to the group of animals normally parasitized. The few forms in marsupials have tended to develop an asymmetry of the bursa, while those in Artiodactyles present a rather close-knit group with numerous minor variations. In South America, the species in edentates and marsupials form a quite natural group, indigenous and peculiar to that continent with their nearest relations in the insectivore *Solenodon* of the West Indies and, curiously, in pangolins of Africa and Asia.

There are two exceptions to this general evolutionary sequence. Two or three blood-sucking forms (*Haemonchus* and *Mecistocirrus*) most atypically, have developed ovarian and testicular tubes which actually spiral round the intestine. The other exception is the genus *Nematodirus* which in morphology, habitat, and biology is markedly different from all the others and may well have developed from a different ancestor than has *Trichostrongylus*. Its completely bilobed bursa suggests a very primitive form.

Another side-line has developed in the bats, with various appendages on the head but with no mouth capsules.

One genus—*Molineus*—while a typical Trichostrongyle has a very interesting host distribution occurring only in the insectivore, lemur, cebid group and in carnivores (in which it is the only Trichostrongyle) in all continents, including Madagascar. It is absent from Old World monkeys.

A great many species are monodelphic and they offer a most complex group for classification; systematists are in complete disagreement as to how to divide them. They are generally known as Heligmosomes and they are essentially parasites of rodents, although other mammalian groups have also been infected. They are extremely labile and many show a tendency to asymmetry and the monodelphic condition is almost certainly a secondary one. Some forty "genera" have been described and many of the species have been at various times placed in several different genera. The classifications have no phylogenetic significance and are merely an aid to identification on *ad hoc* characters such as length of spicules, presence of a rudimentary mouth capsule, longitudinal striations on the cuticle and so on. Their structure suggests that they are derived from the didelphic condition and the host distribution of the species with this characteristic suggests that they also are polyphyletic in origin and that loss of an ovary has occurred several times. The male bursa is often asymmetrical and the fact that this occurs in the Australian species of the didelphic *Trichostrongylus*, reinforces this opinion.

Monodelphism is most frequently seen in the rodent parasites and, from the number of species described, it was probably early established in these animals. Rodents fall into three quite well established groups: myomorph or mice-like, sciuromorph or squirrel-like and hystricomorph which are essentially confined to South America together with the Canadian porcupine and some poorly studied rats in Africa. (The porcupine, which is a recent migrant to the north, has no trichostrongyle parasites.)

A considerable number of species have been recovered from South American hystricomorph rodents, marsupials, and edentates—as well as from the more recent immigrant mice and squirrels. However, there is a certain pattern about those from the native fauna that suggests a monophyletic origin. The genus *Viannella*, for example, which occurs in South American rodents, has a very closely related species in the African reed mouse *(Thryonomys)* which many vertebrate authorities regard as a hystricomorph. This strongly suggests that the original ancestor of the native South American Trichostrongyles arrived from Africa in an early ancestor of the hystricomorph rodents, diversified, and spread to the marsupials and edentates; palaeontology suggests that this would be in the Oligocene or possibly earlier.

With the union of the Americas in the Pliocene, a number of murids and sciurids entered South America and brought other monodelphic species with them and this has undoubtedly added to the present fauna.

The so-called Metastrongyles are really two groups of worms. The first, which live in the bronchi, would seem to be somewhat specialized Trichostrongyles which have migrated to their present position in the bronchi of equines, cattle, sheep and the Malayan tapir and have adopted a specialized life cycle. *Dictyocaulus* larvae develop outside of the body, it is true, but do not feed and can develop adequately in water, while *Metastrongylus* has an external development in earthworms; both have a somatic migration in their hosts.

The second group are true lungworms, with an unusual larval type which requires an intermediate host for its development to an infective stage; this is usually an unspecific gastropod but modifications occur. They live in the lung substance or its blood supply, *not* in the air spaces, and the eggs embryonate and hatch *in situ*, the larvae migrating to the alveoli and so to the exterior. Some species have a rudimentary "bursa" but there is no reason to believe that this has any near relationship to that shown by the trichostrongyles. It is a highly evolved and specialized group, often with extensive developmental migrations in the host and its affinities are far from clear. Dougherty (1949a, b, 1951a, b), on morphological grounds, separated this group into subfamilies which fell quite neatly into three main host-lines—occurring, respectively, in the carnivores, the Cetacea and the Artiodactyles (with a few side forms of minor importance), suggesting a very old origin of the parasites and a possible common origin of the host groups.

The bursate nematodes with true buccal capsules probably rose from *Amidostomum*-like ancestors but they have become much enlarged and have a very elaborate sucking apparatus. There are several major groups of these worms of which four are particularly significant. The oldest is the Sclerostomes—a group of rather stout worms which reaches its greatest abundance in the Perissodactyles, the elephants and the kangaroos. There are very few species in birds but ostriches and rheas (but not the other "ratites") have species quite closely related to those in horses. A rather archaic form occurs in *Testudo*. The entire group not only has highly elaborate mouth capsules and numerous branches to the dorsal ray of the bursa, testifying to their great antiquity, but in each of the mammalian hosts a large variety of very similar species exist. There can be no doubt that about the end of the Mesozoic, Perissodactyles, elephants, kangaroos, ostriches and rheas occupied the same territory. Infections are by ingestion of the third-stage larva which is extremely resistant to desiccation and can live outside of the body in a dormant condition for several years.

This evolution may have started at a very early time in the evolution of land animals, so early that the modern hosts are no longer regarded as close relations. Nevertheless, their ancestors, whether related or not, must have had a common habitat and a common habit in the distant past when they were more susceptible to infection with a more generalized ancestor of their parasites.

Thus, the Sclerostomes of the elephants have a large number of representatives in each species of elephant; a few worms are peculiar to each elephant but most genera have representatives in both.

The same story holds for the horses which have as large a number of closely related forms. Those in the horse are somewhat different from those in the donkey and a little more so from those in the zebras. However, all belong to the same subfamily as those in the elephants, although none are identical.

The rhinoceroses are intermediate between these two groups (although classified with the horses as Perissodactyles) and have species in common with the elephant and the horse.

It follows that both the Paenungulata (the elephants) and the Perissodactyles (the horses and rhinoceroses) must have evolved together in the not too distant past—probably being together about the time grass was spreading over the land in the early Tertiary—and probably, almost certainly, in Africa.

An even more striking case occurs in the Australian kangaroos, which have a very varied fauna of some forty kinds of Sclerostomes related to these two previous groups but much more distantly than those in elephants and horses. Kangaroos are marsupials and the others are placentals and they are regarded as having evolved separately from the mammal-like reptiles in the Cretaceous. However, it seems highly probable that the ancestors of the kangaroos (no fossils of which are known earlier than the Pleistocene) must have been in close association with the ancestors of these ancient primitive ungulates—possibly with the proto-elephants rather than with the equines, and in Africa rather than in Holarctica.

What is even more significant is the presence of species of Sclerostomes in the African ostrich and the South American rhea and their absence in all other living birds—including the Australian ratites. The worms in both are different from each other—that of the ostrich being very close to some in the horses, those of the rhea a little less so. It follows, however, that the ancestors of both must at one time have lived in the same habitat as each other as well as in that of the equines, the elephants and the kangaroos.

The second group—the Oesophagostomes—is of more simple form, is probably younger, and probably arose from a different ancestor. They are (or were) mainly African or Asiatic in their distribution and occur in Artiodactyles and to a lesser degree in primates. There is none in the Perissodactyles or elephants.

The third group, the hookworms, is a somewhat mixed group of tissue-feeding nematodes inhabiting the small intestine of mammals. Many, but not all, succeed in entering the body of the host via the skin; Fülleborn (1921, 1929, 1931) showed that their migration route to the lungs and hence to the intestine, is a purely mechanical one, in no way comparable to the similar migration route of Ascarids; the larva, if swallowed, remains in the intestine to become mature. The hookworms appear to be a relatively new development. No host has the wide variety of species that is seen in the Sclerostomes, but they occur in a great many different kinds of mammals from pangolins to man and ruminants and pigs to carnivores. They are absent from Australia and South America and do not occur in the Perissodactyles or elephants.

The fourth group is found in snakes and consists of the single genus *Kalicephalus*. It is quite cosmopolitan in its distribution occurring *inter alia* in native snakes in Australia and South America and had obviously started to diversify in the Mesozoic. It does not occur in crocodiles or turtles.

There is a striking similarity in the life cycles of all the bursate nematodes (excepting the "Metastrongylidae" which have evolved on different lines). Infection is attained by the entrance of a third-stage larva into the host. This larva has developed *on the ground* from an egg passed in the faeces of the host. The larva has some rather feeble powers of climbing vegetation and a few have developed the ability to penetrate the unbroken skin of the host; this, however, is exceptional and is mainly seen in hookworms of primates and carnivores. Most gain admittance by ingestion. Consequently, *grazing* animals swallowed greater numbers of larvae than browsing animals. It follows, therefore, that the great increase in variety and numbers of the bursate nematodes must have occurred with the development of grass—probably in the early Tertiary.

Before the Palaeocene, mammals were small and relatively primitive and their helminths could be expected to have less rigorous physiological requirements; in other words host specificity would be weak and a variety of hosts could be more easily infected than when evolution has produced more specialized animals. As the hosts diverged from each other and as grazing increased, the intake of worms increased also and minor mutations occurred; the variety of closely related but morphologically distinguishable forms in each species of host would increase. This is apparently what happened in the Perissodactyles and the Paenungulates and in all probability it happened in Africa.

Just how, alone among the mammals, kangaroos also acquired about forty related species, belonging to but quite distinct from the species in equines, rhinoceroses and elephants, is not known. They obviously originated at a relatively primitive stage in the evolution of the marsupials but later than the most primitive when they were essentially insectivorous. The earliest herbivores must have been the original hosts because they are completely absent from the insectivorous forms in both Australia and South America. They originated also *before* grazing replaced browsing in the early Tertiary and they originated at a time when primitive Perissodactyles and Paenungulates occurred in the same locality. It is suggested that this was in the Antarctic Continent in the late Mesozoic or very early Tertiary when that continent was in physical contact with Africa, South America and Australia —either by land-bridges or as palaeomagnetism strongly suggests, before the break-up of a Mesozoic Gondwanaland.

This conclusion is further strengthened by the close relationship of the strongyles of the African ostrich and the South American rhea which alone among birds, harbour bursate nematodes closely related to those in horses. Again there can be no doubt that they not only originated together but that they at one time were in the same areas as the early horses. As South America was separated from Africa during the early Tertiary (at least) and as fossil rheas are known there from the Pliocene, this must have been early in their evolution at a time when the continents were in close juxtaposition.

6. and 7. The remaining large groups of nematodes, the Spirurids and the Filarids have become so specialized in their life cycles and habitats that their evolution is completely obscure. They are not related to each other although both require intermediate hosts.

The Spirurids are not very host specific, are carried by faecal-feeding insects, and infect by being swallowed. They can encyst if swallowed by an unsuitable host and remain viable until swallowed by the correct one. Most inhabit the anterior part of the digestive system but a few live in the orbit. The guinea-worm of Africa and western Asia is an aberrant member of this family.

The Filarids, on the other hand, are all somatic parasites, very host specific and are carried by blood-sucking dipterous insects. While they vary considerably in morphology, the female genital opening is always in the region of the oesophagus—a unique position among nematodes. They may have developed from plant-infecting nematodes in the Mesozoic but how is not obvious.

III. The Evolution of the Vertebrate Hosts

The sequence of the evolution of the hosts of the helminths is much better understood and has been very extensively traced through geological time (Romer, 1945; Watson, 1951). Nothing is known, however, about their very early development in the Proterozoic or in the early Palaeozoic.

The earliest true vertebrates recognized are agnathous types in the *Silurian* but many were in such an advanced state of development that their origin must have been much earlier. In the *Devonian* vertebrate evolution had progressed considerably; jawed fishes appeared, shark-like types followed by bony fishes and towards its close by developments of Amphibia, from a crossopterygian stock. Amphibian development reached its peak, however, in the succeeding *Carboniferous* period towards the end of which the earliest reptiles appeared. In the *Permian* Amphibia decreased while diversification of the primitive reptiles increased.

The diversification of the land tetrapods began with the amphibian-reptiles and evolved along three distinct lines. From the Synapsida reptiles came the therapsids which, in turn, gave rise to the monotremes and mammals (Broom, 1932). From the Diapsida developed the "modern" reptiles (snakes and lizards) and the birds while crocodiles and *Sphenodon* are survivors of the original stem. Tortoises are survivors of the third division—the Anapsida.

There are, accordingly, three different basic evolutionary series of land vertebrates (in addition to the Amphibia) which have evolved separately since the earliest *Mesozoic*.

In the earliest Mesozoic—the Triassic—the typical reptiles of that Era became evident, those from which originated the much later reptiles and birds and which gave rise to the dinosaurs—the archosaurs. Turtles and rhynchocephalians originated then and so did the ictidosaurs which may have included the protomammals. In the succeeding *Jurassic* deposits, fragments of primitive mammals and specimens of feathered thecodont reptiles (the so-called earliest birds) appeared; so also did the earliest teleost fishes and frogs, lizards and crocodiles. In the long *Cretaceous* period which followed, the dominant fauna was reptilian, although most kinds became extinct by its end. However, remains of both marsupial-like and placental-like mammals are found as well as birds with teeth. In the seas, modern sharks developed and teleosts increased.

The earliest remains of placental mammals have been found in the Cretaceous of Mongolia but it is obvious that the absence of skeletons from other areas from an earlier formation does not prove their non-existence there. On theoretical grounds, it is almost certain that mammals, as we define them today, had their origin in the Trias—70 or 80 million years earlier. To a palaeontologist, a mammal is a mammal by definition only and the skeletal differences between mammals and reptiles in the Mesozoic are quite arbitrarily fixed. We have no information whatever as to when warm blood, four-chambered heart, and so on, originated. Nor have we any information that the very specialized foetal development of the marsupial, as we know it now, existed in the earliest marsupial: the evidence is entirely skeletal that "marsupials" and "placentals" were different evolutionary lines and it is quite probable that the biology of parturition in both groups is relatively modern and did not come into existence simultaneously with the change in the jaw-bone and the development of the ear-bones which characterized the evolution of the reptile-mammal into the mammal-reptile. It is also probable that we should consider the monotremes as modified therapsid reptiles which had developed some of the characteristics which we, as biologists, consider distinguish the modern mammals from the modern reptile.

The evolution of birds has progressed along parallel lines but has not diverged so much. Birds are merely feathered reptiles, the other modifications we associate with them being the consequence of their activities. Feathers quite clearly first appeared for a reason analogous to hair in mammals, e.g. to insulate the skin and prevent heat loss. Flight feathers came later—in some birds possibly never, although some, like the kiwi, have obviously lost those which their ancestors possessed. The earliest so-called birds, such as *Archaeopteryx*, are not birds at all, but specialized reptiles with feathers.

The Cenozoic (or Tertiary) saw the development of modern land faunas—especially the true birds and the mammals. The earliest Tertiary—the *Palaeocene*—is very poorly known but its fauna contained opossums, insectivores and early carnivores, primates and ungulates of unfamiliar types. Australia and South America were island continents at this time and although nothing is known of the early development of Australian marsupials, in South America commenced the evolution of local marsupials, edentates and subungulates, a development which continued independently of the rest of the world until almost the end of the *Pliocene*. Rodents and rabbits appeared fully developed at the end of the Palaeocene, and Perissodactyles and Artiodactyles abruptly in the early *Eocene:* bats appeared a little later, about the same time as the whales.

In the *Oligocene*, development was rapid and although little is known of African deposits, it is probable that the elephants as well as the hystricoid rodents—which suddenly appeared in South America in the *Miocene* and formed from now on a considerable proportion of its fauna—developed in Africa just as did the platyrrhine monkeys. This period saw the beginning of our modern families of mammals as well as of birds which appear to have quite suddenly exploded into the innumerable flying families. However, ground birds, such as giant rheas, were present in South America earlier as remains have been found in the Pliocene. These, and the ostrich of Africa, may have been primarily flightless unlike the other "ratite" birds which appear to have become secondarily flightless.

The insectivore-primate group is one of the earliest and still least specialized of the large groups of mammals. The first trace of insectivore remains was about the middle of the Mesozoic. Their main development appears to have been in the Old World and from that group arose the primates—

some of which succeeded in reaching South America by the mid-Tertiary.

It is believed that most of the other major groups of mammals arose from the primitive insectivores.

The ungulates are not a single phylogenetic group. In spite of their name, their basic uniting feature is not their hooves but their herbivorous diet and consequently, their teeth. They are classified nowadays into several quite separate groups, some of which are completely extinct, some nearly so, with only a few flourishing. There is an ancient African group of subungulates—the elephants, the hyraces and the sea-cows—which has almost disappeared. There are the "odd-toed" ungulates—the Perissodactyles—which include the horses, rhinoceroses and tapirs and which also are on the way to extinction. The third persisting group—the "even-toed" or Artiodactyles—is actually three quite distinct types—the camels (which are also disappearing), the pigs and hippopotomi which are much reduced, and the true ruminants, which if one excepts the rodents, today form the dominant group of herbivorous animals.

There is no trace of any ungulate group in the Cretaceous but a number of forms now extinct, were present in the Palaeocene; of these the condylarths (or their ancestors) may represent an ancestral stem of some later forms. There were others, however, well developed at the beginning of the Tertiary, including the highly specialized Notoungulata and Litopterna which flourished in South America almost until its union with the north. In that continent also was a group of elephant-like animals which but for their geographical isolation would be classed with the elephants but which by many palaeontologists (although not all) for this reason are placed in a separate order—the Pyrotheria.

There were no "placental" carnivores in South America in the early Tertiary, their place being taken by a variety of "marsupial" carnivores. The usual palaeontological hypothesis that these South American ungulates, many of which were extremely large animals, disappeared because of the invasion of South America by placental carnivores from the north towards the end of the Tertiary, ignores the fact that they were apparently disappearing long before that event, as well as the equally pertinent fact that the invasion from the north was actually a very meagre one both of carnivores and herbivores, most of which in any event failed to survive. It is quite definitely a *post hoc, ergo propter hoc* argument, based essentially on the completely unproven belief that placental carnivores must, *ipso facto*, be more efficient than the "marsupial" carnivores which lived in South America during the entire Tertiary.

The elephant group—the subungulates or Paenungulata—developed and probably originated in Africa, although they migrated widely to all northern continents and even finally to South America after its junction in the late Tertiary with the north, although they failed to survive there. They were a distinctive group by the Eocene and must have originated early in the diversification of the placentals. The Pyrotheres are quite obviously an early offshoot from this group which developed in the early Tertiary in South America.

The earliest Perissodactyles also originated in the Palaeocene and, by the Eocene, the three modern divisions of horses, tapirs and rhinoceroses were distinct—as well as a number of others now extinct. The horse group, like the elephants, became widely distributed throughout the north from the Eocene onwards as did the tapirs in the Oligocene and the rhinoceroses a little earlier. All three groups developed in Holarctica but their place of origin in the Palaeocene is unknown. The surviving populations of all are, however, essentially tropical.

The somewhat catch-all group we call the Artiodactyles actually consists of three groups, all with traces of origin in the Eocene. The Suina—the pigs and hippopotomi—were well differentiated by the Oligocene. The Tylopoda—the camels—were even earlier and date from the late Eocene while the Ruminantea were (with the exception of *Tragulus*) even later—the first deer being found in mid-Miocene and the Bovidae towards the end of that period.

The place of origin of these groups is unknown but the camels saw their greatest development in North America, the pigs in Holarctica, grazing ruminants in the Old World tropics, and the browsing deer in the temperate north.

Points of interest in the classification of the cud-chewers are that the camels and llamas are unique among mammals in possessing *oval* non-nucleate red blood cells and in possessing a much simpler series of "stomachs" than the deer and the bovines—more analogous to that seen in the quite unrelated kangaroos.

The edentates are South American (with the possible and somewhat doubtful pangolins of Africa and Asia). They are highly specialized animals with skeletal and physiological features which separate them from all other mammals. While a primitive edentate stock has been suggested to have existed in North America early in the Tertiary (the Palaeanodonta) there is no doubt that they were present in isolated South America at the beginning of the Age of Mammals and that they developed there throughout the Tertiary.

The pangolin is probably not an edentate although it may have originated from pre-edentate stock. The aardvark of Africa is certainly not an edentate and may have originated from the condylarths or another early ungulate group.

The rodents are a relatively modern group, originating in the late Eocene or early Oligocene to become the most successful of living mammals. The place of origin is unknown and the earliest fossils are relatively advanced types.

The rabbits and hares are not closely related to the rodents and probably date from the Palaeocene.

The carnivores, now a widely distributed group, were absent from both Australia and South America in the early Tertiary; apart from the dingo, they are still absent from Australia. The absence of placental carnivores from these continents can be most plausibly explained by assuming that the carnivores and the Artiodactyles evolved in Africa *after* the two southern continents had become isolated, and after the commencement of diversification of the marsupials.

The marsupials are believed to have originated more or less simultaneously with the insectivores. They have, however, been isolated in South America and Australia during the Tertiary and in both, considerable diversification of types, including some quite large carnivores, have arisen. Most of these have survived in Australia but, except for a few opossum-like forms, they gradually disappeared in South America. After the junction of North and South America, opossums invaded and became established in North America.

IV. The Development of the Host–Parasite Relationship

It can be accepted that parasites were originally free-living animals and that the ancestors of the modern parasites were able to encounter a biochemical and biophysical environment to which they could adapt relatively simply. This may have occurred in stages; saprobiosis, for example, could easily have preceded actual parasitism. This would occur not only once, but a number of times within the same phylum. It did postulate some degree of propinquity; both parasite and host must have been present at the same place and at the same time. It also postulated relatively wide powers of adaptation on the part of the parasite and a relatively primitive or unspecialized internal environment on the part of the host.

As the host became more specialized, the parasite (at least in most cases) became more and better adapted to its environment, until it became so specialized in its requirements that it could live only in that particular evolving line. In this way, a more or less strict degree of host specificity resulted.

Analysis of host–parasite associations accordingly shows that:

1. Animals with comparable food habits tend to have similar kinds of parasites.
2. Only some groups of parasites show a high degree of host specificity.
3. Old-established hosts tend to have numerous species of parasites related to each other.
4. Related hosts tend to have related parasites.
5. The more highly specialized the host (in an evolutionary sense) the more specialized are its parasites and the more restricted its parasitic fauna.
6. Related parasites have evolved from a common ancestor, whether in related hosts or not.

7. Related parasites in unrelated host groups have developed divergent evolutionary lines and the earlier the original association, the greater the divergence and the more highly specialized the parasite.

8. Unrelated host groups which have numerous parasites related to each other, must in the early days of their evolution, have lived in common with each other.

Palaeontology not only is a science dealing with morphology and classification, but its main interest lies in the fact that it is also the science dealing with the distribution of extinct animals and plants; in other words, it is partly palaeobiogeography. It is true that the materials on which to base its deductions are scanty (often consisting of a few teeth or a fragment of jawbone) and scattered, and that many of the pronouncements of palaeozoogeographers are largely in the realm of educated science fiction. Scott (1913) once described it as the "most inexact of the sciences because it has such a large subjective element and depends so much upon the judgement of the individual naturalist." It can, however, receive considerable assistance from palaeobotany, and in recent years it has had a great deal of suggestive evidence from geophysics, tectonics, palaeomagnetism and palaeoclimatology. If the Theory of Evolution is accepted—and all of palaeontology is based on it—it can also receive assistance from a study of the parasitic worms. Cumulatively, the combined evidence of all these varied disciplines is changing the picture.

The different groups of helminths—ignoring the obvious aberrant species and doubtful identifications—show a definite distribution pattern among the different groups of animals.

In *elasmobranchs*, we find several very characteristic kinds of tapeworms as well as several Ascarid and Capillarid nematodes.

In *Teleosts*, there are representatives of both groups of tapeworms—those with and those without a patent uterus. The dominant groups are the tapeworms and the Ascarids; Capillarids are rare while trematodes of all kinds are common.

In *Amphibia*, the picture changes basically. We find Oxyurids of various kinds and the first of the Trichostrongyles *(Oswaldocruzia)*. *Capillaria* is rare.

In *tortoises*, we meet the primitive anoplocephaline *Oöchoristica*—which is common also in snakes and lizards—but the dominant parasites are Oxyurids. In *Testudo* is an archaic and specialized Sclerostome *(Sauricola)*, and a single genus of Ascarid.

In *crocodiles* there are no tapeworms but a number of Ascarids are present.

In *lizards*, Oxyurids are common but the didelphic trichostrongyle *Oswaldocruzia* has passed over from the Amphibia. Ascarids and Capillarids are rare.

In *snakes*, while Ascarids do occur, the important nematodes belong to the specialized genus *Kalicephalus*. Capillarids and the South American *Cruzia* also appear but are somewhat uncommon.

In *birds*, the predominant parasites are tapeworms, while Ascarids and heterakid Oxyurids are common nematodes. In ostriches and rheas, however, the parasitic fauna is highly specialized and includes members of the genus *Trichostrongylus* and Sclerostomes found in no other birds.

In *monotremes*, the genus *Trichostrongylus* occurs, somewhat modified in the Australian fashion, as are the tapeworms.

In *South American opossums* occur tapeworms of the genus *Linstowia*—quite closely related to *Oöchoristica* but still quite distinct from species in Australia. The pinworm *Cruzia* is common and Capillarids, whipworms and some specialized Trichostrongyles occur.

In *Australian marsupials*, the fauna is markedly different. *Linstowia* also appears, but it is a different subgenus and it is accompanied by another anoplocephaline, *Bertiella*. Some specialized Trichostrongyles are present as well as numerous kinds of Sclerostomes, distantly related to those in ungulates. There are no Oxyurids but *Parastrongyloides* and *Strongyloides* are present.

In *edentates*, *Cruzia* and a number of other Oxyurids appear as well as some specialized Trichostrongyles, while tapeworms of the genus *Oöchoristica* occur.

In *insectivore-primates*, tapeworms are rare. Among the nematodes, however, we find Oxyurids, *Parastrongyloides* and *Strongyloides*, a few rare Trichostrongyles, including, however, *Molineus* which occurs in lemuroids and cebid monkeys.

In *Dermoptera* occurs the tapeworm *Bertiella*, and several unusual Oxyurids.

In *bats* we have Capillarids and a number of highly specialized Trichostrongyles.

In the *New World primates*, tapeworms of the genera *Oöchoristica* and *Bertiella* are present as well as some locally acquired Trichostrongyles. They are hosts to various species of the Oxyurid genus *Enterobius*—a genus (recently reclassified into several genera on very minor distinctions) which like *Bertiella*, is essentially confined to primates.

In the *Old World primates*, *Bertiella* is also found as well as *Enterobius* while in the apes, Oesophagostomes and hookworms occur also.

In *rodents*, which form the dominant mammalian group, the parasites of the hystricomorphs differ markedly from those of the others. In these South American rodents, we find the tapeworm *Oöchoristica* and a group of Trichostrongyles—probably originating from the same stem as the genus *Viannia*. In the Patagonian cavey, however, there is a Trichostrongyle which has its relations in Holarctica and Africa.

The other rodents (and rabbits) are richly supplied with nematodes—Oxyurids, Capillarids and whipworms, Trichostrongyles, and Metastrongyles. A number of tapeworms are present including the genus *Inermicapsifer* which occurs only in Holarctic rodents and in *Lemuroids*.

In *Perissodactyles* occur the tapeworm genus *Anoplocephala*, a very atypical Oxyurid, and a great variety of Sclerostomes—both in equines and rhinoceroses.

Horses as well as tapirs have the same species of lungworm.

In *elephants*, we have essentially the same kind of fauna of Sclerostomes, together with the tapeworm genus *Anoplocephala*, but no Oxyurids have been recorded.

In *hyraces*, we find the tapeworm *Inermicapsifer*, and a pinworm related somewhat distantly to those in Perissodactyles, as well as an unusual Sclerostome.

Among the *Artiodactyles*, the ruminants and camels have an essentially similar fauna represented by the more advanced Anoplocephalines, a great variety of Trichostrongyles (with double ovaries), Oesophagostomes, hookworms, Metastrongyles, Capillarids, and whipworms. Ascarids and Oxyurids are rare or virtually absent.

In *pigs*, the picture is quite different. There are no tapeworms and, while *Ascaris* is common, the evidence suggests it is a recent parasite acquired from man. Trichostrongyles are represented by the single genus, *Hyostrongylus*, but there are a number of Oesophagostomes, hookworms and Metastrongyles as well as whipworms.

In *carnivores*, the tapeworms are essentially of the recent Taenia group. They have several species of related hookworms belonging to a different family from those in ruminants and camels but close to those in pigs. They have a number of Metastrongyles but the Trichostrongyles are almost confined to the single genus *Molineus*, *Ollulanus* being the only other found. The Capillarid group occurs commonly and *Trichinella* is characteristic of the carnivores. Two or three specialized Ascarids occur.

In the *Pinnipedia*, we find Ascarids and hookworms while in the Cetacea, occur Ascarids and a variety of Metastrongyles.

V. Parasites and the Evolution of the Hosts

All the evidence points to the complete isolation of South America from North America, and Australia from Asia at the beginning of the Tertiary. Belief in earlier connexions between South America and North America, and Australia and Asia is based mainly, if not entirely on palaeontology. The tectonic evidence appears to be absent and there is a general belief among palaeographers that the mid-Mesozoic Sea (Tetys) separated the two southern continents from the northern ones during the latter half of the Mesozoic and also, intermittently, separated Africa from Europe. The evidence from palaeomagnetism strongly reinforces this belief.

If we *assume* there was no northern connexion, how can we reconcile this with the available palaeontological evidence? The presence of dinosaurs of various kinds shows quite definitely that *all* the continents were connected by land in the early Mesozoic—and probably earlier (Romer, 1945). If this was the case, the connexion *must* have been via Antarctica and Africa, and at the beginning of the Mesozoic there would be a more or less single continental mass—an epi-continental world consisting of all the pre-Cambrian shields which still form the bases of the present continents. There would be an orientation of North America to enclose the present Arctic Ocean against northern Europe and Asia. Africa would have a connexion with southern Europe (which may have broken earlier than the others), while South America, Antarctica, Australia, Madagascar, and probably India, would huddle together around South Africa. This hypothesis agrees completely with the palaeomagnetic observations made during the International Geophysical Year and with the submarine observations made during the same period. These connexions broke during the latter half of the Mesozoic and under the influence of convection currents in the mantle of the earth, as has been shown by the geophysicists, drifted slowly apart. North America swung around to its present position, South America broke away from Africa and later from Antarctica, and Australia, which moved more erratically than the others, did so also. India moved up to its present position, pushing behind it the enormous mass of the central Asian Mountains. During this time, there was considerable vulcanism, particularly on the moving edge of the continents.

We now have an Arctic Mediterranean surrounded by the huge continental mass of Holarctica and a southern antarctic continent, joined by tenuous undersea ridges to South America, to Australia, and to Africa. The two worlds North and South, are united by the tenuous bonds of "bridges"—sometimes incomplete—across the area of the old mid-Mesozoic Sea.

The original South American fauna offers an excellent set of examples for the elucidation of some of these evolutionary problems.

South America was isolated from the rest of the world through the greater part of the Tertiary—and possibly very much earlier. At the beginning of that era, it was developing a unique fauna of its own, including several groups of primitive ungulates, now extinct, and its marsupials and edentates, small remnants of which survive as oppossums, anteaters, sloths and armadillos. About the middle of the Tertiary, platyrrhine monkeys and hystricomorph rodents appear; they may have been present earlier but there is no trace of them in the fossil fauna. In the Pliocene, junction was made with North America and a limited interchange of fauna took place. South America received some ungulates, carnivores, and rodents, and these brought their parasites with them.

We (Cameron and Myers, 1961) have recently shown that the didelphic Trichostrongyles of the opossums and edentates form a rather compact group distinct from most other genera and present evidence of a local evolutionary sequence. Their ancestors must have been present in South America by the Palaeocene. Their only described relations occur in the Old World pangolins (once considered to be related to the South American edentates) and in the very archaic insectivore *Solenodon* of the West Indies.

An examination of the monodelphic Trichostrongyles in opossums and edentates, shows a somewhat similar picture although the details are more obscure because of the great lability of this group (Yamaguti, 1961; Skrjabin *et al.*, 1952, 1954). Except for one "catch-all" genus (*Longistriata* which has an almost cosmopolitan distribution) all are unique and show distinct relationships to each other. Moreover, the species in the South American rodents are limited to South America and probably belong to a single evolutionary line. It is of great interest to note that one of these genera has a very close relation in an African hystricomorph (*Heterocephalus*).

A number of related species occur in the deer which entered South America in the late Tertiary and which, as they do not occur elsewhere, presumably were picked up from the local mammals.

Outside South America, monodelphic species occur only in rodents, insectivores and bats. In myomorph and sciuromorph rodents, the groups of worms are separate and less diversified than those in hystricomorphs, *Longistriata* occurring in all three groups. While it is hardly conclusive, the evidence points to an early origin of the hystricomorph rodents in *Africa*.

Didelphic Trichostrongyles exist in Amphibia, lizards, snakes and birds. They also occur in Australian marsupials where they quite closely resemble the type genus *Trichostrongylus* but show a tendency to asymmetry of the bursa. There are no monodelphic Trichostrongyles in these marsupials although they occur in American opossums. This suggests that the monodelphic species originated as a mutation or mutations after the Australian marsupials had been isolated from the other continental masses. The didelphic Trichostrongyles were of earlier origin and occur in all continents. Monodelphic Trichostrongyles do occur in Australia but in the native rats—a group of very specialized rodents which entered Australia late in the Tertiary—possibly in the Pleistocene and obviously came overseas from eastern Asia. The appearance of these parasites in the rats reinforces the opinion that the monodelphic species were of relatively late evolution.

The distribution of the Oxyurids is equally significant (Cameron, 1956). Those from equines and hyrax are quite atypical and show no relationship to those in other animals. However, typical species occur in Amphibia, are somewhat rare in lizards and snakes, are absent in crocodiles but extremely common in chelonia. A great many occur in birds of various kinds but none in the ostrich or rhea.

In mammals they have a very patchy distribution. Species have been found in all divisions of the insectivore-primate group; they are quite specialized and archaic in the lower primates but in the monkeys and apes they are congeneric. They have obviously evolved with the group and there is a tendency for all members of a genus of primate to have the same species of

Enterobius. This genus occurs in South American monkeys as well as in African species, apes and man.

In Old World rodents and rabbits, they are common while in the hystricomorphs they are rare. The genus *Cruzia*, however, occurs in a variety of South American animals and is occasionally found in rodents. The genus *Wellcomia* occurs in both South and North American porcupines and in the African *Heterocephalus glabai*, which Simpson (1945) somewhat doubtfully considers to be a hystricomorph.

VI. Conclusions

There seems very little doubt that development of parasitism took place in invertebrates long before the appearance of vertebrates. The proto-cestodes developed originally in Crustacea and later in mites and insects, the proto-digenetic trematodes in molluscs—particularly in gastropods. Some of the Nematoda were also insect parasites—particularly the Oxyuridae but many did not become parasitic until the coming of the vertebrates and a great many of the kinds now existing had to await the appearance of the tetrapods and did not attain their full development until the herbivores changed from browsers to grazers. By this time—probably in the early Tertiary—all the modern groups of tetrapods were well defined and the modern species of helminths were taking form. Their ancestors however lived in the long distant past—probably the Palaeozoic and certainly the Mesozoic.

The present distribution of the parasites, the known past distribution of the hosts and the modern findings of geophysics, palaeoclimates and palaeomagnetism agree well with the polyphyletic origin of mammals from the primitive therapsids. The marsupials, the edentates and the old ungulates originated in Africa—possibly in the Cape-Karroo districts of South Africa where so many mammal-reptiles evolved and probably in association at first with a glossopteris flora (Broom, 1932). Some at least migrated to the old Antarctica—a hypothesis which is more easily understood by observations on a globe rather than on a Mercator-like map. Continental drift in the late Mesozoic distributed the older South American and Australian faunas to develop in isolation during the Tertiary. The newer ungulates, the rodents, the carnivores and the insectivores evolved farther north and subsequent to the separation of the southern continents. This may also have been in Africa but it could equally well have been in Holarctica or Asia.

There are no traces of rodents or primates in South America before the mid Tertiary, after that continent was separated from the rest of the world. There is no evidence, other than palaeontological, that it was ever connected to North America before the Pliocene and the palaeontological evidence can equally well be explained by an early connexion with Africa, and through Africa, with the rest of the world. This is true also of Australia which had no trace of rodents before the Pleistocene and no trace of marsupials before the

Pliocene, although Mesozoic reptiles occur. While it is true that the absence of fossils may merely mean an unsuitable terrain for their preservation, it may also mean that the fauna was not present.

There is little doubt that Australian rodents were rafted from Asia to that continent. This is probably true also of the South American rodents and cebid monkeys. Their parasitic faunas suggest an earlier arrival of the rodents than the monkeys which had obviously been in contact with Old World species before evolving in South America.

These conclusions are admittedly hypothetical. Many more parasites have still to be described from animals in their natural state and native habitats. Much more has to be learned about the bionomics of both parasite and host. The available data however provide important collateral information on the evolution and past migrations of their hosts and the simplest explanation, in the light of our present knowledge, seems to support an African origin of the original forms and a northward and southward distribution of both parasite and host as well as a Gondwana in the early southern hemisphere.

References

Baer, J. G. (1932). *Rev. suisse Zool.* **39**, 195–228.

Baer, J. G. (1934). *Bull. Soc. Neuchâtel. Sci. Nat.* (1933), **58**, 57–76.

Baer, J. G. (1940). *J. Parasit.* **26**, 127–34.

Baer, J. G. (1946). "Le Parasitisme." 232 pp. F. Rouge et Cie, Lausanne.

Baer, J. G. (1947). *Ann. Sci. Franche-Comte. Besançon* **2**, 99–113.

Baer, J. G. (1950). *Rev. suisse Zool.* **57**, 553–59.

Baer, J. G. (1951). "Ecology of Animal Parasites", 224 pp. University of Illinois Press, Urbana.

Baer, J. G. (1957). *In* "First Symposium on Host Specificity among Parasites of Vertebrates", pp. 270–92. Institut de Zoologie, University of Neuchâtel, Neuchâtel, Switzerland.

Basir, M. A. (1956). "Oxyurid Parasites of Arthropoda. A Monographic Study. 1. Thelastomatidae. 2. Oxyuridae." 79 pp. *Zoologica, Stuttgart* Heft (106), **38** 2. Lief.

Baylis, H. A. (1938). *In* "Evolution" (G. R. de Beer, ed.), pp. 249–70. Clarendon Press, Oxford.

Broom, R. (1932). "The Mammal-like Reptiles of South Africa", pp. 376. H. F. and G. Witherby, London.

Cameron, T. W. M. (1956). "Parasites and Parasitism", 322 pp. Methuen, London.

Cameron, T. W. M. and Myers, B. J. (1961). *J. Helminth. R. T. Leiper Suppl.* 25–34.

Chabaud, A. G. (1959). *Bull. Soc. zool. Fr.* **84**, 473–83.

Dougherty, E. C. (1949a). *Parasitology* **39**, 218–21.

Dougherty, E. C. (1949b). *Parasitology* **39**, 222–34.

Dougherty, E. C. (1951a). *Parasitology* **41**, 91–96.

Dougherty, E. C. (1951b). *J. Parasit.* **37**, 353–78.

Fuhrmann, O. (1932). "Les Ténias des Oiseaux", 383 pp. *Mém. Univ. Neuchâtel* **8**.

Fülleborn, F. (1921). *Arch. Schiffs- u. Tropen-Hyg.* **25**, 367–375.

Fülleborn, F. (1929). *J. Helminth.* **7**, 15–26.

Fülleborn, F. (1931). *Naturwissenschaften* **19**, 79–82.

Holmes, A. (1947). *Trans. geol. Soc., Glasg.* **21**, 117.

Janicki, C. von (1930). *Zool. Anz.* **90**, 105–205.

Johnston, T. H. and Mawson, P. M. (1940). *Trans. roy. Soc. S. Aust.* **64**, 363–70.

Joyeux, Ch. and Baer, J. G. (1934). *Biol. med., Paris* **24**, 1–25.

Krabbe, H. (1869). *Dansk. Vidensk. Selsk. Skr. naturvid. math. Afd.* (5), **8**, 249–363.

Le Roux, P. L. (1924). *J. Helminth.* **2**. 111–34.

Leuckart, K. G. F. R. (1866). "Die Parasiten des Menschen und die von ihnen herrührenden Krankheiten." Ein Hand- und Lehrbuch für Naturforscher und Aerzte. 2. Aufl., **1**, 3. Lief. 1. Abt. 855–1000 pp. C. F. Wintersche Verlagshandlung, Leipzig and Heidelberg.

Lubinsky, G. (1960). *Canad. J. Zool.* **38**, 605–12.

Mackerras, M. J. (1959). *Aust. J. Zool.* **7**, 87–104.

North, F. K. (1963). The geological time scale. *In* "Programme Royal Society, Canada, June meeting 1963, Laval University, Quebec," p. 29.

Premvati, (1959). *Canad. J. Zool.* **37**, 75–81.

Romer, A. S. (1945). "Vertebrate Paleontology." 3rd ed. 491 pp. University of Chicago Press, Chicago.

Schad, G. A. (1963). *Nature, Lond.* **198**, 404–406.

Scott, W. B. (1913). "A History of Land Mammals in the Western Hemisphere." pp. 693. Macmillan Co., New York.

Simpson, G. G. (1945). *Bull. Amer. Mus. nat. Hist.* **85**.

Skrjabin, K. I., Shikhobalova, N. P., Schulz, R. S., Popova, T. I., Boev, S. N. and Delyamure, S. L. (1952). "Key to the Parasitic Nematodes. vol. iii. Strongylata." Izdatel'stvo Akademii Nauk SSSR, Moscow. (English translation by Israel Program for Scientific Translation, 1961, 890 pp.)

Skrjabin, K. I., Shikhobalova, N. P., Sobolev, A. A., Paramonov, A. A. and Sudarikov, V. E. (1954). "Descriptive Catalogue of Parasitic Nematodes. vol. iv. Camallanata, Rhabditata, Tylenchata, Trichocephalata, Dioctophymata and a Classification of Parasitic Nematodes under Hosts." (in Russian). 927 pp. Izadtel'stvo Akademii Nauk SSSR, Moscow.

Smyth, J. D. (1962). *Parasitology* **52**, 441–57.

Spasski, A. A. (1951). "Essentials of Cestodology. vol. 1. Anoplocephalate Tapeworms of Domestic and Wild Animals." 783 pp. Izdatel'stvo Akademii Nauk SSSR, Moscow. (English translation by Israel Program for Scientific Translation.)

Stunkard, H. W. (1923). *Bull. Amer. Mus. nat. Hist.* **48**, 162–221.

Stunkard, H. W. (1937). "The Physiology, Life Cycle, and Phylogeny of the Parasitic Flatworms." American Museum Novitates No. 908, 27 pp. American Museum of Natural History, New York.

Stunkard, H. W. (1957). *Z. Tropenmed. u. Parasit.* **8**, 254–63.

Stunkard, H. W. (1962). *Quart. Rev. Biol.* **37**, 23–34.

van Cleave, H. J. (1948). *J. Parasit.* **34**, 1–20.

van Cleave, H. J. and Mueller, J. F. (1934). *Roosevelt Wild Life Ann.* **3**, 161–334.

Vogel, H. (1941). *Zbl. Bakt.* Orig. **148**, 78–96.

Vogel, H. (1942). *Zbl. Bakt.* 1 Abt. Orig. **149**, 319–33.

Watson, D. M. S. (1951). "Paleontology and Modern Biology." Yale University Press, New Haven.

Webster, G. A. (1958). *Canad. J. Zool.* **36**, 435–40.

Webster, G. A. and Cameron, T. W. M. (1961). *Canad. J. Zool.* **39**, 877–91.

Webster, G. A. and Cameron, T. W. M. (1963). *Canad. J. Zool.* **41**, 185–95.

Yamaguti, S. (1958). "Systema Helminthum. vol. 1. Part I. and Part II. The Digenetic Trematodes of Vertebrates." 1575 pp. Interscience Publishers, New York.

Yamaguti, S. (1959). "Systema Helminthum. vol. II. The Cestodes of Vertebrates." 860 pp. Interscience Publishers, New York.

Yamaguti, S. (1961). "Systema Helminthum. vol. III. Part I. and Part II. The Nematodes of Vertebrates." 1260 pp. Interscience Publishers, New York.

Leishmania

S. ADLER

Department of Parasitology, Hebrew University, Hadassah Medical School, Jerusalem, Israel

I. Introduction

The present review deals mainly but not entirely with work carried out during the last twenty-five years. No attempt is made to catalogue the considerable literature which has accumulated during this period. Stress is laid on those problems most likely to engage research workers in the near future because of their intrinsic biological interest or their urgency or both.

The epidemiology of leishmaniasis is replete with problems which call for the attention of the ecologist, pathologist, entomologist, immunologist and protozoologist. It is very probable that much intensive work will be carried out on this subject in the near future in East Africa and in Central and South America.

All parasitic diseases are grist to the immunologist's mill, but few are as promising as the leishmaniases and none pose the basic problems of immunology in so glaring a fashion, involving as they do the lympho-macrophage system *ab initio* and inducing in this system, in the case of visceral infections, a hyperplasia seldom equalled by other parasitic diseases. In the case of cutaneous leishmanias the whole immune process, with its successes and failures, can be followed under the microscope.

Specific identification within the genus *Leishmania* has perplexed many who have attempted it. The complexity of the subject is such that eminent protozoologists have more than once changed their opinions after some years of work. The problem of identification within a group whose members show an almost uniform morphology is of prime interest to the student of micro-evolution. Mesnil in discussing the leishmanias once remarked: "Ils se cherchent encore". Identification must in this case be approached by antigenic analysis and the recognition of slight but significant biological differences. In the evolution of protozoa, antigenic differences are a step in speciation and are established before the appearance of new taxonomic characters. It is therefore impossible to avoid the controversial subject of nomenclature. Hard and fast rules need not be applied. Where antigenic differences are associated with pathological manifestation and other biological differences, as they are in *Leishmania*, it is convenient to acknowledge them by an appropriate nomenclature. There are at least eight distinct serotypes of *Trichomonas vaginalis*, but in this case antigenic differences are, as far as is known, of no pathological or epidemiological significance and need not be given a systematic status.

The leishmanias provide an instructive example of how slight differences between closely related pathogens can be of considerable epidemiological significance.

Workers on leishmaniasis have been so heavily involved in the problem of transmission of species parasitizing man that they have not devoted sufficient time to the study of the leishmanias of lizards. These are of considerable interest in themselves, in addition to which they illustrate, as Hoare (1948) has pointed out: "the probable course of evolution from insect leptomonads to mammalian leishmanias". Hoare's prediction has been justified by the demonstration of antigens common to a lizard *Leishmania* and to *L. donovani*, *L. infantum*, *L. brasiliensis* and *L. tropica*. Moreover this lizard *Leishmania* has produced infections, albeit transient, in man, mouse and hamster. One may hope that parasitologists will, in the near future, pay more attention to this rewarding subject.

The metabolism of the leishmanias is omitted from this review. It is felt that this subject should be dealt with by a biochemist as part of a review on the metabolism of the Trypanosomidae in general; this would form an interesting chapter in comparative physiology.

II. Morphology

In the light microscope no morphological differences can be observed between the intracellular stages of *Leishmania* sp. except in the case of *Leishmania enrietti* where this stage is larger, the flagellar vacuole bigger and the kinetoplast thinner than in other species. Parrot *et al.* (1932) have pointed out that the leishmanoids of *L. infantum* in the dog are smaller than those of *L. tropica* in the same animal (this is also the case in the hamster and mouse). The flagellates in culture are similar, or at most show only minor and inconstant differences, in all mammalian and lizard *Leishmania* species.

The advent of electron microscopy has already provided valuable information, particularly on the structure of the flagellum and kinetoplast but it is too early to predict whether it will establish criteria for specific diagnosis on the basis of ultrastructure.

It is not necessary to repeat the available current descriptions of the leishmanoid and leptomonad stages as seen in the light microscope. A few points of interest which are within the scope of light microscopy and which need further elucidation will be discussed before studies dealing with data accumulated from electron microscopy.

The transformation of a leishmanoid to a leptomonad is of special interest, particularly so as every stage in the development of a flagellum *ab initio* to full functional maturity can be observed directly in living and stained material. This transformation, first observed by Rogers (1904) provided the key to the systematic position and eventually to the solution of the problem of transmission. The process was studied by Leishman and Statham (1905), then by Adler and Adler (1956) and Pulvertaft and Hoyle (1960). Adler and Adler used the following method; heavily infected hamster spleens are ground up and the leishmanoids separated by a method described by Adler and Ashbel (1940). Washed L.D. bodies are sown on Locke-serum-agar (semi-solid) and incubated at 26°C. The inoculum should be sufficiently large to facilitate microscopic examination of loopfuls of material removed before multiplication begins. At various intervals after transfer to 26°C loopfuls of the inoculum are removed for direct examination and for stained preparations.

During the first 20 h at 26°C the organisms go through a stage of growth and internal reorganization. The leishmanoid grows and the anterior end, through which the flagellum will eventually emerge, broadens. The organism tends to assume a piriform shape. The flagellar vacuole increases in size till it is almost as wide as the broad anterior end of the organism and occupies more than half the area between the nucleus and the anterior margin. After about 20 h at 26°C the first sign of a flagellum is seen. It consists of a short stumpy protrusion, widest at its distal end and extending beyond the anterior margin of the organism. When it has grown to about 1 μ in length it performs occasional pendular movements which are sufficient to rotate the organism through a small angle. The flagellar protuberance becomes attenuated during growth and the flagellum finally attains its full size. During the whole of this period it is capable of pendular movement only, with a beat of a few to 130 per min, coupled with bending as the length increases. Full length is attained 4–4½ h after the first appearance of signs of a flagellum. By the time the flagellum has attained full size the body has assumed the leptomonad form.

Shortly after attaining its full size, waves pass along the flagellum but translatory movement may be delayed for a short time. The expanded distal end of the flagellum is often attached to a piece of debris and its activity may cause the whole body of the organism to swing through an angle, or the leptomonad may be unattached and stationary, although rapid waves pass along the flagellum.

The flagellum of *Leishmania* sp. is an organ of attachment as well as an organ of locomotion. This can be clearly seen both in cultures and in the anterior part of the midgut of sandflies, where numerous flagella are seen attached to the rhabdorum and to the oesophageal valve by their expanded distal ends.

Flagellar movement may be accompanied by flexion and extension of the body which contribute to steering; this implies a co-ordinating mechanism, i.e. a primitive neuromotor mechanism, for which no morphological basis has yet been demonstrated. The proximal half of the flagellum may occasionally be held rigid while waves pass along the distal half.

It will be shown elsewhere (Section III) that the development of the flagellum is affected and even inhibited by suitable concentrations of immune serum; this development is therefore controlled by substances which are antigenic.

By the time the flagellum has attained its full size and function the flagellar vacuole has diminished in size and approached its normal dimension. In the normal leptomonad it contracts and dilates at irregular intervals. Occasionally it appears to break up into smaller vacuoles which reunite and reconstitute the organelle. It has also been observed to rotate, in leptomonads taken both from cultures and from the midgut of infected sandflies; the mechanism for this movement is not clear.

When the leptomonad is in the process of division a new flagellum is being formed. The general outlines are as described above, i.e. enlargement of the flagellar vacuole and growth of a flagellum which passes through a stage of pendular movement. In a dividing leptomonad the normal functioning flagellum of the "parent" can be seen together with all stages of the new developing flagellum. Adler and Adler (1955) considered that the development of the flagellum takes place in three stages: (1) the protrusion which attains a length of 1–2 μ beyond the anterior end of the organism; (2) growth of this protrusion till it attains its maximum size; (3) the appearance of waves along the whole length of the flagellum. This division into three distinct stages is not artificial. Between 1927 and 1943 a Chinese strain of *L. donovani* was observed in our laboratory. Short stumpy flagella which did not develop beyond stage 1 were the only ones observed in this strain. Hindle (1930) infected one out of twelve hamsters with this strain.

Pulvertaft and Hoyle (1960) made a cinemicrographic study of *Leishmania donovani* in all stages of development. They describe the process as follows:

"the flagellum is first seen as a tiny stump arising from the vacuole. At first there is an occasional small flicker of movement; as it increases in length, movement becomes more frequent and more rapid but is of a simple pendular type".

The scope of light microscopy is not sufficient to determine the exact anatomical relationship between the flagellar vacuole and the base of the flagellum. The whole series of events from the introduction of the leishmanoid stage into medium at 26°C till the formation of a mature flagellum invites study with the aid of the electron microscope. It is not certain that the flagellum actually arises from the vacuole but the latter is obviously intimately associated with the formation of a flagellum, a fact already noticed by Leishman in 1905. Factors such as immune serum which inhibit the development of the flagellum in cultures induce hypertrophy of the flagellar vacuole (up to 18 μ in diameter—see Section III).

As in the case of other protozoa the vacuole may have an osmo-regulatory function and may possibly have a role in excretion. In stained agglutinated flagellates an eosinophil deposit is often seen near the anterior end of the flagellates. As will be shown in another section, both the leishmanoid and the leptomonad stage of *L. tropica* or *L. mexicana* secrete soluble antigens.

Both volutin and lipid granules are present in the bodies of both leptomonads and leishmanoids, particularly in cultures on media containing immune serum. These granules are not an indication of age or degeneration, because they may be present in young and vigorous cultures particularly in the presence of immune serum.

The structure of the nucleus and nuclear division have recently been studied by Sorouri (1955) in *L. tropica*, Sen Gupta and Ray (1954), Chakraborty *et al.* (1962) in the case of *L. donovani.* According to Sorouri chromatin is distributed on the periphery of the nuclear membrane in the resting nucleus which contains a relatively large endosome. In early prophase the chromatin is condensed in six masses or "chromatids". In late prophase the "chromatids" pair and during pairing move toward the centre. Each pair constitutes a chromosome; therefore, the chromosome number is 3. In metaphase the chromosomes form a metaphase plate. In anaphase daughter chromosome bands migrate towards the nuclear poles, three bodies being recognizable in each band. During telophase the nuclear membrane constricts and two daughter nuclei are formed. Division of the kinetoplast probably commences during prophase, progresses slowly during metaphase and is complete during anaphase. The second flagellum is visible during late anaphase; its development probably commenced during early anaphase. Sen Gupta and Ray maintain that the chromosome number is 6 in *L. donovani.* These authors figure 6 chromosomes at each nuclear pole during nuclear division. By the use of Feulgen, and Gomori's technique for alkaline phosphatase, they concluded that there is a close parallelism between the distribution of DNA and alkaline phosphatase. Chakraborty *et al.* (1962) applied Feulgen staining to suitably fixed material after hydrolysis in HCl. According to these authors eight individual chromosomes can be counted during early anaphase and the kinetoplast divides in late anaphase. The possibility of syngamy is mooted on the basis of a preparation which may be interpreted as the attachment of two gametes and other preparations which resemble first meiotic divisions and eventually haploid nuclei. They make no final decision on syngamy which they leave open for further investigation.

Electron microscope studies on the intracellular leishmanoid were made by Chang (1956) on *L. donovani* and Garnham and Bird (1962) on *L. mexicana*. According to Chang the outer membrane of the cell is composed of a smooth outer layer and inner finely ridged layer. The nuclear wall also consists of a double membrane. The axoneme is seen with or without fibrils according to the level sectioned; it originates from a basal granule. The kinetoplast and flagellar vacuole are not described but the latter is shown in one of the photographs, interpreted as a basal granule. The other description, of *L. mexicana*, coming later than Chang's with the benefit of six years of advances in interpretation, is more complete. The organism is surrounded by a double membrane. Immediately below the inner membrane a row of 130–200 hollow fibrils is found, each fibril being 20 Ä in diameter. No obvious nuclear membrane was seen "perhaps because active division was taking place" (according to Sorouri the nuclear wall does not disappear during division). The flagellum arises from the blepharoplast or cylinder containing nine peripheral double fibrils ending in a plate from which the axoneme begins. The blepharoplast lies on the margin of the flagellar vacuole. Garnham gives a valuable description of the kinetoplast and points out that it is a body of unexpected complexity. The part seen under light microscopy appears in the electron microscope as an electron dense granular band with a distinct and regular fibrillar pattern lying inside an enormous body with a double membrane and having all the characters of mitochondria. All leishmanoids (and leptomonads) have this structure and Garnham notes that it is larger in *L. mexicana* than in *L. donovani.*

Electron microscope studies of the leptomonad forms can be divided into two stages, pre- and post-thin sectioning technique. In the first stage the work of Lofgren (1950), Sen Gupta *et al.* (1951), and Das Gupta *et al.* (1954) sufficed to show that the flagellum is a complex structure containing a number of fibrils but this number was not accurately determined. Das Gupta *et al.* recorded a maximum number of nine and further noted the existence of fibrils associated with the pellicle which they interpreted as myoneme fibrils. These fibrils may be the same structures swollen in the process of preparation which Garnham later found in the leishmanoid. They may explain the movement of flexion and extension in the leptomonad which couple with those of the flagellum and contribute to the steering of the flagellate stage. Thin sections helped to elucidate the structure of the flagellum and showed that the basic pattern is similar to that of flagella and cilia in general.

Inoki *et al.* (1957) described the emergence of the flagellum from a "narrow deep pit" (the reservoir of Clarke and the kinetosome vacuole of other writers). They found no trace of a blepharoplast or of a parabasal but mentioned a kidney-shaped body apart from the base of the flagellum, enclosed in a membrane and having a periodic structure. This obviously refers to the kinetoplast.

Pyne (1958) recorded that the pellicle is about 400 Å thick and is composed of two layers, the inner layer showing a series of fibrils with indications of a regular periodicity (in material treated in distilled water for 24 h after fixation). The nuclear membrane is composed of a thin outer layer 100 Å and a thicker inner layer 400–500 Å in width. A maximum of five osmophilic bodies were observed in the nucleus, possibly chromosomes, but the author did not indicate a chromosome number. Mitochondria are limited by a membrane up to 100 Å thick. The flagellar sheath is continuous with the outer layer of the pellicle and is formed by an invagination of the latter. The number of fibrils is given as ten, one central and nine peripheral; the central fibril is 450 Å thick, the peripheral ones 350×250 Å. The fibrils terminate in swollen granular structures which lie near the parabasal body. The parabasal body (obviously the kinetoplast of other writers) is 0·4–0·7 μ in breadth and 0·1 μ thick. No reference is made to the mitochondrial nature of this body. No eosinophil vacuole (syn. flagellar vacuole) was observed.

Pyne and Chakraborty (1958) studied the basal apparatus of the flagellum in *L. donovani.* They confirmed the "double nature" of both the central and peripheral flagellar fibrils. Each subfibril has a diameter of 200 Å. The flagellum emerges through an invagination of the pellicle. This invagination forms a vacuole for which they propose the name "kinetosomal vacuole". A new name—kinetosome—is also proposed for the base of the flagellum. Both the inner and the outer layer of the pellicle take a part in forming the membrane enclosing the kinetosomal vacuole; this membrane is 400 Å thick. The kinetosome is formed by the flagellar fibrils which run into the body of the flagellate for a distance of 1–2·5 μ. The kinetoplast ($0{\cdot}7 \times 0{\cdot}3$ μ) is a banded structure surrounded by a membrane. No reference is made to its mitochondrial character. The origin of the flagellar sheath is obscure but in the opinion of these authors it is not as they once believed a continuation of the pellicle.

Clark (1959) made the very significant observation that the flagellar vacuole appeared to empty into the reservoir. As will be shown in another section, soluble antigens are secreted both by leishmanoids and leptomonads of *Leishmania.* It is tempting to suggest that these antigens are secreted from the flagellar vacuole.

Clark and Wallace (1960) in a study of four genera of Trypanosomatidae including *L. tropica* refer to the invagination which includes the base of the flagellum as the reservoir; (this is obviously synonymous with the kinetosomal vacuole of Pyne and Chakraborty and the deep pit of Inoki *et al.*). They describe the kinetoplast as an elongated oval body surrounded by a double membrane, and containing a band of electron dense fibres orientated antero-posteriorly and closer to the anterior wall as a constant feature of kinetoplasts. The organization of fibres in the vertebrate parasitizing flagellates differs from that in the purely insect flagellates examined, *Herpetomonas culicis*, *H. muscatum*, *Crithidia fasciculatus*. The posterior portion of the kinetoplast contains structures similar to the cristae of mitochondria. The blepharoplast extends into the reservoir and is separated from the axoneme by a basal plate. The site of the flagellar vacuole is occupied by transparent spaces associated with Golgi material.

It is evident that the electron-microscope study of *Leishmania* is still in its infancy. There is no unanimity as to terminology, details and interpretation. Even so it has provided valuable information on the morphology of the leishmanias, particularly on the structure of the flagellum and kinetoplasts, which could not have been obtained by other means.

III. Identification of *Leishmania* spp. of Mammals and Lizards

Leishmania spp. have so far been recorded from mammals and lizards; none have as yet been recorded from birds.

The *Leishmania* spp. of man were until recently differentiated mainly on the basis of the type of case from which they were isolated, e.g. *L. tropica* from Oriental sore, *L. donovani* from kala azar, and partly on their geographical distribution, e.g. *L. infantum* from Mediterranean kala azar (though many considered *L. infantum* a synonym of *L. donovani*), *L. brasiliensis* for American non-visceralizing leishmaniasis. Russian authors (1946) distinguished urban from rural cutaneous leishmaniasis clinically and epidemiologically and created the names *L. tropica* var. *minor* for the former and *L. tropica* var. *major* for the latter.

It is now clear that the clinical picture is a useful, but by no means infallible guide to diagnosis. The primary lesion which occasionally occurs in East African kala azar months before clinical manifestations of visceral involvement could, on the above system, be attributed to *L. tropica*; moreover in some cases the disease may not progress beyond the primary stage. The clinical picture in active visceral leishmaniasis is identical in all endemic foci; differences in cutaneous manifestations occur before, during and after the active period but they are not constant and are of no help in identifying a culture. In spite of common properties there are minor differences in pathology associated with major epidemiological consequences (as in the case of Mediterranean and Indian kala azar) and other differences such as response to treatment (East African is less responsive to antimonial treatment than Indian kala azar) which must have a biological basis.

The confusion in nomenclature of *Leishmania* spp. associated with Central and South American cutaneous and muco-cutaneous leishmaniasis is not yet cleared up. The name *L. brasiliensis* Vianna 1911 was until recently applied indiscriminately by most authors to organisms from all cases of South American cutaneous and muco-cutaneous leishmaniasis without regard to clinical data; Laveran (1917) maintained that the parasite found in the South American disease is indistinguishable from *L. tropica* and he together with Nathan-Larrier suggested the name *L. tropica* var. *americana* for all strains isolated from non-visceralizing American leishmaniasis. Pessoa and Barreto (1944) in their indispensable monograph on South American cutaneous leishmaniasis, which is a mine of information on all aspects of the subject, stated that the difference between the Old and New World form of disease is not absolute, but that in respect of *L. tropica* and *L. brasiliensis*, they and most authors temporarily considered the two diseases and their causative agents as distinct until future investigations resolve the problem.

Velez (1913) made the first attempts at differentiation on clinical and epidemiological grounds and created the name *L. peruviana* for the uta of Peru, Biagi (1953a-c) suggested the name *L. tropica mexicana* for the cutaneous leishmaniasis of Mexico. The intention in both cases was to distinguish between espundia and non-metastasising cutaneous leishmaniasis.

Floch (1954) proposed the name *L. tropica guyanensis* for the *Leishmania* of French Guiana. He, like Biagi, considered that in the present state of knowledge, cultures, animal experiments and serology are "incapables" of differentiating *Leishmania* sp. In French Guiana the majority of cases of cutaneous leishmaniasis do not metastasise in the nasal mucosa but a minority do; therefore the disease differs from chiclero ulcer in which involvement of the nasal mucosa is unknown and from espundia in which the latter is common. He agrees with the view that all cutaneous leishmaniases are caused by one species, *Leishmania tropica*, and he recognized three subspecies in South America, *L. tropica brasiliensis* for the causative agent of espundia, *L. tropica mexicana* for that of chiclero ulcer and *L. tropica guyanensis* for the cutaneous leishmaniasis which occasionally metastasises in the nasal mucosa in French Guiana.

Medina and Romero (1959) created the name *L. brasiliensis pifanoi* for the parasite of a rare form of cutaneous leishmaniasis which Convit designated as leishmaniasis tegumentaria diffusa.

Garnham and Lewis (1959) isolated a strain of *Leishmania* in British Honduras which they identified with the *L. tropica mexicana* of Biagi; they found that this strain produced local tumours and visceral infections in hamsters; in the local tumours the parasites were distributed on the periphery of the histiocytes, a condition not observed in infections of *L. tropica* in hamsters. Garnham (1962) raised this strain to specific rank, i.e. *L. mexicana.* Muniz (1953) had demonstrated a similar distribution of parasites in infected histiocytes both in the cutaneous lesions of a Bolivian case and in subcutaneous tumours and viscera of hamsters inoculated with cultures from this case. The clinical description of the case was very different from chiclero ulcer or bay sore.

Experiments on laboratory animals have given variable results. Wenyon (1911-12), Aragao (1927), Pedroso (1923), Geiman (1940) reported successful inoculations in dogs; Montenegro (1923) failed to infect dogs.

The results recorded in experiments with hamsters are also variable. Fuller and Geiman (1942) reported local lesions in hamsters (*L. tropica* produces heavy visceral infections in hamsters); we have mentioned the findings of Garnham, Lewis and Muniz. Guimaraes (1951) using a strain designated as *L. brasiliensis* from Amazonia produced visceral infections in hamsters. Coelho (personal communication) produced local lesions with a strain isolated from a case of espundia. Negative results have also been recorded.

Experiments on mice have also given non-uniform results varying from negative in the experiments of Mazza (1926), local lesions and metastases in those of Adler and Theodor (1930) to visceral involvement in those of Guimaraes (1951) (local lesions, metastases round the anus and visceral involvement may occur in the experience of the reviewer in mice infected with cultures of *L. tropica*). The most constant results so far have been obtained in monkeys, e.g. Amaral (1941) infected monkeys *Macacus rhesus* with *L. brasiliensis* and produced local lesions.

The results of inoculation into man have been equally contradictory; Montenegro (1923) successfully infected a new site in one active case by injecting material from an infected site, but obtained negative results with a similar procedure in six other cases. Pessoa and Arantes (1944) failed to infect seven lepers by injection of rich cultures of *L. brasiliensis*. On the other hand Trejos and Echandi (1951) infected two volunteers by injection of cultures of a form which they considered to be *L. brasiliensis*. Adler and Gunders (1964) infected a volunteer by injecting material from a culture of *L. mexicana*, recently isolated from a hamster.

Muniz (1953) summing up the situation regarding the South American cutaneous and muco-cutaneous leishmaniases wrote "up to now it is well established that all the procedures used for the differentiation of the species of *Leishmania*, such as reaction of immunity, aspects in culture and pathogenicity for certain animals are not available. The *L. tropica* for instance was distinguished from *L. brasiliensis* because the latter did not give systemic lesions in hamsters. However, it is known today that *L. brasiliensis* is able to produce systemic infection in that animal. Variation in the behaviour of a strain cannot be a sure basis for the distinction of species chiefly if one considers that such variation depends upon the parasite, the host and epidemiological factors."

The only instance of unanimity with regard to animal inoculation in a

South American leishmaniasis is the case of *L. enrietti* Muniz and Medina (1948) which in the laboratory was found by Muniz and Medina (1948), Torres *et al.* (1948), Paraense (1953), Adler and Halff (1955) infective for guinea pigs and not for other adult laboratory animals. Adler and Halff (1955) recorded transient infections in mice inoculated during the suckling period. The natural host of this species remains to be discovered. The reviewer found that this species shares some antigens with *L. tropica, L. mexicana* and *L. brasiliensis* but can be distinguished from them.

The confusion and contradiction between the results reported by different authors is in the opinion of the reviewer due to two factors. In the first place the cutaneous and muco-cutaneous leishmaniases of Central and South America are caused by a number of distinct species, each with its own spectrum of host infectivity. Secondly, in the experience of the reviewer some *Leishmania* sp. lose their infectivity even for highly susceptible laboratory animals such as the hamster after periods of culture in the laboratory. The different species vary in this respect. *L. tropica,* as Parrot (1929) first showed, and the reviewer repeatedly confirmed, loses its infectivity for mice after eighteen months or so on culture media. It also loses its infectivity for hamsters but not for man. Adler and Zuckerman (1948) found that a strain maintained during twenty-two years in culture was still infective for man (long after it had lost infectivity for mice and hamsters). In the experience of the writer strains of *Leishmania* isolated from cases of kala azar in Kenya lose their infectivity for hamsters, which are highly susceptible to recently isolated strains, after a few months in culture. On the other hand Adler (1961) found that a strain of *L. infantum* isolated in Malta proved infective for hamsters after twenty-seven years in culture. Strains of *L. mexicana* lose their infectivity for hamsters after about five months in culture. The writer has not detected antigenic differences between the infective and non-infective Kenya strains. Adler (1940) found it difficult to infect human beings with Indian *L. donovani* by injection of rich cultures and material from heavily infected hamsters (one case out of five became infected). On the other hand, Manson-Bahr (1959) found that cultures of a Kenya strain of *Leishmania* isolated from cases of kala azar are infective for man. It appears that in general *Leishmania* spp. maintain indefinitely their infectivity for susceptible hosts by cyclical passages through sandflies.

There has been a tendency of late among South American workers towards the opinion that *L. brasiliensis* is not a homogeneous species. Pifano (1960) regards *L. brasiliensis* as a complex of varieties, species or races absolutely different from each other. Forratini (1960) adopts the view that *L. brasiliensis* contains races and subspecies in the process of speciation. Cultures which he isolated from three species of wild rodents were not infective for other animals and the organisms therefore seem to be restricted to specific hosts. Findings of this kind must be interpreted cautiously. We have already stressed the fact that strains may lose their infectivity for some hosts a short time after isolation in culture. Hertig *et al.* (1954) isolated a *Leishmania* from spiny rats in Panama; cultures were not infective for the original host.

The most recent classification of the parasites of American cutaneous and muco-cutaneous leishmaniasis is that of Pessoa (1961), who was at first inclined to think that a single organism, *L. brasiliensis*, was the cause of all forms of South American cutaneous and muco-cutaneous leishmaniasis but later, on the basis of geographical distribution and clinical data, finally (1961) recognized five distinct varieties. He adopted the trinomial system of Hoare (1955) and accepted the following:

1. *L. brasiliensis brasiliensis* Vianna 1911 (metastases in oro-pharynx in 80% of cases);
2. *L. brasiliensis guyanensis* Floch, 1953; *L. brasiliensis guyanensis* (metastases in nasal mucosa in 5% of cases);
3. *L. brasiliensis mexicana* Biagi, 1953 (syn. *L. tropica mexicana* Biagi);
4. *L. brasiliensis peruviana* Velez, 1913 (for uta);
5. *L. brasiliensis pifanoi* Medina and Romero, 1957 (for chronic diffuse cutaneous leishmaniasis with a characteristic clinical picture described by Convit *et al.*, 1959).

Pessoa, one of the leading investigators of South American leishmaniasis, has, after many years of study, adopted a pluralist attitude with which the reviewer is in complete agreement. Nevertheless, clinical data cannot be applied for specific determination of an individual strain which may have been derived from 20% of cases of *L. brasiliensis brasiliensis* sensu Pessoa or 95% of the cases of *L. brasiliensis guyanensis* which do not metastasise in the nasal mucosa. As in all protozoal infections, a single species need not give a uniform clinical picture.

In the absence of morphological differences between alleged species of *Leishmania* (except in the case of *L. enrietti* in the guinea pig), the limitations of clinical data and the variable results of animal inoculations, attempts to differentiate *Leishmania* sp. by serological methods are the only resort.

Noguchi (1924, 1926) was the first to differentiate *Leishmania* sp. by serological tests. He cultivated leishmanias on semi-solid medium containing various concentrations of homologous and heterologous sera (1924). He noted homogeneous growth in semi-solid medium containing normal serum and granular growth on media made up with homologous serum. (Study of a micro-organism growing in medium containing homologous serum was introduced by Charrin and Rogers (1889) in the case of *Bacillus pyocyaneus* and this method is probably the oldest of all serological tests.) Noguchi (1926) also used agglutination tests (0·05 ml immune serum to 1 ml suspension of flagellates examined immediately, after 30 min, 18 h and 24 h) and complement fixation in a study of *L. donovani*, *L. infantum*, *L. tropica* and *L. brasiliensis*. On the basis of the above tests Noguchi concluded that

L. donovani, *L. brasiliensis*, and *L. tropica* can be differentiated and that *L. donovani* and *L. infantum* are closely related.

Noguchi's opinion was confirmed by a number of workers, among them Kligler (1925), Fonseca (1932) and Cunha and Chagas (1937).

Khodukin *et al.* (1936) stated that they could distinguish strains "plus ou moins" by their titre but "the possibility of error was not always excluded"; spontaneous agglutination was not rare. After absorption by one species sera failed to agglutinate other species. *L. brasiliensis* was the only strain which could be differentiated by complement fixation. No single method (agglutination, growth on medium containing immune serum, complement fixation, Rieckenberg phenomenon) was found to be decisive.

Cunha (1942) revised his opinion and maintained that it was not possible to differentiate *Leishmania* sp. by agglutination tests because all species have a common antigen which constitutes the major part of the agglutinogens in recently isolated cultures. Chang and Negherborn (1947), using suspensions of flagellates in saline (100×10^6/ml) and examining 15–30 min after adding serum, concluded that it was possible to distinguish *L. tropica*, *L. brasiliensis*, *L. donovani* and *T. cruzi* by agglutination although in some experiments "non-specific clumping creates confusion". Rosette formation occurs after 2 h in controls. Cross reactions are weak.

Kirk (1950) summed up the situation as follows: "attempts to differentiate *Leishmania* by bacteriological technique have given extremely contrary results." Sen and Mukherjee (1961) stated that they could distinguish between organisms isolated from post-kala azar dermal leishmaniasis and those isolated from active kala azar, by agglutination and absorption tests.

There are a number of reasons for the discrepancies between the findings of various workers and even the same workers in their attempts to differentiate *Leishmania* spp. by serological methods. In the first place, all strains from South American cutaneous and muco-cutaneous leishmaniasis were considered to belong to one species. *L. brasiliensis*. It is now clear that a number of distinct species were included under this name. In the second place, agglutination of the flagellate stage of leishmanias may be complicated by a number of factors such as spontaneous clumping and rosette formation. Immobilization tests do not give uniform results because the phenomenon does not occur simultaneously in a population of flagellates. The action on flagella is also not simultaneous; a number of flagella may be contorted, swollen and motionless, others are unchanged and active. What is more important is the fact that the antigenic composition of *Leishmania* is complex and *a single reaction such as complement fixation or agglutination, cannot possibly reveal the action of the various antibodies evoked by corresponding antigens.*

The reviewer found the following method satisfactory. Medium (Locke-serum-agar) is distributed in Kahn tubes, 2 ml per tube. Immune serum prepared in rabbits by the usual routine is added to make the required dilutions (from 1:5 to 1:20,000). Most sera are effective up to 1:1,000 but good sera may give a titre of 1:20,000, exceptionally 1:50,000. *L. brasiliensis* (from cases of espundia) is a particularly good antigen and often produces sera with a titre of 1:20,000. After adding specific serum in various concentrations, tubes are sown from rich cultures with homologous and heterologous strains; loopfuls of fresh material from cultures or immune serum are examined at intervals of several days under a coverslip with all powers of the microscope and Giemsa stained preparations made. (It is interesting to note that cultures made up on medium containing immune serum live considerably longer than those made up on normal serum.) Growth on dilutions of 1:5 is occasionally retarded temporarily; a few sera completely inhibit growth in this dilution.

It is advisable to examine the cultures by the above method every 3 or 4 days throughout the whole life of a culture and collect data on flagellar development and types of colonies produced, etc. A suitable serum shows the following effects in culture of a homologous organism.

1. Growth of the organism in the form of syncytia which may attain a size of 450 μ, many of them with bizarre shapes. The formation of syncytia indicates inhibition of cytoplasmic cleavage without interference with nuclear multiplication and increase in cytoplasm.

2. Inhibition of growth of flagella which may be complete in the highest concentrations. In concentrations in which inhibition of flagellar development is not complete, a number of immature flagella can be observed which are capable of pendular movement only. The margins of some flagella are distorted and show local bulges.

3. Differentiation of both aflagellar and flagellar forms on the margins of syncytia (even of the same syncytium).

4. Enlargement of flagellar vacuoles. This is a striking phenomenon. In the normal non-dividing organism the vacuole is 1 μ or less in diameter. In the higher concentrations of immune serum the vacuole may attain a diameter of 18 μ. In some individual aflagellar forms the flagellar vacuole enlarges till only a thin rim of cytoplasm and finally no cytoplasm remains. In some of the syncytia uniformly glassy structureless areas are seen; they appear to be associated with hypertrophy of flagellar vacuoles and accompanying elimination of cytoplasm. The hypertrophy of the flagellar vacuole may be due partly to the absence of an adjoining flagellar complex. According to Clark and Wallace (1960), the vacuole of the normal flagellate secretes into the reservoir near the vacuole and in the absence of this structure the vacuole undergoes hypertrophy.

5. In cultures 1–3 weeks old, fibrils are deposited in the matrix of syncytia. Between the compartments formed by these fibrils flagellar and aflagellar forms differentiate; necrotic material is also formed.

6. Colonics of active flagellates are formed in higher dilutions. As the dilution increases the number of free active flagellates and non-agglutinated flagellates increase. The final titre is the dilution in which all the flagellates are free and no agglutinated clumps are found.

Stained specimens show enlarged aflagellar forms with several nuclei. Errors of development such as aflagellar forms with one or more nuclei and no kinetoplasts occur.

In the higher concentrations of specific serum the whole symmetry of normal leptomonad and leishmanoid disappear but this does not interfere with nuclear multiplication or growth. The morphological aberrations mentioned above invite study with the aid of the electron microscope in the expectation of throwing light on normal structure and function, particularly of the flagellum in all stages of development and of the flagellar vacuoles. The phenomena mentioned above are obviously artifacts which have no counterpart in the normal life of *Leishmania* spp. but they are interesting in so far as they show the action of antibodies on the living organism in culture and are useful in differentiating *Leishmania* spp. It is also clear that the titre, i.e. the lowest dilution in which no agglutination occurs, is by no means the only important factor in the study of specific serum. (Strictly speaking we are not dealing with agglutination because with the method described above the flagellates grow in agglutinated masses, i.e. they do not separate after division and there is no agglutination of individual flagellates which were previously unagglutinated.)

Agglutination titres are useful in comparing two organisms when the differences are considerable; differences in the other phenomena, e.g. formation of syncytia, type of colony, etc., may occur in cases where there is no marked difference in titre (this may partly account for the discrepancies noted in the usual agglutination tests).

The above method can be used with sera which have been absorbed by heterologous *Leishmania* sp., e.g. *L. brasiliensis* grown on its homologous serum incorporated in Locke-serum-agar, after absorption by *L. mexicana* produced syncytia in dilution of 1:500 (Fig. 1a, b). It is very useful to select a suitable serum, i.e. one which produces syncytia and aflagellar forms in titres of 1:5 to 1:20 or more with homologous strains. With this method it was very easy to show that *L. brasiliensis*, *L. mexicana*, *L. tropica* and *L. donovani* could be differentiated from each other. Furthermore *L. tropica minor* could readily be differentiated from *L. tropica* and the former should be given specific rank.

Strejan (1963) found that *L. mexicana* and *L. tropica minor* will not grow on Citri and Grossowicz's semi-defined medium on which *L. tropica* grows profusely. *L. tarentolae* is so far the only species which grows on a fully defined medium elaborated by Trager (1957).

It is important to note that serological differentiation of species does not

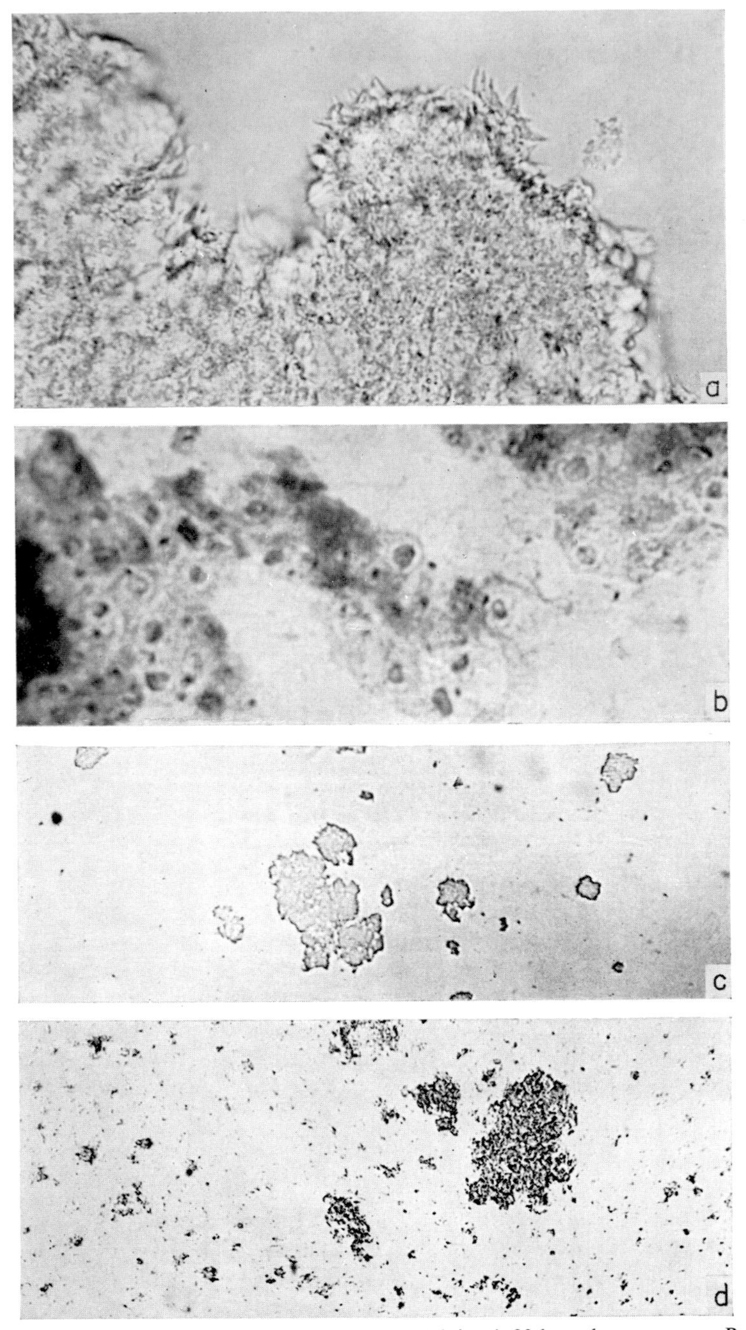

FIG. 1. (*a*) *L. brasiliensis* in Locke-serum-agar containing 1:20 homologous serum. Part of a large syncytial mass. Leptomonads and aflagellar forms have differentiated on the margin. Note many large flagellar vacuoles in syncytia. From cultures 15 days old. Fresh material. × 630.

(*b*) *L. brasiliensis* in Locke-serum-agar containing 1:20 homologous serum. Stained material of part of a syncytium showing differentiation into aflagellar forms. From cultures 17 days old. × 1,700.

(*c*) *L. donovani* (Indian strain) grown on Locke-serum-agar containing 1:20 homologous serum. The culture contains syncytial masses and no free flagellates. Fresh material from cultures 6 days old. × 100.

(*d*) *L. infantum* (Maltese strain) grown on Locke-serum-agar containing 1:20 anti-*L. donovani* serum (same serum as in *c*). The culture consisted of clumps of flagellates and many free active flagellates. × 100.

necessarily parallel cross-immunity, e.g. individuals immune to *L. tropica* after spontaneous cure are also immune to *L. mexicana* though the two species are distinct.

All the *Leishmania* species of mammals we have tested have antigens in common manifested by cross agglutination. In the case of South and Central American strains only *L. mexicana* and *L. brasiliensis* have so far been examined by the above method. It is now important to determine whether or not the cases of espundia from different foci of South America are caused by a homogeneous single species.

The same problem presents itself in the case of the non-metastasising leishmaniases, i.e. whether the organisms from cases of uta can be distinguished from those of chiclero ulcer and from the cutaneous leishmaniasis of Panama. Furthermore, a serological examination of strains isolated in the early stages of infection when the lesions are localized, may prove valuable for prognosis and treatment by distinguishing non-metastasising strains or species from those liable to metastasise in the nasal mucosa.

In the case of the Old World leishmaniases it is important to compare strains from different epidemic foci of visceral leishmaniasis and to determine the status of espundia in the Sudan. It may be noted that Indian strains of *L. donovani* can be distinguished by the above method from Mediterranean strains of *L. infantum* (Fig. 1c, d).

The above findings were made with sera prepared from cultures. Adler and Adler (1955) found that sera prepared by injection of L.D. bodies extracted from the spleens of hamsters heavily infected with a strain of *Leishmania* isolated by Dr. Heisch from a case of kala azar in Kenya, gave agglutinating titres of 1:10 to 1:50 with the homologous leptomonad strain maintained in cultures on Locke-serum-agar. Sera prepared with the latter gave a titre of 1:500. This indicates differences in antigenic structure between the leishmanoid and the flagellate stage. These differences cannot be ascribed entirely to the absence of a flagellum in the leishmanoid.

Aflagellar forms of *L. adleri* were produced by growth on medium containing 10% immune serum. After repeated washing the material was used for preparing sera in rabbits; these sera gave titres no higher than 1:20 to 1:50. Sera produced by the leptomonads of the same strain produced titres up to 1:20,000. The low titre produced by the aflagellar forms may have been due to antibodies firmly fixed to the organisms and not removable by repeated washing and centrifuging. It is therefore possible that the low titre produced in rabbits by the leishmanoids from the spleen of infected hamsters may have been in part due to antibodies which they absorbed in the spleen.

Note on Nomenclature

Hoare (1955) proposed a trinomial system for the genus *Leishmania* acknowledging biological differences and clinical data between strains, each strain constituting a nosodeme. The reviewer prefers a binomial system which was employed in the early period of research on leishmaniasis by those who considered *L. donovani* and *L. tropica* as distinct species. Speciation on the basis of marked difference in antigenic structure and biological characters is convenient in microbiology. Garnham (1962) prefers a binomial system as one "less likely to create anomalies and create confusion in the future".

A. LEISHMANIAS OF LIZARDS

Nine species of *Leishmania* have been described from lizards, *chamaeleonis*, *henrici* and *zmeevi* as species of *Leptomonas*. With their hosts and geographical localities they are:

L. adleri Heisch, 1955: *Latastia longicauda revoili* (Kenya).

L. agamae David, 1929: *Agama stellio* (Tiberias).

L. ceramodactyli Adler and Theodor, 1929: *Ceramodactylus dorias* (Baghdad).

L. chamaeleonis Wenyon, 1921: *Chamaeleon vulgaris* (Egypt, Israel), *C. ellioti* (Uganda), *C. pardalis*, *C. oustalei*, *C. verrugosus*, *C. lateralis*, *C. brevicornis* (Madagascar).

L. gymnodactyli Khodukin and Sofiev, 1947: *Gymnodactylus caspius*, *Agama sanguinolenta*, *Pharinocephalus heliodactylus*, *P. myctaceus* (U.S.S.R.).

L. hemidactyli Mackie, Gupta and Swaminath, 1923: *Hemidactylus gleadovii* (India).

L. henrici Leger, 1918: lizards of genus *Anolis* (Martinique).

L. tarentolae Wenyon, 1921: *Tarentola mauritanica* (N. Africa, Sicily).

L. zmeevi Andrushko and Markov, 1955: Leptomonads in two (unnamed) species of lizards (Karakum desert).

Species of *Leishmania* have not been recorded from the American continent or from Australia. Dr. Bruce MacMillan (personal communication) did not find any in geckoes examined in New Guinea.

The leishmanias of lizards deserve further study because they are of interest in themselves and, as Hoare (1948) indicated, because of the light they may throw on the evolution of the Leishmanias in general. *L. chameleonis* is the most primitive of known species and is restricted to the cloaca of chameleons; Frenkel (1941) recorded invasion of epithelial cells of the large intestine in a chameleon infected experimentally by oral administration of infected faeces. Cultures from heart's blood were negative. *L. henrici* represents a further stage in the evolution of leishmanias; flagellates occur in the cloaca and occasionally invade the bloodstream. All the other above mentioned *Leishmania* sp. of lizards are blood and tissue parasites which cannot tolerate bacterial contamination, whereas *L. chameleonis* and *L. henrici* obviously live in an environment containing a rich bacterial flora.

No detailed studies have been made on the antigenic relationship between the *Leishmania* spp. of lizards nor have experiments been carried out on immunity and cross-immunity. The gecko *Tarentola mauretanica* was at one time suspected to be a reservoir of human leishmaniasis because in culture its parasites are morphologically indistinguishable from the human leishmanias. Adler and Theodor (1931) found that this gecko is not susceptible to *L. infantum*. Mohiuddin (1959) has shown that *L. adleri* is infective for lizards of diverse genera and geographic distribution. He infected lizards belonging to the genera *Malesia*, *Lacerta*, *Acanthodactylus* and *Agama*. In all cases, including its original host, the infection is cryptic and can be demonstrated by blood culture only. Manson-Bahr and Heisch (1961) inoculated four volunteers; three developed nodules at site of infection. L.D. bodies were demonstrated in one case up to 5 days after injection but not later. The writer received a strain of *L. adleri* from Dr. Heisch soon after its isolation and inoculated hamsters with it. No parasites were found in the hamsters by microscopic examination and it appeared that the organism is non-infective for the animals. Later results showed that *L. adleri* produces in hamsters transient cryptic infections lasting up to 5 weeks. Prolonged microscopic examinations revealed no L.D. bodies but cultures from the spleen were positive. Transient infections lasting up to 5 weeks were also produced in mice inoculated intradermally during the suckling period; L.D. bodies could be demonstrated by microscopic examination of connective tissue adjacent to the skin up to 10 days after injection and cultures made from the dermis were positive up to 5 weeks after injection. Inoculation of aflagellar forms of *L. adleri* from cultures grown in 1:5 immune serum gave the same results as cultures of active flagellates when inoculated into baby mice. *L. adleri* is so far the only known species of *Leishmania* which produces infections in both lizards and mammals, cryptic and of long duration in the former, cryptic and transient in the latter. Heisch (1958) found a sandfly *Phlebotomus clydei*, which feeds on lizards and mammals, naturally infected with a *Leishmania* probably *L. adleri*. Like mammalian *Leishmania* spp., the flagellates in the sandfly adopted an anterior position. It is therefore interesting to note that Adler and Adler (1955) found that *L. adleri* shared antigens with *L. donovani*, *L. infantum*, *L. brasiliensis* and *L. tropica*. The general biological characters, particularly the antigenic relationship with mammalian leishmanias, indicate that *L. adleri* presents an interesting phase in the evolution of *Leishmania*, i.e. a transition from a purely reptilian to a mammalian parasite.

Further studies on the leishmanias of reptiles are called for. The examination of reptiles for *Leishmania* sp. by culture of heart's blood should be rewarding, particularly in Africa, which contains many species of *Phlebotomus* that feed on lizards. It would be interesting to determine whether there are tissue and blood *Leishmania* sp. in South American reptiles and whether *L. adleri* is infective for American reptiles. The presence of a lizard *Leish-*

mania in Kenya with easily demonstrable antigens shared with *L. donovani*, *L. infantum*, *L. brasiliensis* and *L. tropica*, is of great interest. *Leishmania* is obviously an ancient genus, long antedating the entry of man into America, as shown by the presence of autochthonous American rodent *Leishmania* sp. differing antigenically from Old World species, though infective for some Old World mammals. It is not too far-fetched to speculate that the current mammalian species of *Leishmania* originated from a type similar to *L. adleri*. The only *Leishmania* sp. recorded from a reptile outside the Old World is *L. henrici* from Martinique. This species represents a transition phase from an intestinal to a blood parasite. A study of its antigenic relationship (if cultures were available) would be most interesting.

IV. Maintenance of Strains

It is easy to maintain strains in culture. Many satisfactory media have been described and used. It is important to bear in mind that strains change after various periods in culture, particularly in their infectivity for laboratory animals. As previously stated, strains of *L. mexicana*, Kenya strains of *L. donovani* and strains of *L. tropica* lose their infectivity for mice and hamsters.

It is therefore advisable to maintain strains by continuous passage through hamsters and in the case of *L. enrietti* through guinea pigs.

Cultures should also be transferred to deep freeze (after adding glycerine up to 10%) soon after isolation. Dr. Ann Foner (1964) has recovered twenty-nine strains after 13 months at –60°C.

V. Immunology

The leishmaniases are of particular interest to immunologists because of the striking increase in immunologically competent cells throughout the course of all clinical types, generalized in visceral and local in cutaneous and mucocutaneous.

The series of changes in the population of these cells can be conveniently studied and their significance evaluated in the case of Oriental sore. The first stage is characterized by a proliferation of histiocytes in which L.D. bodies multiply. Infected histiocytes are more active than normal ones in all clinical types of leishmaniasis. Infiltration by histiocytes is followed by a secondary local infiltration of mononuclears, lymphocytes and plasma cells. This secondary infiltration eventually limits the proliferation of histiocytes and reduces the number of parasites till spontaneous cure is established (Adler, 1948). The mechanism by which the round cells and plasma cells reduce the number of parasites is not clear. It is not possible to detect with certainty classical antibodies in the circulating blood of immunes. Whether or not the lymphocytes and plasma cells produce antibodies locally *ad hoc* remains to be investigated. Histological examination shows that parasites are less numerous in the immediate proximity of plasma cells and round cells than in other parts of the same lesion. It is easy to produce antibodies in rabbits by the usual procedures but these do not appreciably inhibit the growth of *Leishmania* sp. in cultures.

Immunity to *L. tropica* follows spontaneous cure and experimental attempt at reinfection by inoculation of flagellates from cultures gives negative results. The injected parasites are not destroyed immediately. Adler and Zuckerman (1948) inoculated a spontaneously cured volunteer with 6 million parasites; after 24 h a culture from the same site gave a positive result, after 48 h a negative result. Reinfection in spontaneously cured cases living in endemic foci are rare: a few instances have been reported five and more years after initial cure.

The duration of the various stages in the above mentioned sequence of cellular changes varies in different cases. In a few cases the period of histiocytic proliferation is brief, and round and plasma cell infiltration together with giant cells is established several weeks after the lesion is apparent. In these cases the lesion resembles lupus vulgaris histologically. Immunity against *L. tropica* is not established till the lesion has gone through the whole gamut of the above described changes. Marzinowsky and Schorenkova (1924) reported experimental reinfection after surgical removal of an active lesion. Senekji and Beattie (1941) showed that new lesions can be produced experimentally by inoculation of flagellates from cultures into patients with active lesions. Moschkowsky (1941) maintained that lesions produced by super-infection commence at the stage achieved by the current lesions. Dostrowsky *et al.* (1952) proved that lesions produced experimentally in infected individuals correspond to the current condition of the original lesions.

Dostrowsky described and named the condition known as leishmaniasis recidivans; this consists of areas of infiltration adjacent to the scars of old lesions which had apparently healed spontaneously. The condition may appear years after apparent cure. Parasites are extremely rare in these lesions and in nearly all cases can only be demonstrated by inoculating a number of tubes of suitable medium. Clinically and histologically the disease resembles lupus and has not infrequently been diagnosed as such by clinicians. This condition can be regarded as an incomplete immune response which reduces the number of parasites to a minimum without eliminating them completely. The very few remaining parasites stimulate the continued local mobilization of round cells and plasma cells. In these cases the incomplete immune process is associated with a pathological condition which persists for years. The above described phenomena indicate that the acquisition of immunity to *L. tropica* is not a one step process.

The sequence of local cellular changes characteristic for Oriental sore also occurs in South American cutaneous and muco-cutaneous leishmaniasis.

Weiss (1943) found that in cases of uta in Peru cure is usually followed by a lasting immunity. The majority of cases of chiclero ulcer are also immune but reinfections are more common in Mexico than elsewhere. Biagi (1953a-c) has recorded twelve reinfections in thirty-seven Mexican cases residents of endemic foci, 6–21 years after spontaneous cure.

It has already been indicated that round cells and plasma cells play an essential role in acquired immunity. This view gains support from a study of the condition described by Convit (1958) as leishmaniasis tegumentaria diffusa. In this condition extensive areas of skin are involved. The dermis is packed with heavily infected histiocytes which display a very marked physical activity, forcing their way into the dermis where many of them die and leave a deposit of L.D. bodies. Invasion of the nasal mucosa is slight. Cultures can be made from the circulating blood but there is no multiplication of parasites in the viscera. The Montenegro test is negative. There are no clinical indications of any immune response. The number of cases reported in Bolivia, Venezuela and Brazil is small and there is no information on the immunological status of infected individuals regarding other antigens. Examination of skin biopsy material from the only case seen by the writer (demonstrated by Professor Guimaraes of Salvador) showed that the lesion consisted of dense masses of histiocytes without any secondary infiltration of lymphocytes and plasma cells. A strain obtained from this case was examined by the writer and found to be an excellent antigen in the rabbit although it was not antigenic in the patient from whom it was isolated. The reviewer found that this strain could readily be distinguished serologically from *L. tropica* and *L. mexicana* and was closely related to *L. brasiliensis*. The absence of lymphocyte and plasma cell infiltration in the above case and their association with eventual spontaneous cure in other cases are evidence for immunological role of lymphocytes and plasma cells in leishmaniasis.

Muco-cutaneous leishmaniasis presents special features; the initial cutaneous lesion usually heals spontaneously as in the case of Oriental sore but metastatic lesions may occur in the mucosa and cartilage of the nose, mouth and naso-pharynx. Infection of the nasal mucosa may occur while the initial lesion is still active, according to Pessoa and Barreto (1944), or they may appear at various intervals, even years after the initial lesion has healed. Lindsay (1935) observed the appearance of metastases years after the healing of the primary skin lesions in Paraguayan soldiers subjected to the stress of war. Whatever immunity has been acquired following spontaneous cure of the initial skin lesion does not protect the cartilage and mucosa of the naso-pharynx once they have been invaded. It is difficult to account for this phenomenon but the following finding may be relevant. When the cartilage is invaded by infected histiocytes, the latter multiply and form nests of infected cells. In a biopsy of cartilage seen by the author, a zone of lysis of cartilage was found surrounding nests of infected histiocytes. There was no secondary invasion of round cells and plasma cells and this may explain the absence of any local protective mechanism in the invaded cartilage. The destruction of cartilage is not specific for *L. brasiliensis*; it occurs frequently in infections with *L. mexicana* (chiclero ulcer), and occasionally in *L. tropica* when the primary lesion is on the external ear and invasion of the adjacent cartilage by hyperactive infected histiocytes occurs. Metastatic lesions of the nasal mucosa from which infected histiocytes have easy access to the adjacent cartilage is typical for *L. brasiliensis.*

Cases of Oriental sore show a delayed hypersensitivity test (Montenegro reaction, leishmanin reaction) in response to the intradermal inoculation of dead and washed leptomonads. The reaction can be evoked by leptomonads of *L. tropica* or any other known species of *Leishmania, L. mexicana, L. donovani, L. brasiliensis* and even by a lizard *Leishmania, L. gymnodactyli.* It can also be evoked by antigens prepared from *Trypanosoma cruzi* and according to Depieds *et al.* (1958) by antigens prepared from *T. equiperdum.* This delayed hypersensitivity is in itself no indication of immunity; it first appears in the earliest stage of the disease while the parasites are still multiplying (Adler and Zuckerman, 1948).

The reaction is positive in all subsequent stages and persists long after spontaneous cure and probably throughout life (the writer has found the reaction positive 35 years after spontaneous cure). It can be elicited by 10^5 washed and phenolysed flagellates in 98% of cases of active Oriental sore; we elicited a Montenegro reaction by intradermal injection of 10^3 flagellates in a hypersensitive immune. The reaction in itself is no index of immunity. It constitutes a mechanism for mobilizing immunologically competent cells to a site containing antigens of *Leishmania* sp. or closely related organisms; the cells are mobilized early in the infection but they are effective only after immunity has been established; the specific mobilization of cells in response to antigen antecedes the establishment of immunity. In the absence of a mechanism for mobilizing effective immunologically competent cells the immune process could not be implemented.

The delayed skin reaction which can be conveniently studied in the cutaneous and muco-cutaneous leishmaniases is one of the crucial problems of current immunology; it has accordingly evoked a vast amount of experimental work and theoretical discussion. Immunologists are not yet agreed as to whether or not circulating classical antibodies play an essential role in this reaction.

Cutaneous leishmaniasis is a localized process but the immune mechanism is effective over the whole area of the body. This implies a response of the immunological mechanism to antigens derived from the parasites and liberated from the parasitized histiocytes, i.e. liberation of soluble antigens. Adler and Gunders were able to demonstrate the presence of such antigens by the following methods (1964, in press). *L. tropica* or *L. mexicana* is cultured in a tube of N.N.N. medium to which 4 ml saline is added. After 4–6 days the fluid is centrifuged in order to remove leptomonads and then filtered through a Zeiss; 0·1 ml of the filtered fluid is injected intradermally into an immune volunteer and into a control. An immediate reaction may appear both in the immune volunteer and in the control but this disappears within a few hours; a delayed reaction occurred within about 12 h in an immune but not in controls. This reaction differed from that of the Montenegro reaction in so far as it consisted of an area of erythema 2×4 cm surrounding a papula. The erythema lasted a month or so but the papula corresponding to the Montenegro reaction persisted for several months. There are apparently two components involved in this reaction, one corresponding to the antigen demonstrated by the Montenegro test and another responsible for the relatively extensive erythema. A specific reaction in a spontaneously cured volunteer can also be produced by the following method; to 1 g of hamster spleen heavily infected with *L. mexicana* add 3 ml saline and allow to stand overnight in the refrigerator at 4°C; the fluid is withdrawn, centrifuged and the supernatant is filtered through a Zeiss; 0·1 ml is inoculated intradermally into a volunteer and a reaction similar to that described above is obtained. The filtered serum of a hamster heavily infected with *L. mexicana*, but not that of a normal hamster, also gave a delayed local erythema which lasted about 3 weeks in one volunteer but the central papula was not observed; controls were negative. In addition to soluble antigens which enter the general circulation where they can be demonstrated by skin tests, L.D. bodies of *L. tropica* have occasionally been reported in lymph glands draining the infected area.

Dostrowsky and Sagher (1957) have shown that the delayed reaction in the case of Oriental sore can be completely inhibited if cortisone is injected locally together with the test antigen.

The delayed skin reaction is positive in all the South American cutaneous and muco-cutaneous leishmaniases. As in the case of *L. tropica* this reaction is positive in the earliest stages of the disease, within 8 days according to Echandi (1953). It can be evoked with antigen prepared from *L. tropica, L. brasiliensis, L. mexicana, L. enrietti* and *Trypanosoma cruzi.*

As previously stated, classical antibodies have not yet been demonstrated with certainty in Oriental sore. In the case of muco-cutaneous leishmaniasis contradictory results have been reported and the position is obscure. Moses (1919) reported positive complement fixation reactions with a homologous antigen in 80% of cases. Positive reactions with *L. brasiliensis* as antigen have been reported by Cunha and Dias (1939) in all forms of leishmaniasis and also in Chagas' disease. Mazza (1935), Cunha and Dias (1939) and Pessoa and Cardoso (1942) reported positive complement fixation in cutaneous leishmaniasis by using antigen prepared from *Trypanosoma cruzi.* Martins (1941) claimed that complement fixation is valuable for diagnosis though he pointed out that the preparation of a suitable antigen is difficult and false positives are common.

The immunity acquired to *L. tropica* somewhat resembles the homologous skin graft reaction in so far as it is implemented by lymphocytes and plasma cells; classical antibodies cannot as yet be demonstrated with certainty as participants in this immunity. The acquisition of immunity requires more time to become established than in the case of the homograft reaction—usually from 3 to 18 months.

A. VISCERAL LEISHMANIASIS

The cellular reactions in the visceral leishmaniases are similar to those found in cutaneous leishmaniasis, viz. proliferation of histiocytes and secondary infiltration by round cells and plasma cells, but they are generalized throughout the viscera particularly the spleen, liver, bone marrow and lymphatic glands. The invasion of lymphatic glands is more pronounced in East African than in other visceral leishmaniases and can be exploited for diagnostic purposes by gland puncture. In the dog and in rodents, infected histiocytes are always distributed throughout the dermis where they are accessible to sandflies. Infection of the dermis is a variable factor in human visceral leishmaniasis; its epidemiological significance will be discussed in another section.

No other protozoal infection in man, dog or hamster shows a similar increase in histiocytes, round cells and plasma cells. This cellular reaction can be interpreted as part of a protective mechanism (as is obviously the case in cutaneous leishmaniasis) which in the case of man, hamster and spermophile is unsuccessful; most untreated human infections and all hamster and Macedonian spermophile infections end fatally. In contrast to infections of cutaneous leishmaniasis, cases of human and canine visceral leishmaniasis definitely show circulating antibodies demonstrable by complement fixation.

Positive results have been obtained with antigens prepared from infected tissues of hamsters, e.g. Hindle *et al.* (1926) from infected human spleens, Chang and Chang (1951), Weng *et al.* (1953).

Positive results in Indian kala azar have been recorded by Lowe and Greval (1939), Greval *et al.* (1939), Sen Gupta (1943, 1945), Sen Gupta and Adhikari (1952) using antigens prepared from tubercle bacilli or from Kedrowsky's bacillus. Similar antigens prepared from acid-fast bacilli are frequently used in Brazil for the diagnosis of human and surveys of canine kala azar and the results have been summarized by Nussenzweig (1956). Brazilian workers favour antigens prepared from tubercle bacilli, in preference to homologous antigens. Pellegrino *et al.* (1958) carried out complement fixation with an antigen prepared from *Mycobacterium butyricum* and found it satisfactory for the diagnosis of human and canine kala azar. Mayrink (1961) made a careful study of complement fixation with antigens prepared from *M. butyricum*. He found the test positive in most cases of kala azar but false positives rendered it unreliable.

One of the drawbacks of the complement fixation test for leishmaniasis is the occurrence of positive results in Chagas' disease both with homologous and heterologous antigens.

A pronounced hyperglobulinaemia with an increase in globulin/albumen ratio is a feature of visceral leishmaniasis in man and dog. Ada and Fulton (1948) found an increase in beta and gamma globulins in infected hamsters, Rossan and Stauber (1959) an increase in globulins also in infected cotton rats, chinchillas, mice and Mongolian gerbils.

In a series of experiments Wertheimer and Stein (1944) demonstrated that the increase in globulins in canine kala azar is maintained in the presence of protein hunger combined with extensive blood loss. The globulins in these circumstances are not produced at the expense of reserve proteins. It therefore appears that they are preferentially synthesized in spite of protein deficiency. It would be interesting to determine whether this also applies to specific antibodies.

Viera da Cunha *et al.* (1959) have shown that there is no quantitative correlation between serum globulins and antibodies in South American kala azar; cases with the highest serum globulin values may show a low complement fixation titre. The antibodies produced obviously have no protective value. It thus appears that the proliferating immunologically competent cells, stimulated by the parasites, secrete an excess of gamma globulin but fail to stamp the molecules with the seal of specificity.

The increased physical activity of parasitized histiocytes characteristic of all the leishmaniases is particularly striking in visceral leishmaniasis; in sections of the spleen infected cells can be seen forcing their way through the wall of the sinuses into the portal circulation.

The immunological value of the hypertrophied R.E. system and the concomitant increase in immunologically competent cells (except polymorphs) in visceral leishmaniasis is not clear; as regards the *Leishmania* it is obviously an ultimate failure. There is no blockade of the R.E. system if phagocytosis is considered an index of activity in this system. Adler (1940) found indiscriminate phagocytosis by histiocytes in a case of experimental visceral leishmaniasis in man. He also showed (1954) that hamsters infected with *L. donovani* are far more resistant to *Plasmodium berghei* than normal hamsters; their histiocytes are loaded with malaria pigment no less than those of controls. In addition to phagocytosis direct destruction of parasites *P. berghei* can be observed in the circulating blood of hamsters infected with *L. donovani.* Similar findings were made later in the case of hamsters with visceral infections produced by *L. tropica.*

Paradiso (1926) observed that malaria was very rare in cases of kala azar in children in Sicily. This observation was made at a time when malaria was very common in the island. Napier *et al.* (1933) concluded that there was an antagonism between kala azar and malaria. Sen Gupta (1944) found that in the few untreated cases of kala azar which showed malaria parasites their number was small; co-existing malaria in kala azar is usually latent and becomes patent as a result of treatment with antimonials. Henderson (1937) found malaria parasites in 30% of cases of kala azar in the Fung province of the Sudan (20% *Plasmodium falciparum* and 10% *P. vivax*). Blanc *et al.* (1959) suggested that the reported rarity of malaria in cases of kala azar may be due to defects in diagnosis. Sati (1962) described an outbreak of severe kala azar, fatal within a few weeks, in the Southern Fung in the Sudan; 42% of the cases were simultaneously infected with malaria.

It is highly improbable that the above discrepancies are due to defects or errors in diagnoses of malaria, particularly as the data in India and Sicily were based on examinations in institutions with a vast experience in laboratory diagnosis of both malaria and kala azar. The explanation is to be sought, among other factors, in differences between leishmanias characteristic for each endemic focus.

It is possible that the enormous increase in immunologically competent cells and the hyperglobulinaemia associated with visceral leishmaniasis may confer a degree of non-specific protection against some pathogens. The capacity of man and animals infected with visceral leishmaniasis to form antibodies against bacterial and other antigens remains to be investigated.

According to Napier (1946) individuals cured of kala azar by chemotherapy are immune and do not become reinfected while residing in endemic foci. He also stated that parasites may be detected in the viscera shortly after clinical cure not followed by relapses; this indicates that the acquired immunity to Indian kala azar subsequent to the destruction of a large number of parasites is an instance of premunition. Post-kala azar dermal leishmanoid may be regarded as an example of premunition in which the immunity maintained by residual parasites is sufficient to protect the host against visceral involvement without inhibiting their multiplication in restricted areas of the dermis.

Prata (1957) noted that parasites were found in eleven out of thirteen Brazilian cases of kala azar 1½–11 months after clinical cure had been established; these cases did not relapse. This is an obvious example of premunition. In cases of Mediterranean infantile visceral leishmaniasis we have observed relapse some months after apparent cure.

Adler and Tchernomoretz (1946) noted that specific treatment of canine visceral leishmaniasis is followed by temporary clinical improvement but the infection is not eliminated; infected histiocytes remain in the dermis and corneo-sclerotic junction and the visceral infection recurs. It is therefore doubtful whether an effective immunity to *L. infantum* occurs in dogs.

Kirk (1938) has suggested that the cases of espundia occasionally observed in the Sudan represent the tertiary forms of local kala azar. These cases show no symptoms typical of visceral leishmaniasis; they differ from South American espundia in so far as no scars referable to a primary infection are present. If Kirk's hypothesis is correct the immunological status of these cases is a special type of premunition in which the residual parasites are associated with protection against the manifestations of visceral leishmaniasis though they cause a serious local condition. The problem can only be solved by isolating strains of *Leishmania* from cases of espundia in the Sudan and comparing them serologically to strains isolated from local cases of visceral leishmaniasis.

B. SKIN REACTION IN VISCERAL LEISHMANIASIS

Sen Gupta and Mukherjee (1962) found that cases of active Indian kala azar do not show a Montenegro reaction when challenged with 0·05 ml of a suspension containing 5×10^6 parasites. (In the case of Oriental sore the reaction is positive in 98% of cases after inoculation of 10^5 parasites suspended in 0·1 ml phenol 0·5%.) Alencar (1958) found that Brazilian cases of kala azar also gives a negative Montenegro reaction. Few cases of Mediterranean kala azar have been tested and they were negative. Sen Gupta and Mukherjee (1962) examined seventy-three cases of post-kala azar dermal leishmanoid: twenty-six were positive and forty-seven were negative. Out of twenty-five controls twenty-four were negative and one positive.

The responses vis-à-vis the Montenegro reaction in East African kala azar are quite different from other types so far studied. Working in Kenya, Manson-Bahr *et al.* (1959) found the Montenegro test positive in three out of four cases of post-kala azar dermal leishmaniasis and in fifteen cases of active and cured kala azar. Manson-Bahr (1961) made a more detailed study of the Montenegro test in Kenya. The test was negative in sixteen cases of early kala azar but became positive 6–8 weeks after treatment. In 119 normal volunteers inoculated with a strain from a gerbil ninety-five converted a negative to a positive Montenegro after 6–8 weeks; seventy of these volunteers were examined 24 months later and none were found infected (six were challenged with a human strain and none became infected). Of those tested for the Montenegro reaction forty-seven were positive and eleven reverted to negative.

C. IMMUNITY TO HETEROLOGOUS STRAINS. IMMUNIZATION

As previously mentioned, persons living in an endemic focus who have recovered from a non-metastasising form of cutaneous leishmaniasis do not as a rule become reinfected. (They are apparently less rare in infections with *L. mexicana* than in the case of *L. tropica*.) Acquired immunity has been exploited in order to protect the face from cutaneous leishmaniasis by inoculating flagellates from cultures of L.D. bodies from hamsters into the buttock. In view of the fact that all the mammalian leishmanias have common antigens in addition to specific antigens it is interesting to determine whether closely related strains can cross-immunize.

Dr. Gunders and the writer found that two volunteers previously infected with *L. tropica* (one experimentally with a Jericho strain derived from a sandfly *P. papatasi* and one naturally infected in Baghdad) were both immune to *L. mexicana* (in press). A control volunteer became infected. This result is interesting in view of the fact that *L. tropica* and *L. mexicana* can be distinguished by serological methods. The converse experiment, i.e. inoculation of *L. tropica* into a volunteer recovered from *L. mexicana*, has not as yet been carried out. The above result suggests the desirability of determining whether *L. mexicana* or *L. tropica* can provide partial or complete protection against *L. brasiliensis*, the causative agent of espundia, by far the most serious form of leishmaniasis in man and one which is more refractory to treatment than any other. *L. brasiliensis* can easily be distinguished from *L. tropica* and *L. mexicana* serologically but shares some antigens with both. Adler and Theodor (1927) inoculated a volunteer (spontaneously recovered from an infection with *L. tropica*) with *L. brasiliensis* from a sandfly *P. papatasi* infected from cultures on one site and with cultures of the same strain on another site. The result was negative. The strain had been on culture media for 2½ years before the experiment. (The strain was isolated by Noguchi and Lindenberg on 2 March 1924 in Sao Paulo.)

Senekji (1943) injected cultures of *L. brasiliensis* into a man presumably immune to *L. tropica*; the result was negative. Unfortunately no conclusions can be drawn from the above experiments because in other experiments injections of cultures of *L. brasiliensis* into non-immunes have given negative results.

The case of *L. tropica* var. *major* (rural type of cutaneous leishmaniasis) and *L. tropica* var. *minor* (urban type) are of particular interest. These two types of leishmaniasis are clinically distinct. Russian authors (1944) originally considered that there is no cross immunity between these two types. The evidence on this point is somewhat ambiguous. Ansari and Mofidi (1950) inoculated *L. tropica* var. *major* into two volunteers who had previously recovered from *L. tropica* var. *minor*; the result was negative. Rodiakin (1957) states that *L. tropica* var. *major* protects against both types and proposes the use of this strain for vaccination against both. Kosevnikov (1957) commenting on Rodiakin's findings states: "We do not regard the problem as finally solved but the categorical statement that there is no cross-immunity should be revised". (We are indebted to Dr. C. A. Hoare, F.R.S., for drawing our attention to this work.) The writer found that *L. tropica* var. *minor* can be distinguished serologically from *L. tropica* (in press).

Manson-Bahr (1959) has carried out experiments with various *Leishmania* spp. on volunteers recovered from kala azar in Kenya. Some were inoculated with an Indian strain of *L. donovani* and others with a Mediterranean strain of *L. infantum*. In all cases the result was negative. He concluded that the Kenya strain immunizes against other forms of visceral leishmaniasis. These findings are difficult to assess; Adler (1940) found that it is not easy to infect adults by inoculation of cultures of Indian strains of *L. donovani*. Adler and Theodor (1931) failed to infect a normal adult by inoculating flagellates of *L. infantum* from an infected sandfly *P. perniciosus*. Sen Gupta (1943) inoculated Indian *L. donovani* into a patient with leukaemia and produced a local nodule but no visceral infection.

Manson-Bahr (1961) produced Oriental sore in cases recovered from kala azar in Kenya by the inoculation of material from cultures of *L. tropica*. This experiment clearly proves that *L. donovani* and *L. tropica* do not cross-immunize. He also found (1961) that volunteers inoculated with a strain of *Leishmania*, previously isolated by Heisch from a gerbil, developed a leishmanioma at the site of injection but no signs of visceral leishmaniasis followed; when challenged later by inoculation of a strain isolated from a human case of kala azar they did not become infected. Serological comparisons between the human and gerbil strain have not as yet been carried out.

Later Manson-Bahr *et al.* (1963) inoculated four volunteers with a strain of *Leishmania* found by Heisch *et al.* (1962) in a sandfly *Phlebotomus martini*; three of them were Montenegro negative with no previous history of leishmaniasis and one had been inoculated with the gerbil strain 4½ years previously. All four volunteers developed nodules from which cultures of *Leishmania* were obtained. Two of the Montenegro negatives developed kala azar, the third developed a nodule from which cultures were isolated during a period of 11 months. In the volunteer previously inoculated with the gerbil strain a nodule developed from which cultures were obtained during a period of 7 weeks, after which it faded rapidly. This volunteer did not develop kala azar. Manson-Bahr's experiments *in toto* justify his contention that immunization against East African kala azar in Kenya can be achieved by injections of cultures of the gerbil strain.

Nevertheless, vaccination of half the population of an endemic area in Kenya by injection of living cultures of a gerbil strain, the remaining half serving as a control, gave no appreciable protection; almost equal numbers of cases appeared in both groups (Manson-Bahr, personal communication). Further work is necessary to explain the striking discrepancy between laboratory experiments and field trials.

Successful vaccination in man has hitherto been achieved by the inoculation of living infective organisms, e.g. *L. tropica* in the case of Oriental sore and the gerbil strain of Heisch in Kenya in the case of visceral leishmaniasis.

Inoculation of dead flagellates into susceptible animals has not given the same uniform results. Pessoa and Barreto (1944) protected four out of six monkeys *Macacus rhesus* against *L. brasiliensis* by repeated inoculation of dead flagellates.

Coutinho (1954) inoculated fourteen guinea pigs with *L. enrietti* which had previously been injected with dead flagellates. All the animals became infected and seven died.

It would be interesting to carry out experiments on suitable animals with living non-infective strains. Adler and Zuckerman (1948) failed to protect hamsters against an infective strain of *L. tropica* by previous intrasplenic and intraperitoneal injections of a non-infective strain. The hamster is, however, not a suitable animal for experiments of this type because, in our experience, it does not develop an effective immunity to any Old World species of *Leishmania* to which it is susceptible.

VI. Epidemiology

Endemic foci of leishmaniasis are scattered over wide areas of the globe from China across Asia, India, Persia and Afghanistan, the Caucasus, the Middle and Near East, the Mediterranean basin, Portugal, Abyssinia, East and West Africa, the Sudan and China in the Old World and from Mexico to the northern part of the Argentine in the New. Throughout this wide region there is no territorial continuity between endemic foci which may be separated from each other by extensive lacunae: some like Bengal and Assam occupy a large territory, others are limited in size. Some are permanent, others transient. The climatic conditions of the various foci range from arid to tropical humid, their terrain varies from semi-desert to forest and jungle, their altitude from below sea level (Jericho) to 2,500 m above sea level, their flora and fauna vary accordingly. Regional epidemiological peculiarities associated with the above factors together with differences among species of *Leishmania* and the bionomics of local vectors, are only to be expected. In spite of these differences a number of general principles can be profitably utilized in the study of all types of foci.

The epidemiology of leishmaniasis whether visceral, cutaneous or mucocutaneous is in every case determined by a reservoir of infection (animal, human or both) from which specific vectors, i.e. one or more local Phlebotomine sandflies, infect themselves by ingesting leishmanoids from blood or infected tissues. The leishmanias go through a cycle of development in the local specific vector or vectors which transmit by depositing flagellates from a contaminated proboscis into the skin of susceptible hosts. The various leishmanias are perpetuated by cyclical transmission to and from a reservoir. The known distribution of leishmaniasis corresponds to that of sandfly vectors no less than that of malaria to Anophelines. Any other mode of transmission except by sandflies, e.g. the unique case of marital transmission described by Symmers (1960), transmission by blood transfusion (Chang *et al.*, 1948; André *et al.*, 1957) and intra-uterine infection (Low and Cook, 1926) are rarities of no epidemiological significance.

The most important advance in the elucidation of the epidemiology of the leishmaniases during the last two decades is undoubtedly the increasing recognition of the importance of animal reservoirs in most endemic foci, i.e. that leishmaniasis is a zoonosis (with the possible exception of Indian kala azar). The evidence for this is both direct, by discovery of the reservoir host or hosts in endemic foci, and indirect by an analysis of the aetiology in the various foci, e.g. the infection of human beings during transit through uninhabited territory, such as deserts in Turkistan, forests and jungle in South America, in circumstances which preclude a human source of infection. Formerly only one instance of a proved animal reservoir was known, i.e. the dog in endemic foci of visceral leishmaniasis in the Mediterranean basin, Caucasus, Turkistan, China north of the Yangtze and Brazil. The evidence for implicating the dog as reservoir was clear and can be summed up briefly. The infection in these foci occurs both in man and dog. Inoculation of parasites isolated from man into dogs produces an infection in these animals indistinguishable from the natural one. In all the above endemic foci the infection is far more common in dogs than in man. Yakimoff (1915) found 35% of dogs infected in Samarkand. In Malta the infection rate in dogs was found to be 12% in 1911 (Critien 13% in 1911, Wenyon 10% in 1914, Adler and Theodor 10% in 1931 and 1932), while the number of human cases was about 90 per annum in a population of 300,000. In Emek Jezreel in Israel there have been two human cases during the last thirty years though the infection rate in dogs was 20% during the 1930's. In Malta the infection rates in sandflies *P. perniciosus* fed on dogs approached 100% in heavily infected animals, whereas the infection rate produced by feeding on severe human cases of infantile kala azar was negligible (Adler and Theodor, 1935). Clearly the disease is disseminated by sandflies feeding on infected dogs and human cases are much less important as a source of infection. Deane (1956) has reported an infection rate of 15% in sandflies *P. longipalpis* fed on a human case in Brazil as compared to higher rates in the same species fed on dogs (24%) and local foxes (*Lycalopex velutus*, up to 100%). In Brazil human cases may therefore play a role as a source of infection though they are much less important than dogs and local foxes.

In Baghdad and Teheran cutaneous leishmaniasis was common in both man and dogs. The inoculation of a strain isolated from a dog and inoculated into man produced a lesion indistinguishable from naturally acquired Oriental sore (Adler and Theodor, 1930). The disease in Baghdad cannot, however, be considered strictly as a zoonosis, for human cases could readily serve as a source for the infection of both human and canine hosts via the vector (high infection rates occur in sandflies *Phlebotomus sergenti* fed on human lesions). Urban Oriental sore in Baghdad may be considered as an amphixenosis, a term introduced by Hoare (1962) to denote infections interchangeable between man and other vertebrates; in this case man and dog can both serve as links in the chain: "vertebrate host—sandfly—vertebrate host". Man probably plays a more important role than the dog in this respect because canine infections in Baghdad appear in October and attain their maximum incidence in the winter months when sandflies are not prevalent, whereas human cases with lesions suitable for infecting *P. sergenti* are present throughout the whole year.

Until Latyshev and Kryukova working in Central Asia (Hoare, 1944) had definitely incriminated a wild animal, i.e. gerbils, as a reservoir of one particular strain of *L. tropica*, the dog living in close association with man was the only proven reservoir for a *Leishmania* infective for man, i.e. *L. infantum*. With the realization of the importance of wild animal reservoirs, a considerable amount of work has been carried out in Central and South America, Kenya, the Sudan and Tadzikistan on the subject. This will be discussed later.

In view of the fact that research on the epidemiology of leishmaniasis will for some time be concentrated on the problem of animal reservoirs as a source for human infections, it is useful to examine the criteria essential for incriminating any particular vertebrate for this role. These are: (1) distribution of the suspected reservoir in an area where human infections are acquired; (2) frequent association between sandfly vector and the suspected reservoir incidental to the bionomics of both; (3) the occurrence of natural infections of a type sufficient to propagate the infection in the reservoir via the sandfly vector. Russian workers (1944) found all the above criteria satisfied in Turkestan where they demonstrated that the following rodents *Rhombomys opimus* (the most important), *Meriones erythorous*, *M. meridianus* and *Spermaphilopsis leptodactylus* are naturally infected with *L. tropica* var. *major*, that sandflies *Phlebotomus caucasicus* and *P. papatasi* live in the burrows of these gerbils and perpetuate the disease among them by cyclical transmission. In Tadjikistan the gerbils *Pallasiomys erythronrus* and *P. meridianus* were found to act as reservoirs. It was found that the parasite of the gerbil was infective for man and vice versa. In Persia, Ansari and Faghik (1953) noted *Rhombomys opimus* as reservoir. Human beings passing through territory occupied by gerbils acquire the infection from a sandfly previously infected on a gerbil. Human cases though they may be infective for sandflies, are not an essential link in the chain of factors which perpetuate the parasite. This chain can be broken only by destroying one of the essential links. It is interesting to note that the Russian investigators considered that in Turkistan *P. papatasi* played a more important role than *P. caucasicus* in infection of humans passing through gerbil territory because the former sandfly is more anthropophilic than the latter. In Persia *P. caucasicus* is decidedly anthropophilic.

In Indian kala azar man himself satisfies all the above criteria of a reservoir of natural infection, i.e. close association with the vector sandfly, *P. argentipes*, a high infection rate in the latter fed on human cases, and finally transmission to man by the bite of infected sandflies (Swaminath *et al.*, 1942). Man is an efficient reservoir of Indian kala azar because of two characteristics of the disease, viz. the presence of parasites in the peripheral blood in sufficient numbers to produce a high infection rate in the vector *P. argentipes* (Knowles *et al.*, 1924) and the condition known as post-kala azar dermal leishmaniasis in which localized areas of infected skin, accessible to the vector, appear after cure. (L.D. bodies can be detected in blood smears in the majority of cases in contrast to Mediterranean and East African kala azar.)

In the search for an animal reservoir the following precautions should, as far as possible, be observed:

1. A specimen inhabiting an endemic focus found susceptible to infection in the laboratory is not necessarily a reservoir, e.g. the spermophil *Citellus citellus* in Macedonia is highly susceptible to *L. infantum*. Laboratory infections are always heavy and numerous parasites are found in the peripheral blood. The Chinese hamster *Cricetus barabensis graseus* is susceptible to visceral leishmaniasis. High infection rates are found in vectors fed on individuals with laboratory infections and these animals are accordingly ideal for experiments on transmission. Neither of these species are reservoirs of leishmaniasis probably because there is no ecological association between them and the local sandfly vectors.

2. Finding sporadic natural infections in a species inhabiting an endemic focus is strong presumptive evidence but not a conclusive proof. The animal in question may be a blind alley for the parasite if the degree of infestation in the peripheral blood or skin is not sufficient to produce adequate infection rates in sandflies, e.g. most human cases in the Mediterranean basin in the case of *L. infantum*. The parasites may be numerous in tissues inaccessible to the vector, e.g. nasal cartilage and mucosa in espundia after the initial skin lesion has healed.

Information on infection rate in sandflies fed on natural and experimental infections and the distribution of flagellates in the infected sandflies is desirable in all cases of suspected vectors.

A. THE EXAMINATION OF ANIMALS FOR LEISHMANIASIS

A sandfly vector becomes infected either from the peripheral blood, e.g. *P. argentipes* fed on cases of kala azar, on a local skin lesion, e.g. *P. sergenti*, *P. papatasi* fed on an Oriental sore, or by feeding on the unbroken skin of animals in which the dermis is uniformly infected, e.g. *P. perniciosus*, *P. major*, *P. chinensis*, *P. longipalpis* fed on dogs naturally infected with visceral leishmaniasis or on hamsters with laboratory infections of visceral leishmaniasis.

An animal should be examined by smears and cultures taken from the viscera, culture of heart blood, cultures and smears from skin lesions when present, cultures and smears from dermis of the unbroken skin. The author has found the following method useful in examining the dermis in animals which showed no external signs of skin lesions. Normal skin is incised and saline injected locally through the incision into the dermis through a fine capillary; the saline is withdrawn and re-injected several times with the same capillary and then sown on suitable medium. The incision in the skin is then extended, small pieces of connective tissue adjacent to the skin removed, teased out on a slide to make a thin layer, allowed to dry and stained with Giemsa. Cultures made from the dermis reveal infections of the dermis too slight to be detected by microscopy.

Heisch (1954) has used the following method: material from spleens of animals caught in the wild is injected into hamsters. With this method he succeeded in isolating *Leishmania* from a gerbil *Tatera vicina* and a ground squirrel *Xerus rutilus*. *L. donovani*, *L. infantum*, *L. tropica* and *L. mexicana* all produce visceral leishmaniasis in hamsters. Leishmanias isolated by this method cannot be identified without serological investigation.

A slight visceral infection in an animal demonstrable by Heisch's method and negative on microscopic examination, or by culture of spleen juice such as Heisch observed in gerbils in Kenya does not *necessarily* invalidate the animal as a reservoir. Working with the strain isolated by Heisch *et al.* (1962) from a sandfly *P. martini* in Kenya, the reviewer injected baby mice intradermally. During an observation period of 96 days, cultures were frequently obtained from the dermis while cultures made from the spleen were negative after more than 2 days subsequent to inoculations. A specific vector might well become infected by feeding on an animal with an infection such as described above, although this infection might well be overlooked during an examination of the viscera both by culture and by smear. The importance of examining the dermis of suspected vectors by the above method where laboratory facilities are available, is obvious.

Hertig (1962) has recently introduced xenodiagnosis for examining animals caught in the wild. Positive results obtained by this method should be the most instructive of all.

The above brief outline of the problem of animal reservoirs may serve as a preliminary to discussing the various types of foci.

B. VISCERAL LEISHMANIASIS

There are three distinct types of foci of visceral leishmaniasis.

1. *Indian Kala Azar*

This type is unique in so far, as already stated, man himself fulfils all the conditions of an ideal reservoir of infection. Dogs have not been found infected in Bengal and Assam though hundreds have been examined at the height of major epidemics. There are many records of experimental infections of dogs and jackals both by inoculation of L.D. bodies and of cultures (e.g. Patton, 1913; Laveran, 1917). The vector, *P. argentipes*, feeds readily on man and cows in nature; it would be interesting to determine if it feeds on dogs in the laboratory.

The fact that man is an efficient carrier within the zone of distribution of *P. argentipes* does not exclude the possibility of another natural host. Attention has been focused mainly on dogs and less on the rodents as potential reservoir. The writer at one time (1947) thought on epidemiological grounds that East African visceral leishmaniasis is closer to the Mediterranean than to the Indian type but a serological examination of strains from Kenya (unpublished) indicates that the East African type is closer to the Indian *L. donovani* than to the Mediterranean *L. infantum* (from which it is readily distinguished by serological methods). In view of the accumulating evidence of a rodent reservoir in East Africa, it would seem more profitable to look for natural infection in Indian rodents rather than in canines. (*P. argentipes* has been found to feed readily on rodents in the laboratory.) Another possibility presents itself, namely, that Indian kala azar originally derived from an animal source from which it eventually became independent; this would imply that Indian kala azar according to Pavlovsky's conception is an anthroponosis, i.e. an infection restricted to man but originally derived from lower animals (Hoare, 1962). At present man himself is an excellent reservoir.

All age groups are susceptible but the age groups 10–20 years provide almost 60% of the total number of cases. The periodic flare up of the disease to epidemic proportions involving large populations has been characteristic of Indian kala azar. There is a continuity in the progress of the disease during epidemics, i.e. house to house and quarter to quarter. This is in keeping with transmission by a sandfly *P. argentipes* which breeds in close proximity to human habitations, has a short range of flight and easy access to a reservoir of infection, on which it feeds readily.

2. *Visceral Leishmaniasis Associated with a Canine Reservoir*

This epidemiological type has a wide but discontinuous distribution. It includes the whole Mediterranean basin, Portugal, the Atlantic coast of North Africa in the west, the Caucasus, Central Asia and China north of the Yangtze in the Far East and parts of South America, particularly Brazil. Recently Taj el Deen and Aloasi (1954) have found the disease in Iraq. In the Mediterranean basin 80% of all cases occur in children under five and approximately 94% in children under ten. The relative rarity of the disease in adults is not, in the opinion of the reviewer, due to immunity acquired by exposure to the disease in infancy, for the number of cases occurring in adults visiting endemic centres is not considerable. Taub (1956) noted that the serum of human adults contains a factor (a euglobulin) which destroys *L. infantum* and that this factor is absent in children up to the age of six. This finding tallies with the age distribution of the disease in endemic foci. Maggiore (1923) failed to infect adults by the inoculation of infected bone marrow and Adler and Theodor (1931) failed to infect an adult volunteer by the inoculation of *L. infantum* from an experimentally infected sandfly, *P. perni-*

ciosus, though susceptible animals were readily infected by this method. The annual incidence of the disease was fairly constant in endemic foci in the Mediterranean basin. The number of cases has diminished since the introduction of modern insecticides. Prior to this period there appeared to be a fairly stable equilibrium between reservoir, human cases and vector in endemic foci. In striking contrast to Indian visceral leishmaniasis, large scale epidemics were unknown (small outbreaks in what were regarded as new foci in the Mediterranean region did not involve more than twenty to thirty cases).

In Turkistan and the Caucasus the epidemiology resembles that found in the Mediterranean basin and differences such as the relative incidence of human and canine infections can be ascribed to differences in the bionomics of the local vectors. In parts of Asiatic Russia man becomes an incidental host if he invades uninhabited territory occupied by a reservoir of infection and a vector. This occurred in Tadzhikistan in 1950 and 1951 where outbreaks occurred in workmen opening up new territories. Symptomless infections were found in wild jackals. The symptomatology of the disease in dogs in the Mediterranean basin is variable (Adler, 1936); the common symptoms are seborrhea due to infiltration round sebaceous glands, partial depilation, ulceration and emaciation, but an infected animal may be symptomless. Severe cases improve clinically if they are kept on a diet of fresh meat without any other treatment (Adler *et al.*, 1938). It is therefore probable that *L. infantum* was less pathogenic for the wild dog living as a hunting animal than for his domesticated descendants. The epidemiology of visceral leishmaniasis in South America resembles that of other foci where canines are reservoirs but has several interesting peculiarities of its own, related to the distribution of the vector *P. longipalpis*. Although a few autochthonous cases had been previously recorded by Migone (1913) and Mazza and Cornejo (1926), it was only after Penna (1934) had recorded forty-seven instances of visceral leishmaniasis diagnosed *post mortem* during the examination of liver sections from 47,000 viscerotomies performed for the purpose of yellow fever investigations, that the importance of the disease in South America was realized. Subsequently extensive epidemiological studies initiated by the late Evandro Chagas have been carried out by Brazilian scientists who mapped out the distribution of the disease in Brazil and showed that it was mainly rural and sylvestran. Chagas *et al.* (1938) found that *P. longipalpis* was abundant in foci of the disease in man and dog. Chagas (1939, 1940) infected *P. intermedius* and *P. longipalpis* in the laboratory, Deane (1956) working in Ceará demonstrated L.D. bodies both in the dermis of dogs and the local fox *Lycalopex vetulus* naturally infected with visceral leishmaniasis. Deane (1956) infected *P. longipalpis* by feeding on naturally infected humans, dogs and foxes (raposas) *Lycalopex vetulus*. The course of infection in *P. longipalpis* was similar to that recorded for *L. tropica* in

P. papatasi, *L. donovani* in *P. argentipes* and *L. infantum* in *P. perniciosus*. Histological examinations by Deane (1956) of skin biopsies from natural infections of visceral leishmaniasis gave the following very significant result: parasites were detected in the dermis of 5 out of 27 human cases, in 38 out of 49 dogs and in 3 out of 4 foxes. These findings indicate that some human cases can serve as a source of infection for *P. longipalpis* and that the disease in Brazil is to some extent an amphixenosis (between man and dog) although man, as previously stated, plays a much smaller role than the dog or fox in perpetuating the parasite. The distribution of the disease among different age groups somewhat resembles that of Mediterranean kala azar but there is a tendency to attack children of a somewhat higher age group than in Mediterranean countries. Thus Alencar (1958) analysing more than a thousand cases found 60% were in children up to the age of five and 80% up to the age of ten. Deane (1956) emphasized the fact that visceral leishmaniasis in Brazil occurs only where *P. longipalpis* is found in association with human habitations. De Leon and Figuerra (1958) found the same relationships in Guatemala. The distribution of *P. longipalpis* in South America is not known completely, but it is already clear that the area of distribution is not continuous but interrupted by extensive territorial lacunae in which the insect has not been found. *P. longipalpis* has recently been found in Venezuela and Hertig (1962) has recorded a single specimen from Panama. Sandflies have a limited range of flight and do not tend to occupy new territories. Their present distribution may be explained by the interesting suggestion of Professor Oliviero Mendes da Costa of the Oswaldo Cruz Institute that *P. longipalpis* is an old species and once occupied a large area which became broken up into territories of various size as a result of climatic and geological episodes (personal communication). It may be noted that *P. longipalpis*, which is a domestic species in Ceará in North West Brazil, is common in Lapinha near Belo Horizonte where it is restricted to caves. Mr. Alberto Falcao of the Centro de Pesquisas of Belo Horizonte informs me that in this locality *P. longipalpis* has never been found in human habitations. Human cases of visceral leishmaniasis are rare or absent in this area. Although the sandfly does not enter human dwellings in the neighbourhood of Lapinha, it feeds on man in the laboratory (also on birds, dogs and rodents). No morphological differences were found between specimens of *P. longipalpis* from Ceará and from Lapinha. (It would be interesting to compare infection rates after feeding on naturally infected dogs in *P. longipalpis* from Lapinha to those from Ceara where this sandfly has become domestic. The difference in bionomics between populations of the same species in two areas is most striking.)

3. *Epidemiology of Visceral Leishmaniasis in East Africa*

The epidemiology of the disease in these two territories presents unique features differing from those of the Indian focus of kala azar, and from all foci (including North and North West Africa) with a proved canine reservoir. In spite of these differences the general epidemiological principles previously enunciated are fully applicable.

Before World War II visceral leishmaniasis was rare in Kenya. Since the outbreak of World War II an increasing number of cases have been reported. Fendall (1950) reported fifteen cases from the Kitui district which quickly became an important endemic and epidemic focus. Heisch (1954) reported on an epidemic of 3,000 cases in Kenya. Southgate and Oriedo (1962) pointed out that the disease is still spreading in the Kitui district of Kenya.

Kala azar has been known in the Sudan since Neave (1906) described the first case, but for some years the disease was considered not to be an immediate major problem though a possible menace for the future; it did in fact become a major problem in the form of recent fulminating epidemics.

Both in Kenya and the Sudan occasional outbreaks were noted among *personnel passing through uninhabited territory*. A typical instance was recorded by Cole *et al.* (1942) and Cole (1944), viz. an outbreak of kala azar (sixty cases) in a battalion of the King's African Rifles operating on the borders of northern Kenya, Abyssinia and the Sudan. This finding immediately calls to mind Pavlovsky's conception of natural nidi, i.e. "sections of territory with a definite geographical landscape the biotopes of which form a definite biocenosis". Within this territory parasites (some of which may incidentally be infective for man) are perpetuated among local animals by arthropod vectors. Man is liable to become infected if he encroaches on this territory. This situation raises a series of epidemiological problems relevant to the Sudan and Kenya and, as will be indicated later, to Central and South America. Having become infected by encroaching on, or passing through, a nidus (*sensu* Pavlovsky), can man serve as a reservoir from which the disease can be introduced in settled or nomadic human populations? The issue depends on the type of infection produced in man, i.e. infection of the dermis or circulating blood sufficient to make leishmanoids available to sandflies (as is obviously the case in Kenya, where Minter *et al.* (1962) infected twenty-eight out of fifty-eight sandflies *P. martini* by feeding them on cases of kala azar), and on the presence in or near human habitations of a suitable vector possibly, *but not necessarily*, the same species which propagates the parasite in the nidus. (It is important to realize that some species of *Phlebotomus* contain populations which are domestic and populations which are feral, e.g. *P. papatasi*, *P. caucasicus*, *P. longipalpis*. Furthermore a *Leishmania* sp. may be transmitted by more than one vector, e.g. *L. tropica* by *P. sergenti*, *P. papatasi*, *P. caucasicus*, *L. infantum* by *P. perniciosus* and *P. major*.)

The history of kala azar in Kenya and the Sudan is in full accord with the above conception. It resembles the Indian rather than the Mediterranean form in its age distribution. Henderson (1937) found that 66% of the cases occurred in adults. Sati (1962) states that the disease attacks all age groups but is commoner in adults than in children. As in the case of Indian kala azar, naturally infected dogs have not been found in endemic foci.

The two striking epidemiological contrasts to Indian kala azar are the discontinuity in the distribution of cases and the appearance of new foci and/or the recrudescence of old ones, both in Kenya and in the Sudan. Thus Garnham (1954) in discussing the outbreak in Kitui (Kenya) thought it most unlikely that kala azar had previously existed in this district for a long time. Henderson (1937) in discussing kala azar in villages along the Blue Nile thought that a severe transient epidemic had occurred fifty years previously. Outbreaks may occur in foci old or new separated by considerable distances and intervals of time. Thus the epidemic described by Henderson (1937) occurred in the province along the Blue Nile and Stephenson (1940) reported another epidemic along the White Nile.

Kala azar in the Sudan and Kenya is not static, so much so that it is impossible to predict the course of events particularly with respect to invasion of new territories and the establishment of new foci. This poses the problem of possible outbreaks of kala azar in African territories like Chad where cases of kala azar have been found but the disease is still as rare as it was in Kenya before World War II. The situation will remain obscure till more knowledge accumulates on the ecology of rodents and the distribution and bionomics of sandflies in East and Central Africa.

The occurrence of an animal reservoir in East Africa was suggested by Henderson (1937), the late Dr. R. Kirk (1942) who made fundamental contributions both practical and theoretical to the study of leishmaniasis in the Sudan, Kirk and Lewis (1955), Heisch (1958), and Hoare (1954). By 1941 only a single instance of a natural infection of visceral leishmaniasis in an animal was known in the Sudan, i.e. in a monkey *Cercopithecus ethiops*. Kirk (1956) found one fox *Vulpes pallida* in the Blue Nile endemic area with an ulcer on the face but there was no certainty that it contained *Leishmania*. Since then Heisch, working in Kenya (1957, 1958), has isolated *Leishmania* sp. from the gerbil *Tatera vicina*, from a ground squirrel *Xerus rutilus*, and *L. adleri* from a lizard *Latastia caudata*.

As already indicated, slight differences occur in the pathological and clinical picture between various foci of visceral leishmaniasis. These differences have epidemiological implications in so far as they concern infections in areas of skin accessible to sandflies: they do not apply to individual cases, but they are of statistical significance. They involve: (a) local infection of the skin in the early stages of infection before the advent of symptoms associated with visceral infection; (b) skin and blood infection during manifestation of visceral leishmaniasis and of post-kala azar dermal leishmaniasis.

Having failed to find an animal reservoir Kirk (1956) concluded that post-kala azar dermal conditions would go far to explain the epidemiology of visceral leishmaniasis in the Sudan. He added: "This conclusion greatly diminishes the strength of the arguments for the existence of an animal reservoir host, as it provides an alternative explanation for most of the observations on which they are based". Infections of the dermis in any stage of visceral leishmaniasis are irrelevant to the problem of an animal reservoir but they are, as Kirk emphasized, of prime epidemiological importance. This relative epidemiological significance in other endemic areas has already been indicated; they are of special interest in East African kala azar and will be discussed briefly.

Napier and Krishnan (1931) suggested that primary skin lesions, papulae too small to attract clinical notice, developed at the site of inoculation by sandflies prior to the active phase of Indian kala azar. Kirk (1938) observed one and Kirk and Sati (1940) a number of cases of leishmanioma preceding the febrile stage of kala azar in the Sudan. Kirk (1942) suggested that a leishmanioma is the primary stage, the febrile period the secondary and the occasional cases of espundia the tertiary stage of Sudan kala azar. The work of Manson-Bahr (1959, 1961, 1963) corroborated Kirk's hypothesis and observations on the primary leishmanioma. This author inoculated four volunteers with *Leishmania* originally derived from a wild sandfly *P. martini* which Minter *et al.* (1962) found infected with leptomonads and inoculated into a hamster which developed a visceral infection. (Dr. Heisch sent this strain to the author who found it to be serologically indistinguishable from a human strain isolated in Kenya.) Three of the volunteers developed leishmaniomas on the site of the inoculation. Kirk and Sati (1940) and Cole (1944) recorded skin eruptions containing *Leishmania* which appeared during treatment. These eruptions may occasionally continue for some time after cure but they disappear spontaneously within 6 months of their first appearance. Manson-Bahr (1959) found by biopsies that eleven out of fifty-four Kenya cases had infections of the dermis easily detected by histological examination. As already stated Minter *et al.* infected sandflies by feeding on cases of kala azar in Kenya. It is not clear whether the insects were infected from the blood or dermis or both but circulating blood was regarded as the source of infection.

Marozian (1941) recorded the presence of a few papules initially pin-head in size but growing to the size of lentils during the prefebrile stage of visceral leishmaniasis in Asiatic Russia; they disappear within 4 months and obviously have not the epidemiological significance of the leishmaniomas during the prefebrile period and the skin infections during the febrile period of East African kala azar.

Indian kala azar is almost unique in so far as the post-kala azar dermal leishmanoid appears 1–2 or more years after cure in 10% of cases. These lesions contain numerous L.D. bodies in the dermis and are readily accessible to sandflies and are therefore of prime epidemiological importance. The condition is not common in East Africa and rare in China from where Chen *et al.* (1953) reported two typical cases one of which appeared 3 years after cure of the visceral manifestations.

C. CENTRAL AND SOUTH AMERICAN CUTANEOUS AND MUCO-CUTANEOUS LEISHMANIASIS

The epidemiology of cutaneous and muco-cutaneous leishmaniasis in Central and South America will tax the ingenuity and industry of investigators for a considerable time to come. Foci are scattered over an area extending from approximately 22° north to 30° south of the equator. The full extent of the distribution within this vast area is not known. Biagi *et al.* (1957) writing on the disease in Mexico states that the probable area of distribution is much greater than the known. From time to time new foci of the disease are discovered, e.g. Forratini *et al.* (1959) discovered a new focus in the Amapa region north of the delta of the Amazon and new foci will no doubt be found. The disease is acquired in almost every conceivable terrain ranging from the Cordilleras up to an altitude of 2,500 m to the swamps and forests of Amazonia. By and large it is a non-urban disease commonest in men who work in forests or the borders of forests and jungles and in rural settlements. It can be confidently predicted that in the future as in the past outbreaks will occur in groups of men engaged in opening up new territories.

The leishmanias of mammals have speciated more in the New than in the Old World and each species may well have its own vectors, reservoir and spectrum of host infectivity. Speciation in the genus *Phlebotomus* has also been more extensive and more complex than in the Old World (French Guiana alone boasts of sixty-two species). The bionomics of forest species among which vectors are to be expected will not be easy to elucidate. A few important species appear to have no particular food preferences, e.g. *P. longipalpis*, the vector of kala azar, feeds readily on birds, rodents, equines, dogs and man.

The leishmaniases of the New World being obviously zoonoses, the epidemiology can be formulated in terms of reservoir–vector–man. This simple formulation does not make the subject less difficult. It is relatively easy to find a reservoir among domestic animals when such exists, as the dog for South American and Mediterranean kala azar, but the problem is more complex in the cutaneous and muco-cutaneous leishmaniases of South America. In the latter the search among animals associated with human habitations has not given impressive results. Mazza (1926) found a few infected dogs in the Argentine. Mello (1940) found an infected cat near Belem in Brazil; in this case *Leishmania* was demonstrated by culture. Dr. Alencar of Fortelezza showed the reviewer a case of cutaneous leishmaniasis in a donkey; the diagnosis was confirmed by demonstration of the parasites.

Alencar *et al.* (1960) recorded a *Leishmania* which they thought was probably *L. brasiliensis* in a rat. In any case the presence of rare natural infections in domestic animals does not solve the problem of sylvestran reservoirs.

Shattuck writing in 1938, impressed by the failure to find a reservoir, concluded that "in the present state of knowledge it cannot confidently be stated that any animal, insect or plant serves as a reservoir for the leishmaniae of man". Nevertheless the epidemiological evidence for an animal reservoir was sufficiently convincing to stimulate further search among the rich fauna of South America though it might appear, in the light of previous disappointments, tantamount to seeking for a needle in a haystack. Hertig *et al.* (1954) opened up a new chapter by finding natural infections in two South American rodents, the spiny rat *Proechymis semispinosus* and *Hoplomys gymnurus* in Panama. The infections were demonstrated by culture of heart's blood but no parasites were found in the viscera; 10·5% of trapped spiny rats were found infected. Inoculation of cultures into clean spiny rats gave negative results although cultures were infective for a volunteer and one strain after 4½ years in culture was infective for a hamster. The failure to transmit the infection to spiny rats in the laboratory is one of a number of surprises encountered by workers on leishmaniasis. Furthermore the infection apparently disappeared locally from spiny rats; the natural infections recorded by the authors may have been a transient episode which could easily have misled less cautious observers. It would be interesting to determine whether the connective tissue in infections of this type contains parasites, as is the case in mice infected with the Kenya strain of *Leishmania* isolated from *P. martini* (1964). In leishmaniasis tegumentaria diffusa, parasites of man can be cultured from the circulating blood, the viscera are not parasitized and parasites are numerous in the dermis.

Forratini (1960) recorded natural infections in three rodents, *Kannabateomys ambyonyx*, *Cuniculus paca* and the agouti *Dasyprocta cyarae*; the latter two species had previously been suspected. (Some thirty species of sandflies have been found associated with *Cuniculus paca*; this illustrates the complexity of the problem in South America.)

Lainson and Strangways-Dixon (1962) found natural infections in the following rodents in British Honduras: (1) the tree rat *Ototylomnis* sp. (6 out of 13 specimens examined); (2) the white footed mouse *Nyctomis sumichrasti* (1 out of 7 specimens examined); (3) the spiny pocket mouse *Heteromys* sp. (3 out of 44 specimens examined). In all cases the infection was restricted to local lesions in the tail. The lesions contained leishmanoids. Cultures inoculated into hamsters and mice produced lesions indistinguishable from those produced by *L. mexicana*. Cultures made from an infected *Nyctomis sumichrasti* were inoculated into a volunteer and produced a characteristic lesion containing numerous leishmanoids.

This important discovery encourages the hope that reservoirs will be found in other forest regions of South America.

D. TRANSMISSION

Brief allusions to sandfly vectors were made in the preceding section. The behaviour of *Leishmania* sp. in sandflies and the bionomics of the latter are obviously inseparable from the epidemiology of leishmaniasis, but in view of the complexity of the subject and the problems raised by recent work, particularly that carried out in the Gorgas Memorial Laboratory (1961, 1962), it is convenient to treat them in a separate section.

Vectors of leishmaniasis have been indicated in the past on the basis of a local distribution corresponding to that of the disease (the classical example is Sinton's demonstration that the distribution of *P. sergenti* corresponds to that of Oriental sore and *P. argentipes* to that of kala azar in India), on the discovery of sandflies with natural infections of a *Leishmania* infective for man (e.g. Adler and Theodor, 1925), on infection of sandflies by feeding on human cases or experimental animals, and finally on a few occasions by transmission by the bites of infected sandflies.

A sandfly survey is an essential part of every epidemiological investigation. The survey should include human habitations, stables, barns, tree holes, rodent burrows in the vicinity and any natural structures typical for the environment. Heisch (1954) found that in Kenya sandflies occur in the ventilation shafts of termite hills, where they have ample opportunities for feeding on small mammals and lizards which reside there either temporarily or permanently. Heisch *et al.* (1956) captured specimens of *P. martini* in the vicinity of termitaries. In South American forests, this calls for methods of collecting sandflies at various elevations from ground level to the forest canopy. Human and animal baits are a valuable help in capturing sandflies; they also facilitate study of food preferences. Collection should be carried out both during the day and after dark throughout the whole sandfly season. In addition to the examination of preserved material, dissections of fresh material should be made in a search for natural infections. During dissection particular attention should be paid to the accessory glands of females with an empty alimentary tract. The presence of granules in the accessory glands indicates that the insect had digested a blood meal, laid eggs and was ready for another meal; their absence indicates a female which had not yet had a blood meal. In Lapinha near Belo Horizonte, Dr. Mayrink Wilson and the writer found that 60% of wild female *P. longipalpis* with an empty alimentary tract had laid eggs and were ready for renewed feeding.

A judicious appreciation of the systematic affinities of local species in the light of knowledge acquired in other foci may give a useful clue as to local vectors, e.g. Kirk and Lewis (1951, 1955) on the basis of a sandfly survey in the Sudan suggested *P. orientalis*, *P. martini* and *P. lesleyae* as possible vectors—a suggestion which was justified by subsequent work in the case of *P. orientalis* and *P. martini*. The association of *Leishmania* sp. and their known vectors in other foci may be of help in indicating that the known sandfly fauna does not contain vectors and in predicting and searching for the presence of other species. Thus Adler (1936) predicted the presence of a sandfly of the major group in the Sudan before the discovery of *Phlebotomus orientalis*. A critical estimate of a local collection is essential for selecting the probable vector or concluding that the vector has not yet been encountered. After making a sandfly survey in a focus of visceral leishmaniasis in Iraq Pringle (1956) suspected that *P. papatasi* was the local carrier. To the writer it seems improbable that *P. papatasi* transmits kala azar. The most likely carrier is a species of the major group (Pringle has recorded three such species from Iraq: *P. wenyoni*, *P. kandelaki*, and *P. chinensis*). Latyshev *et al.* recorded *P. papatasi*, *P. sergenti*, *P. caucasicus* and *P. chinensis* from Tadzjikistan and considered them all as vectors of cutaneous leishmaniasis; while there is evidence for the first three as vectors on the basis of experimental infections in other foci, there is no evidence from other sources for *P. chinensis* as a vector of cutaneous leishmaniasis. *P. sergenti* and *P. caucasicus* have been rightly incriminated as vectors of cutaneous leishmaniasis in Persia, the latter species because it lives in burrows of gerbils *Rhombomys opimus* among which Ansari and Faghik (1953) found a high infection rate with *L. tropica*. (Both *P. sergenti* and *P. caucasicus* have been infected experimentally.)

Larivière *et al.* (1961) equated the distribution of Oriental sore and *P. dubosci* in West Africa by careful mapping. In five West African foci the distribution of *P. papatasi* var. *bergeroti* corresponded to that of Oriental sore. Gherman *et al.* (1957) indicated *P. perfilewi* as vector of a small outbreak (24 cases) of kala azar in the Crajova region of Romania, Marashvilli (1961a, b), indicated *P. kandelaki* and *P. chinensis* as vectors in Georgia on the basis of distribution. Similarly Rés (1957) found that the distribution of kala azar in Portugal coincided with that of *P. perniciosus*. Examples from the Old World can be multiplied.

It is much more difficult to equate distribution of leishmaniasis with that of known species of sandflies in South America because of the large numbers of species found within a single focus. Nevertheless Deane (1956) showed that *P. longipalpis* is present in every focus of visceral leishmaniasis in South America. Pessoa and Barreto (1944) record a distinct correlation between the distribution of cutaneous leishmaniasis and *P. whitmani*, *P. pessoai* and *P. migo nei* in the state of Sao Paulo. Floch (1954) is of the opinion that the distribution of *P. anduzei* corresponds to that of leishmaniasis in French Guiana. The distribution of uta in Peru coincides with that of *P. noguchi*. Rodrigues and Avilés (1953) list the species possibly associated with leishmaniasis in Ecuador as: *P. dysponatus*, *P. camposi*, *P. leopoldoi*, *P. apicalis*, *P. gomezi*, *P. shannoni*. This brief account illustrates the difficulty of incriminating a particular species as a vector in South America on the basis of distribution apart from the particular case of *P. longipalpis* and kala azar where the evidence is satisfactory.

A species incriminated on justifiable evidence as a vector may have a more extensive distribution than that of the disease, e.g. *P. anduzei* is considered by Floch to be the vector of cutaneous leishmaniasis in French Guiana. This sandfly is common in Cayenne where the disease is not known to occur. A possible explanation of this phenomenon is the absence of a reservoir in Cayenne. A similar state of affairs exists in Italy with regard to *P. perfilewi* the suspected vector of Oriental sore in the Abruzzi region which has a wider distribution than the disease in man. It may be taken as a general rule that whenever a vector associates with man and has a topographically wider range than human infections, an animal reservoir with a distribution more restricted than that of the vector is indicated (Adler, 1962a, b).

Natural infections with leptomonads have been recorded in *P. papatasi*, *P. sergenti*, *P. major*, *P. arpaklensis*, *P. clydei*, *P. argentipes*, *P. martini*, *P. migonei*, *P. anduzei*, *P. panamensis*, *P. longipalpis*, *P. whitmani*, *P. pessoai*, *P. intermedius*, *P. ovallesi* and *P. cruciatus*, The natural infections recorded by workers at the Gorgas Memorial Institute will be dealt with separately. A leptomonad found in a sandfly can be confidently accepted as a stage in the life history of a vertebrate parasite but specific diagnosis is possible only by direct inoculation experiments in animals or by isolation in cultures which can be used for animal experiments and serological investigations.

In describing natural (and experimental) infections it is essential to note the distribution of parasites in the alimentary tract, i.e. whether the infection is anterior or in the hindgut or both.

A survey should obviously be accompanied by or followed by experimental work, i.e. feeding wild or laboratory bred sandflies on naturally or experimentally infected man and animals, or on cultures, and re-feeding on healthy susceptible animals or man. It is advantageous to use wild females with an empty alimentary tract wherever these are available in sufficient numbers; they usually feed more readily and the results are as valid as those obtained with laboratory bred ones. In some cases, e.g. *P. major* the vector of visceral leishmaniasis in Greece, it has not yet been possible to reproduce the physiological status of wild sandflies in laboratory bred ones. Nevertheless work should also be done with laboratory bred sandflies; transmission of kala azar was achieved by Swaminath *et al.* (1942) with laboratory bred sandflies *P. argentipes* and of *L. tropica* by Adler and Ber (1941) with laboratory bred *P. papatasi.* It is moreover possible to breed sandflies on a large scale and thus be independent of seasonal fluctuations.

Transmission experiments have been carried out by collecting wild sandflies, triturating them and inoculating material thus obtained into laboratory animals (e.g. Strangways-Dixon and Lainson, 1962). This procedure should be avoided when possible. At most it merely proves that one of a number of species harbours a *Leishmania*; the identity of the vector, the distribution of flagellates in the alimentary tract and the mode of transmission are not determined and much valuable information is lost. The preliminary work necessary to initiate transmission experiments is very considerable and it is a pity to miss full benefit from it by short cuts which inevitably fail to give essential data. With a little practice a worker can easily dissect and examine 50–100 sandflies a day. Direct inoculation experiments should be attempted only with leptomonads from sandflies specifically identified.

Sandflies are usually infected in the laboratory by feeding on laboratory animals with a heavy infection or by feeding on rich cultures. Under these circumstances very high infection rates (approaching 100%) are produced in vectors, but non-vectors may also be infected although the infection rate in the latter will be low. The results must be assessed critically. The factors which determine the behaviour of a *Leishmania* sp. in a sandfly are to some extent quantitative and a non-vector may be infected by a large dose of parasites. Not every species which has been infected in the laboratory is necessarily a vector, e.g. *P. mongolensis* is listed as one of the vectors of kala azar in China (e.g. Heyneman, 1961), because Young and Hertig (1926) infected 16 out of 661 sandflies by feeding on Chinese hamsters in contrast to 195 out of 384 sandflies *P. chinensis*. *P. mongolensis* should not be considered a vector of kala azar in China.

The behaviour of *Leishmania* in sandflies and its relevance to the problem of transmission was discussed by Adler (1956) and reviewed by Adler and Theodor (1957). The reader is referred to these two publications. The following discussion will be confined mainly to work carried out since 1957; previous work will be mentioned and analysed only in so far as it has a bearing on the problem as it stands at present.

It has been accepted on experimental grounds that leptomonads of mammalian *Leishmania* sp. adopt an anterior position and that few parasites are sufficient to produce a high infection rate in the specific vectors. This generalization still stands for the Old World mammalian *Leishmania* sp. and for visceral leishmaniasis in South America. It was also accepted that lizard leishmanias adopted a posterior position in sandflies and that a natural infection with a posterior station in a sandfly indicated that the vertebrate host was a reptile. David (1929) found that *L. agamae* adopts a posterior position in *P. papatasi*, and Adler and Theodor (1929) noted that *L. ceramodactyli* also adopts a posterior position in the same sandfly. In the case of *L. tarentolae* Adler and Theodor (1935) recorded an anterior position in *P. minutus* while Parrot (1935) recorded a posterior position in *P. minutus parroti*. Heisch (1955, 1958), working in Kenya, found natural infections in *P. clydei*; the flagellates adopted an anterior position in naturally infected sandflies and Heisch was inclined to think that *P. clydei* might be a carrier of kala azar in

Kenya but later concluded that the flagellates found in this sandfly were stages in the life cycle of a lizard *Leishmania* probably *L. adleri* (*P. clydei* feeds both on lizards and on man). It therefore became clear that the position adopted by leptomonads in a sandfly is no certain clue to their vertebrate host, though all the mammalian leishmanias were expected to adopt an anterior position.

The work carried out at the Gorgas Memorial Institute of Tropical and Preventive Medicine, Incorporated, has opened up the whole subject of distribution of leptomonad stages of *Leishmania* sp. in sandflies anew. The salient facts garnered from the reports of the above Institute (1962, 1963) and the note of Johnson *et al.* (1962) are as follows: "Natural infections of wild caught Panamanian *Phlebotomus* sandflies with leptomonad flagellates, consistent morphologically with *Leishmania*, were first found in January 1961, although dissections for this purpose had been made from time to time during the previous year" (quoted from Annual Report of 1962). Natural infections were found in six species: *P. trapidoi*, *P. ylephileptor*, *P. gomezi*, *P. sanguinarius*, *P. panamensis* and *P. shannoni*. Out of 3,112 females dissected 198 were found infected (infection rate of 6·4%). The infection rate increased with the onset of the rainy season, in the case of *P. trapidoi* from 5·8% to 16·5% and in the case of *P. ylephileptor* from 5·1% to 15%.

"*The great majority of the infections in our Panamanian sandflies were in the hindgut with few or no flagellates in the midgut*" (quoted literally from report of 1962). There were, however, two heavy infections of the midgut, one in *P. shannoni* the other in *P. trapidoi*, "In the latter the flagellates were not only in the mid- and hindgut but virtually filled the pharynx" (quoted from report of 1962). Pure cultures were recovered from naturally infected sandflies and one strain isolated from *P. trapidoi* produced a typical lesion in a hamster. The cultures made from sandflies indicate that more than one species of *Leishmania* is involved. (This opinion is based on morphological evidence. It would be interesting to compare the strains serologically.) A total of 244 sandflies fed on infected hamsters. Eighty-eight became infected: *P. sanguineus*, 55 out of 146; *P. gomezi*, 33 out of 98. "Flagellates of a Guatemalan strain of *Leishmania* tend to assume an anterior position in the gut" (quoted from report of 1962). "With the Panamanian and Peruvian strains there is a tendency for the flagellates to assume a more posterior position" (quoted from report 1962). The behaviour of a given strain is the same in both, *P. gomezi* and *P. sanguineus*. In the 1963 report the presence of *P. longipalpis* in Panama (taken at a height of 2,000 feet in Cerro Campo) is reported. One female only was caught and found infected. The distribution of the flagellates in this specimen is not recorded.

It should be noted that *P. longipalpis* has been infected with the Brazilian strain of kala azar (Deane, 1956) and with *L. mexicana* (Coelho and Falcao, 1962); both species adopt an anterior position in this sandfly.

The workers of the Gorgas Memorial Laboratory could not have achieved their results had they not developed methods of capturing forest sandflies at all elevations from the ground to the forest canopy. They would also have missed fundamental biological data had they contented themselves with triturating masses of sandflies instead of dissecting and studying each specimen individually.

One of the striking features of the work in Panama is the high infection rate in wild sandflies which has not been equalled in intense foci of kala azar in India or in foci of Oriental sore such as Jericho and Baghdad. The only endemic focus approaching Panama in infection rates in wild sandflies is Venezuela where Pifano (1958) found 6·94% *P. panamensis* infected with leptomonads. On the basis of their findings the workers at the Gorgas Memorial Laboratory suspect an arboreal reservoir. Their findings show the complexity and difficulties of the problem of South American cutaneous and muco-cutaneous leishmaniasis. It is not merely a question of feeding and re-feeding sandflies and succeeding in transmission in one or more experiments but of unravelling a whole series of ecological problems involving the association of animal reservoirs (still to be discovered) and sandflies and the biology of individual species of *Leishmania* and sandflies.

The frequency of the posterior station in sandflies in Panama is most remarkable. Struck by the above results we re-examined the data published by Adler and Theodor (1935) on feeding experiments with *P. perniciosus* on animals infected with *L. infantum*. It was noted that hindgut infections are comparatively rare; in 30% of sandflies *P. perniciosus* infected by feeding on animals with a Maltese strain of *L. infantum* the infection was confined to the stomach or the stomach and lower part of the cardia, i.e. it was not anterior. Furthermore the distribution of the flagellates in the sandfly did not change appreciably after 4 days, i.e. there was the same percentage of stomach infection without forward progression after 10 days as after 4 days. The condition in the hindgut was not recorded. No particular conclusions were drawn from these observations because it was considered that the 70% anterior infections were sufficient to account for transmission by bite. When *P. perniciosus* fed on a Sicilian strain of *L. infantum* the flagellates almost uniformly adopted an anterior position. There was no difference in the results obtained with sandflies *P. perniciosus* from Malta and those from Catania (Sicily). The above work was carried out with local strains of *L. infantum* and the local carrier, *P. perniciosus*. In the case of infections of *L. tropica* in both *P. papatasi* and *P. sergenti* we invariably found flagellates in an "anterior position" after 3 days; only one strain of *L. tropica* after 22 years on culture media, long after it had lost its infectivity for mice and hamster (although it retained infectivity for man), produced mainly stomach infections in *P. papatasi* (Adler and Zuckerman, 1948). Shortly after its isolation this strain behaved like a typical *L. tropica* and produced anterior infections in sandflies *P. papatasi*.

The findings in the Gorgas Memorial Laboratory were not entirely uniform. The majority of natural infections were in the hindgut but there were two instances of heavy infections in the midgut; one in *P. shannoni* and one in *P. trapidoi.* The variations in behaviour of a single strain of *Leishmania* in the same sandfly and the differences between strains in the same sandfly may suggest that the pattern of behaviour of a *Leishmania* species in sandflies is determined by genetic factors present in both.

Inevitably one asks: "What, if any, is the role of the large numbers of hindgut infections in Panamanian sandflies in the life history of local *Leishmania* spp.? Obviously they cannot be transmitted by bite. Are they blind alleys from which the flagellates cannot escape? Some of them at least are mammalian *Leishmania* spp. Are there other haemoflagellates apart from mammalian *Leishmania* spp. among these natural infections? Is transmission limited to the small minority of sandflies with an anterior infection?".

It is instructive to compare the findings in Panama with those in Brazil and British Honduras. Pessoa and Coutinho (1941) fed sandflies *P. whitmani, P. fischeri* and *P. arthuri* on a lesion in a *Macacus rhesus* infected with *L. brasiliensis*; only two sandflies became infected, one out of forty-six specimens of *P. whitmani* and one out of 246 *P. fischeri*, i.e. the infection rate was less than that found in wild sandflies in Panama or Venezuela. Either the Panamese sandflies are more sensitive to the local leishmanias than the above three species to the strain of *L. brasiliensis* to which they were exposed, or they have access to heavier doses of parasites than those provided by the infected macaque in Pessoa and Coutinho's experiment. Lainson and Strangways-Dixon (1962) working in British Honduras dissected 334 sandflies; flagellates were found in one specimen of *P. ovallesi* (located in the hindgut) and in one specimen of *P. cruciatus*, a suspected vector in Mexico. The distribution of flagellates in the infected *P. cruciatus* was not determined. Three hundred sandflies caught in a given area of forest were triturated in sterile Locke solution and the suspension was inoculated into the back of a hamster which subsequently developed a lesion containing leishmanoids at the site of inoculation.

One hundred females fed on the lesion of a hamster; fifteen females were dissected 5 days after feeding and *all showed massive infections in the foregut*: seven females *P. ylephileptor*, seven females *P. geniculatus* and one female *P. paraensis* re-fed or probed in an attempt to re-feed a second time; five females *P. ylephileptor* re-fed a third time; one female *P. ylephiletor* re-fed a fourth time, and when dissected a day after re-feed the mouth parts contained active leptomonads. A volunteer on whom the single specimen of *P. paraensis* probed for a period of 30 sec without ingesting blood, became infected. Transmission was accomplished by this single sandfly 4 days after it had fed on an infected hamster.

Analysing the above experiments we note: sandflies fed on the lesions of a hamster infected with *L. mexicana*. These lesions contain enormous numbers of leishmanoids. Individuals of three species of sandflies exposed to *L. mexicana* all developed pronounced anterior infections; at least two out of fifteen sandflies eventually developed infections extending into the proboscis. One of the species with infections in the anterior station was *P. ylephileptor* in which natural infections in the "posterior station" were found in Panama. The serological relationship between Panamanian *Leishmania* and *L. mexicana* is not known and therefore it cannot be certain that comparable data are being analysed, but the striking difference between the findings in British Honduras and those in Panama inevitably leads to the question whether the quantity of leishmanoids ingested by a sandfly and the corresponding increase in infection produced, influence the distribution of flagellates in the case of South American cutaneous leishmaniasis. In Old World vectors of *L. tropica* it certainly does not; *L. tropica* assumes an anterior position in *P. papatasi* and *P. sergenti* no matter how slight the infection in the sandfly. In the case of South American visceral leishmaniasis we have noted anterior infections in *P. longipalpis* even when the infection in the sandfly did not exceed twenty flagellates. In view of the surprises encountered during research on leishmaniasis (and the findings in Panama are among the most striking) it is not, in the present state of knowledge, advisable to apply data on the life cycle of *L. tropica* in Old World sandflies to strains which produce cutaneous and muco-cutaneous leishmaniasis in the New World. The findings of Strangways-Dixon and Lainson on *L. mexicana* in sandflies of British Honduras and those of Coelho and Falcao (1962) in Belo Horizonte are nevertheless very similar to those observed repeatedly in the case of *L. tropica* in *P. papatasi* and *P. sergenti*.

Coelho and Falcao (1962) infected wild *P. longipalpis*, caught in Lapinha, by feeding on a lesion in a hamster produced by a strain isolated from a case of leishmania tegumentaria diffusa. The authors refer to this strain as *L. brasiliensis*, though Medina and Romero (1959) named the organisms isolated from this type of case *L. brasiliensis pifanoi*. The sandflies became infected with leptomonads which adopted an anterior position. Five sandflies were triturated and inoculated into hamsters which subsequently became infected. Sandflies *P. longipalpis* and *P. renei* (also from Lapinha) fed on the above hamsters and became infected. In both species flagellates were found in the stomach and anterior part of the cardia. On re-feeding on normal hamsters both species transmitted by bite.

It is noteworthy that *P. renei* had never before been named as a probable vector of cutaneous leishmaniasis. Both *P. longipalpis* and *P. renei* in Lapinha are cave dwellers and have little opportunity of infecting man. (An entomologist who frequently collects sandflies in the caves became infected. As cutaneous leishmaniasis is not acquired in the town of Belo Horizonte he was probably infected while collecting in the caves.) If these two sandflies transmit in nature it can only be to animals visiting or living in caves. In the above experiments the frequency of "anterior station" infections in *P. longipalpis* and *P. renei* fed on lesions in hamsters, is again in striking contrast to the distribution of *Leishmania* in Panamanian sandflies, both naturally and experimentally infected. Natural infections of *P. longipalpis* with leptomonads have been recorded several times; Pifano (1943) found natural infections in Venezuela and attributed them to *L. brasiliensis* though both cutaneous and visceral leishmaniasis occur in Venezuela. Deane (1956) found three naturally infected specimens out of 1,017 caught in Ceará, an important endemic focus of kala azar (this slight natural infection rate in a focus of visceral leishmaniasis is in contrast to the findings in Panama).

Apart from the above experimental transmissions by bite Pifano (1958) recorded transmission by naturally infected sandflies. Seventeen female *P. panamensis* fed on a man while he was serving as bait for attracting sandflies; he developed nodules containing leishmanoids apparently on sites on which the insects had fed. This experiment proves transmission by sandflies, most probably by bite. Unfortunately the insects were not dissected.

In East Africa the search for the vector of kala azar has recently given important and interesting results. Kirk and Lewis (1955) made the first important step in the Sudan by allowing *P. orientalis* (closely related to *P. perniciosus*) to feed on a case with post-kala azar dermal leishmaniasis which appeared after successful treatment of a visceral leishmaniasis; five sandflies out of sixty-six fed at random became infected, one with a marked anterior infection ("many flagellates attached anteriorly to head") and four with infections confined to stomach. They also fed *P. papatasi* and *P. clydei* directly on infected skin of a similar case; three out of forty *P. papatasi* and one out of seventy-three *P. clydei* became infected. The leptomonads assumed an anterior position in *P. clydei*. (Wild sandflies were used in these experiments. The infection in *P. clydei* may have been a natural one acquired during a previous feed on a lizard.)

Heyneman (1963), working in an endemic area in central Sudan, found 22 out of 1,171 wild specimens of *P. orientalis* infected with leptomonads. These natural infections showed masses of elongate flagellates in the anterior and posterior proventriculus.

Heisch *et al.* (1962) after years of search found a sandfly probably *P. martini* naturally infected with leptomonads in an anterior position. These leptomonads proved infective for a hamster and were finally shown to be infective for man. As already stated this strain is serologically indistinguishable from the human Kenya strain. Following on this important find Minter *et al.* (1962) fed sandflies *P. martini* on three cases of kala azar with positive blood cultures. Of fifty-eight sandflies which survived 72 h after feeding on the above cases, twenty-eight had leptomonads in the anterior part of the midgut. It is not quite certain whether the sandflies were infected from the skin, blood, or both. In any case this experiment proves that man can act as a reservoir of kala azar in Kenya should an efficient vector form a part of a stable human environment, or nomadic populations containing suitable cases pass through territories inhabited by a vector. In the latter case an animal reservoir is more important than human cases.

The evidence to date incriminates two species as vectors in East Africa, viz. *P. martini* and *P. orientalis*. Further work including feeding wild and laboratory bred sandflies on naturally and experimentally infected animals (including suspected reservoirs) will be followed with interest.

There are several points of general interest in the development of *Leishmania* sp. in sandflies. In the first place there is no evidence of differentiation into non-infective followed by metacyclic infective forms as in the case of trypanosomes of the *brucei* group in *Glossina*. All stages in the sandfly appear to be infective for susceptible animals. If leishmanoids are introduced into a non-vector sandfly as in the case of a Sudan strain of *L. donovani* into *P. papatasi* the resulting leptomonads are infective for hamsters (Adler, 1947). Cultures retain infectivity for susceptible animals for a number of months or even years, in marked contrast to those of *Trypanosoma rhodesiense* which quickly differentiate into non-infective forms (within 48 h). An infected vector always contains leptomonads capable of further multiplication; thus sandflies *P. papatasi* can be infected by ingestion (through a membrane) of leptomonads which have developed for a period of 8–12 days in sandflies (Adler, 1928).

There is a slight indication of differentiation towards forms which are probably more easily transmitted by bite than other forms in the case of the "short forms", i.e. small thin leptomonads with a flagellum longer than the body, which are found in some infections of *P. sergenti* with *L. tropica* and *P. perniciosus* with *L. infantum*. These forms appear to invade the proboscis more readily than others but they are often absent.

The life cycle of *Leishmania* spp. in sandflies is not a highly specialized one. The only signs of specialization are a selective infectivity for vectors in which, in contradistinction to non-vectors, high infection rates are produced by relatively few parasites and the anterior position adopted by all Old World mammalian species and some New World species in their vectors.

In so far as ingested *Leishmania* sp. are eliminated from non-vectors, it is justifiable to consider the latter as immune but this immunity can be broken down by relatively large doses of parasites as in the case of Sudan *L. donovani* and *P. papatasi*.

There is no inevitability about the inoculation of leptomonads from a heavily infected sandfly into a vertebrate host as in the case of trypanosomes of the *brucei* group or sporozoites of malaria from the salivary glands of their respective vectors. There are many records of feeding experiments with heavily infected sandflies on man and susceptible animals which have given a negative result. Facts such as the above evoked Mesnil's judgment on the leishmanias "Ils se cherchent encore".

While there is almost general agreement on the transmission of leishmaniasis by sandflies and there can be no doubt whatever about the infectivity of *Leishmania* sp. for particular sandflies there are differences of opinion about the mode of transmission. Napier thought that leptomonads are ejected from heavily infected sandflies in an attempt to ingest blood while Adler and Ber (1941) considered that leptomonads are deposited directly from the proboscis if the latter is invaded and that infection of the proboscis is a *sine qua non* for transmission by bite. Adler and Theodor (1935) maintained that the sandfly has no apparatus for ejecting flagellates from the buccal cavity. In the transmission experiments of Adler and Ber (1941) motionless and apparently dead as well as living flagellates were deposited in a puncture wound made by a sandfly. Some of the sandflies involved in transmission probed a number of times in different sites before feeding. In the successful transmission experiment of Strangways-Dixon and Lainson (1962) a single sandfly probed without ingesting blood. In this experiment a number of other heavily infected sandflies fed but did not transmit. The conditions which favour descent of leptomonads down the proboscis are not known but it is significant that slight changes in the maintenance of infected sandflies *P. argentipes* and in the infecting feed of *P. papatasi* facilitated transmission by bite.

Experiments on the mechanism of biting and ingesting blood and tissue fluids such as Gordon and Willett (1956) devised in the case of *Glossina* would be helpful in the case of *Phlebotomus*. Sandflies ingest tissue fluid as well as blood (on one occasion we observed a sandfly which had gorged itself on clear fluid without visible signs of a blood meal). This explains the high infection rates in vectors fed on animals in which the dermis of the unbroken skin is infected.

Adler (1962a) pointed out that leptomonads of *Leishmania* carry a negative charge; the direction of their movement would therefore be influenced by the charge in their immediate vicinity. The pH in the midgut would be determined by the remains of the blood meal and metabolites of the flagellates, anterior to the midgut by fluids (? plant juices) into which the sandflies may have probed after the digestion of a blood meal. These considerations may have a bearing on the distribution of flagellates in the alimentary tract, including the proboscis.

The possibility of a *Leishmania* spp. establishing itself in a new vector with which it previously had no ecological contact is a problem of practical as well as theoretical importance. The adaptation of *Leishmania* spp. to their local vectors shows considerable refinement, so much so that strains of a single species from different foci differ in their infectivity for the same vectors. Thus *L. tropica* from Crete produces a low infection rate in *P. papa-*

tasi and a high infection rate in *P. sergenti* while Jericho strains produce a high infection rate in both. The reviewer did not think it probable that a strain adapted to its specific vectors would become adapted to other sandfly hosts. It is reasonable to assume that these adaptations are associated with genetic factors in the *Leishmania* and in the vectors prevalent in a given focus. Should a *Leishmania* sp. be transported to a new focus with its own distinctive sandfly fauna, would the genetic constitution of any species in the new environment render it a suitable host for the *Leishmania*, e.g. would any South American sandfly prove a suitable host for *L. tropica* or *L. infantum*? It is not difficult to infect a sandfly with an alien species of *Leishmania* if infective doses of leishmanoids or flagellates are sufficiently large, larger than those encountered under natural conditions. The crucial test is the infection rate and distribution of flagellates in sandflies exposed to reasonably small doses of parasites such as are available under natural conditions. The problem is now well within the range of experimental investigation by workers in endemic foci.

Addendum

After the above review was written further information on infections of Panamanian sandflies with *Leishmania* appeared. Hertig and McConnell (1963) reported on experimental infections, Johnson *et al.* (1963) on natural infections of Panamanian sandflies and animal inoculations with leptomonads from the above source.

Johnson *et al.* (1963) point out that the entire host range of the common man-biting sandflies of Panama is not yet known. *Phlebotomus gomezi* and *P. sanguinarius* feed in limited numbers on hamsters but feed readily on opossums, climbing rats (*Tylomys*), squirrels and kinkajous. *P. trapidoi* the most likely suspect (as vector) is most difficult to breed and feed in the laboratory. Collections and dissections indicate that the most likely reservoirs are to be found among arboreal animals.

Leishmaniasis had been contracted by Gorgas Memorial personnel collecting biting insects in the forest canopy in Almirante. The following species *P. gomezi*, *P. sanguinarius*, *P. trapidoi* and *P. ylepheptor* were captured both at ground level and a 36 ft tree platform, while the large majority of *P. panamensis* were captured at ground level. The infection rates in sandflies captured 36 ft above ground level showed a rather higher infection rate than those captured at ground level.

The infection rate in wild sandflies was significantly higher during the rainy season (10·6%) than during the dry season (4·1%). The infection rate among species biting man was 8·5%; the highest infection rate (15·4%) was found in *P. trapidoi*. *Leptomonads were always present in the hindgut of infected females*. They also occurred occasionally in the posterior part of the midgut, rarely in the cardia and only twice in the foregut (oesophagus and pharynx). In 2% of natural infections leptomonads were present in the Malpighian tubules. Lesions produced in hamsters by two sandfly strains were indistinguishable from those produced by Panamanian human strains. Hertig and McConnell (1963) fed over 800 sandflies on cultures of *Leishmania* by the Hertig pipette method and found an infection rate of 81%. After 3 days the anterior part of the hindgut was invaded and by the 4th and 5th day the surface of this part of the hindgut (the part immediately behind the origin of the Malpighian tubules) was often completely covered; very few flagellates were attached elsewhere in the hindgut. These authors state: "at 3 days (2 days in *P. panamensis*) flagellates were beginning to appear in the cardia, anterior to any blood remains, with a tendency to become concentrated near or attached to the proventricular valve". Occasionally at 3 days and often at 4 or 5 days this region was packed and even distended with flagellates. At times the cardia was entirely free of flagellates even though they occurred elsewhere. Infections of the pharynx were also found particularly in *P. gomezi* (6 out of 47 in which the pharynx was dissected out).

McConnell (1963) cultured leptomonads from wild sandflies; of thirteen strains inoculated into hamsters only two (isolated from *P. trapidoi*) produced lesions. Cultures obtained from human lesions invariably produced infections in hamsters. All strains isolated from sandflies contained flagellates with a "greatly elongated posterior end". The number of these forms increase with the age of a culture. An agar-diffusion test indicated that a human strain and two isolates are different. It is uncertain whether there is any relationship between the isolates containing long forms and human leishmaniasis. Granules in the accessory glands occur both in fed and unfed Panamanian laboratory bred sandflies.

Hoogstraal *et al.* (1963) isolated cultures of *Leishmania* from the spleens of two rodents, *Rattus rattus* and a spiny mouse *Acomys* sp. caught in the town of Malakal in the Sudan. Cultures produced visceral infections in hamsters. Human cases do not occur in this urban centre from which man-biting sandflies are absent. It is therefore probable that a non-anthrophilic sandfly associated with rodent burrows maintains the infection in these animals. Hoogstraal and Dietlein (1963) produced visceral infections in hamsters by inoculating triturated sandflies, *P. orientalis*, caught in an endemic area.

Heyneman (1963) recorded twenty-two natural infections with leptomonads among 1,171 specimens of *P. orientalis* caught in an endemic area, the cardia was infected in twelve of the above sandflies.

As previously stated, Kirk and Lewis indicated *P. orientalis* as one of the probable vectors of visceral leishmaniasis in the Sudan. The above findings of Hoogstraal *et al.* and of Heyneman provide factual evidence that this species is indeed a local vector. It is of interest to note that *P.*

orientalis is closely related to *P. perniciosus*, one of the vectors of *L. infantum* in the Mediterranean area; the females of the two species resemble each other very closely but the males can be distinguished by relatively slight differences in their terminalia.

Acknowledgement

The work in this review was aided by grant no. AI-05524-01 TMP from the National Institutes of Health, U.S. Public Health Service.

References

Ada, G. and Fulton, J. D. (1948). *Brit. J. exp. Path.* **29**, 524.
Adler, S. (1928). *Trans. roy. Soc. trop. Med. Hyg.* **22**, 177.
Adler, S. (1936). *In* "Third International Congress of Comparative Pathology", Vol. 1, p. 3.
Adler, S. (1940). *Trans. roy. Soc. trop. Med. Hyg.* **33**, 419.
Adler, S. (1947). *Trans. roy. Soc. trop. Med. Hyg.* **40**, 701.
Adler, S. (1948). "Proc. 4th Congr. trop. Med. Malar. Washington", Vol. 2, p. 1119.
Adler, S. (1954). *Trans. roy. Soc. trop. Med. Hyg.* **48**, 431.
Adler, S. (1956). *Rev. bras. Malar.* **8**, 29.
Adler, S. (1961). *Bull. Res. Counc. Israel* **9**E.
Adler, S. (1962a). *Sci. Rep. Ist. Super. Sanita* **2**, 143.
Adler, S. (1962b). *Rev. Inst. med. trop. Sao Paulo* **4**, 61.
Adler, S. (1964). *Bull. Res. Counc. Israel* (in press).
Adler, S. and Adler, J. (1955). *Bull. Res. Counc. Israel* **4**, 396.
Adler, S. and Adler, J. (1956). *Bull. Res. Counc. Israel* **6**E, 77.
Adler, S. and Ashbel, R. (1940). *Ann. trop. Med. Parasit.* **34**, 207.
Adler, S. and Ber, M. (1941). *Ind. J. med. Res.* **29**, 203.
Adler, S. and Gunders, A. E. (1964). *Trans. roy. Soc. trop. Med. Hyg.* (In press).
Adler, S. and Halff, L. (1955). *Ann. trop. Med. Parasit.* **49**, 37.
Adler, S. and Tchernomoretz, J. (1946). *Ann. trop. Med. Parasit.* **40**, 320.
Adler, S. and Theodor, O. (1925). *Ann. trop. Med. Parasit.* **19**, 365.
Adler, S. and Theodor, O. (1927). *Ann. trop. Med. Parasit.* **21**, 62.
Adler, S. and Theodor, O. (1929). *Ann. trop. Med. Parasit.* **23**, 1.
Adler, S. and Theodor, O. (1930). *Ann. trop. Med. Parasit.* **24**, 197.
Adler, S. and Theodor, O. (1931). *Proc. roy. Soc.* B **108**, 447.
Adler, S. and Theodor, O. (1932). *Proc. roy. Soc.* B **110**, 402.
Adler, S. and Theodor, O. (1935). *Proc. roy. Soc.* B **116**, 516.
Adler, S. and Theodor, O. (1957). *Annu. Rev. Ent.* **2**, 203.
Adler, S. and Zuckerman, A. (1948). *Ann. trop. Med. Parasit.* **42**, 178.
Adler, S., Theodor, O. and Witenberg, G. (1938). *Proc. roy. Soc.* B **125**, 491.
Alencar, J. E. (1958). *Int. Congr. trop. Med. Malaria, Lisbon* **3**, 718.
Alencar, J. E., Pessoa, E. de P. and Fontana, Z. F. (1960). *Rev. Inst. trop. Med., Sao Paulo* **2**, 347.
Amaral, A. D. F. (1941). *Rev. Med. Cir., Sao Paulo* **6**, 113.
André, R., Brumpt, L., Dreyfus, B., Pasalaceq, A. V. and Jacob, B. (1957). *Bull. Mem. Soc. Hôp., Paris* **25/26**, 854.
Ansari, N. and Faghik, N. (1953). *Ann. Parasit. hum. comp.* **28**, 241.
Ansari, N. and Mofidi, Ch. (1950). *Bull. Soc. Pat. exot.* **43**, 601.
Aragao, H. B. (1927). *Bras. Med.* **36**, 129.
Biagi, F. F. (1953a). *Med. Mex.* **33**, 401.
Biagi, F. F. (1953b). *Med. Mex.* **33**, 435.

Biagi, F. F. (1953c). *Rev. Med.* **34**, 385.
Biagi, F. F., Maroquin, F. and Armando, G. (1957). *Rev. Med. Mex.* **37**, 444.
Blanc, F., Merlehot, J. and Aubert, L. (1959). *Bull. Soc. Pat. exot.* **52**, 630.
Chagas, A. W. (1939). *Hospital* **14**, 1.
Chagas, A. W. (1940). *Mem. Inst. Osw. Cruz* **35**, 327.
Chagas, E., Ferreira, L. C., Deane, G., Deane, L., Guimaraes, F. N., Paumgarten, M. J. and Sa, B. (1938). *Mem. Inst. Osw. Cruz* **33**, 89.
Chakraborty, J., Guha, A. and Das Gupta, N. N. (1962). *J. Parasit.* **48**, 131.
Chang, P. C. H. (1956). *J. Parasit.* **42**, 126.
Chang, H. L. and Chang, N. C. (1951). *Chin. med. J.* **69**, 3.
Chang, S. L. and Negherborn, W. O. (1947). *J. infect. Dis.* **80**, 172.
Chang, H. L., Chow, H. K. and Lu, J. P. (1948). *Chin. med. J.* **66**, 325.
Charrin, A. and Rogers, G. H. (1889). *C. R. Soc. Biol., Paris* **41**, 667.
Chen, T., Chen, P. and Li, L. (1953). *Chin. med. J.* **71**, 334.
Clark, T. B. (1959). *J. Protozool.* **6**, 227.
Clark, T. B. and Wallace, F. G. (1960). *J. Protozool.* **7**, 115.
Coelho, M. V. and Falcao, A. (1962). *Rev. Inst. Med. trop. Sao Paulo* **4**, 159, 220.
Cole, A. C. E. (1944). *Trans. roy. Soc. trop. Med. Hyg.* **37**, 409.
Cole, A. C. E., Cosgrove, P. C. and Robinson, G. (1942). *Trans. roy. Soc. trop. Med. Hyg.* **36**, 25.
Convit, J. (1958). *Rev. Sanid. Asist. Soc.* **1–2**, 28.
Convit, J., Alarcon, C. J., Medina, R., Reyes, O. and Kerdeli, V. F. (1959). *Arch. Venez. Pat. trop.* **3**, 218.
Coutinho, J. D. (1954). *Folia clin. biol., S. Paulo* **21**, 321.
Critien, A. (1911). *Ann. trop. Med. Parasit.* **5**, 37.
Cunha, A. M. da (1942). *Mem. Inst. Osw. Cruz* **37**, 35.
Cunha, A. M. da, and Chagas, E. (1937). *Hospital* (*Rio de J.*) **11**, 148.
Cunha, A. M. da, and Dias, E. (1939). *Bras. Med.* **53**, 89.
Das Gupta, N. N., Guha, A. and De, N. (1954). *Exp. Cell Res.* **6**, 353.
David, A. (1929). *Ann. Parasit. hum. comp.* **7**, 3.
Deane, L. M. (1956). "Leishmaniose Visceral no Brasil," Servicio Nacional de Educacao Sanitaria Rio de Janeiro.
De Leon, J. R. and Figuerra, L. N. (1958). *Int. Congr. trop. Med. Malaria, Lisbon* **3**, 747.
Depieds, R., Collomb, H., Mathurin, J. and Ranque, J. (1958). *Bull. Soc. Pat. exot.* **51**, 50.
Dostrowsky, A. and Sagher, F. (1957). *J. invest. Derm.* **29**, 15.
Dostrowsky, A. Sagher, F. and Zuckerman, A. (1952). *A.M.A. Arch. Derm. Syph.* **66**, 665.
Echandi, C. A. (1953). *Rev. Biol. Trop. San José, Costa Rica* **1**, 173.
Echandi, C. A. (1954). *Trop. Dis. Bull.* **51**, 784.
Fendall, N. R. E. (1950). *E. Afr. med. J.* **27**, 291.
Floch, H. (1954). *Arch. Inst. Pasteur de la Guyara Française*, Publication, No. 330.
Foner, A. (1964). *Bull. Res. Counc. Israel* (In press).
Fonseca, F. (1932). *Amer. J. trop. Med.* **12**, 433.
Forratini, O. P. (1960). *Rev. Inst. Med. trop. Sao Paulo* **2**, 195.
Forratini, O. P., Juarez, E., Bernardi, L. and Dauer, C. (1959). *Rev. Inst. trop. Med. Sao Paulo* **1**, 11.
Frenkel, J. (1941). *Ind. J. med. Res.* **29**, 811.
Fuller, H. S. and Geiman, Q. M. (1942). *J. Parasit.* **28**, 429.
Garnham, P. C. C. (1954). *Trans. roy. Soc. trop. Med. Hyg.* **48**, 464.
Garnham, P. C. C. (1962). *Sci. Rep. Ist. Super Sanita* **2**, 76.
Garnham, P. C. C. and Bird, R. G. (1962). *Sci. Rep. Ist. Super Sanita* **2**, 83.
Garnham, P. C. C. and Lewis, D. J. (1959). *Trans. roy. Soc. trop. Med. Hyg.* **53**, 12.

Geiman, Q. M. (1940). *J. Parasit.* (Suppl.) **26**, 22.
Gherman, I., Zaharia, C. A., Birzu, I., Popescu, G. and Tieranu, E. (1957). *Microbiol. Parasitol. si Epidemiol. Bucharest* **2**, 556.
Gorgas Memorial Laboratory (1962). Thirty-third Annual Report of the Work and Operations of the Gorgas Memorial Laboratory, Fiscal Year 1961, Washington, D.C.
Gorgas Memorial Laboratory (1963). Thirty-fourth Annual Report of the Work and Operations of the Gorgas Memorial Laboratory, Fiscal Year 1962, Washington, D.C.
Gordon, R. M. and Willett, K. C. (1956). *Ann. trop. Med. Parasit.* **50**, 426.
Greval, S. D., Sen Gupta, P. C. and Napier, I. E. (1939). *Ind. J. med. Res.* **27**, 181.
Guimaraes, F. N. (1951). *Hospital* (*Rio de J.*) **40**, 43; **41**, 395.
Heisch, R. B. (1954). *Trans. roy. Soc. trop. Med. Hyg.* **48**, 449.
Heisch, R. B. (1955). *Nature, Lond.* **175**, 433.
Heisch, R. B. (1957). *E. Afr. med. J.* **34**, 183.
Heisch, R. B. (1958). *Trans. roy. Soc. trop. Med.* **48**, 449.
Heisch, R. B., Guggisberg, C. A. and Teesdale, C. (1956). *Trans. roy. Soc. Med. Hyg.* **50**, 209.
Heisch, R. B., Wijers, D. J. B. and Minter, D. M. (1962). *Brit. med. J.* **1**, 1456.
Henderson, H. (1937). *Trans. roy. Soc. Trop. Med. Hyg.* **13**, 179.
Hertig, M. (1962). In "Report of the Work and Operations of the Gorgas Memorial Laboratory." Fiscal Year 1961, Washington, 1962.
Hertig, M. and McConnell, E. (1963). *Exp. Parasit.* **14**, 10.
Hertig, M. Fairchild, G. B. and Johnson, C. M. (1954). "Report of the Work and Operations of Gorgas Memorial Laboratory," Washington, 1954.
Heyneman, D. (1961). *E. Afr. med. J.* **38**, 196.
Heyneman, D. (1963). *Amer. J. trop. Med. Hyg.* **12**, 725.
Hindle, E. (1930). *Trans. roy. Soc. trop. Med. Hyg.* **24**, 97.
Hindle, E., Hou, P. C. and Patton, W. S. (1926). *Proc. roy. Soc.* **B 100**, 368.
Hoare, C. A. (1944). *Trop. Dis. Bull.* **41**, 331.
Hoare, C. A. (1948). *Proc. int. Congr. Trop. Med. Malar., Washington* **2**, 1110.
Hoare, C. A. (1954). *Trans. roy. Soc. trop. Med. Hyg.* **48**, 465.
Hoare, C. A. (1955). *Refuah Vet.* **12**, 263.
Hoare C. A. (1962). *Acta tropica* **19** 281.
Hoogstraal, H. and Dietlein, D. R. (1963). *Amer. J. trop. Med. Hyg.* **12**, 165.
Hoogstraal, H., van Peerer, P. F. O., Reid, T. P. and Dietlein, D. R. (1963). *Amer. J. trop. Med. Hyg.* **12**, 175.
Inoki, S., Nukanishi, K., Nukayashi, T. and Ohno, M. (1957). *Med. J. Osaka Univ.* **7**, 719.
Johnson, P. T., McConnell, E. and Hertig, M. (1962). *J. Parasit.* **48**, 158.
Johnson, P. T., McConnell, E. and Hertig, M. (1963). *Exp. Parasit.* **14**, 10.
Khodukin, N. G., Soffieff, M. S. and Keverkoff, N. P. (1936). *Ann. Inst. Pasteur* **57**, 487.
Kirk, R. (1938). *Trans. roy. Soc. trop. Med. Hyg.* **32**, 271.
Kirk, R. (1942). *Trans. roy. Soc. trop. Med. Hyg.* **35**, 257.
Kirk, R. (1950). *Parasitology* **40**, 58.
Kirk, R. (1956). *Trans. roy. Soc. Trop. Med. Hyg.* **50**, 169.
Kirk, R. and Lewis, D. J. (1951). *Trans. roy. Soc. Ent. Lond.* **102**, 383.
Kirk, R. and Lewis, D. J. (1955). *Trans. roy. Soc. trop. Med. Hyg.* **49**, 229.
Kirk, R. and Sati, M. H. (1940). *Trans. roy. Soc. trop. Med. Hyg.* **33**, 501; **34**, 213.
Kligler, I. J. (1925). *Trans. roy. Soc. trop. Med. Hyg.* **19**, 330.
Knowles, R., Napier, L. E. and Smith, R. O. A. (1924). *Ind. med. Gaz.* **59**, 593.
Kosevnikov, P. V. (1957). *In* Introduction to Rodiakin (1957).
Kryukova, A. P. (1951). *Probl. reg. gen. exp. Parasit. Med. Zool. Moscow* **7**, 70.
Kryukova, A. P. (1954). *Trop. Dis. Bull.* **51**, 40.
Lainson, R. and Strangways-Dixon, J. (1962). *Brit. med. J.* **1**, 1596.

Larivière, M., Abonone, E. and Kramer, R. (1961). *Bull. Soc. Path. exot.* **54**, 1031.
Latyshev, N. I. and Kriukova, A. P. (1941). *Trav. Acad. Milit. Méd. Armée Rouge U.R.S.S. Moscow* **25**, 229.
Latyshev, N. I. and Kryukova, A. P. (1943). *Trop. Dis. Bull.* **40**, 24.
Latyshev, N. I., Kryukova, A. P. and Povalishina, T. P. (1951). *Probl. reg. gen. exp. Med. Zool. Moscow* **7**, 35.
Latyshev, N. I., Kryukova, A. P. and Povalishina, T. P. (1954). *Trop. Dis. Bull.* **51**, 37.
Latyshev, N. I., Shoshina, M. A. and Polyakov, A. (1951). *Probl. reg. gen. exp. Parasit. Med. Zool. Moscow* **7**, 63.
Latyshev, N. I., Shoshina, M. A. and Polyakov, A. (1954). *Trop. Dis. Bull.* **51**, 39.
Laveran, A. (1917). "Leishmanioses." Masson, Paris.
Leishman, W. B. and Statham, J. C. B. (1905). *J. R. Army med. Cps* **4**, 321.
Lindsay, J. W. (1935). *Trans. roy. Soc. trop. Med. Hyg.* **18**, 539.
Lofgren, R. (1950). *J. Bact.* **60**, 617.
Low, G. C. and Cook, W. E. (1926). *Lancet* **2**, 1209.
Lowe, J. and Greval, S. D. S. (1939). *Ind. J. med. Res.* **26**, 833.
McConnell, E. (1963). *Exp. Parasit.* **14**, 123.
Maggiore, S. (1923). *La Pediatria* **33**, 169.
Manson-Bahr, P. E. C. (1955). *Trans. roy. Soc. trop. Med. Hyg.* **49**, 304.
Manson-Bahr, P. E. C. (1959). *Trans. roy. Soc. trop. Med. Hyg.* **53**, 123.
Manson-Bahr, P. E. C. (1961). *Trans. roy. Soc. trop. Med. Hyg.* **55**, 550.
Manson-Bahr, P. E. C. (1963). *Brit. med. J.* **1**, 1209.
Manson-Bahr, P. E. C. and Heisch, R. B. (1961). *Ann. trop. Med. Parasit.* **55**, 381, 550.
Manson-Bahr, P. E. C., Heisch, R. B. and Garnham, P. C. C. (1959). *Trans. roy. Soc. trop. Med. Hyg.* **53**, 380.
Manson-Bahr, P. E. C., Southgate, B. A. and Harvey, A. E. C. (1963). *Brit. med. J.* **1**, 1208.
Marashvilli, G. M. (1961a). *Med. Parasit., Moscow* **30**, 188.
Marashvilli, G. M. (1961b). *Bull. trop. Dis.* **59**, 21.
Marozian (1941). *Med. Parasit., Moscow* **10**, 101.
Martins, A. F. (1941). *Rev. Inst. Adolfo Lutz* **1**, 55.
Marzinowsky, E. J. and Schorenkova, A. (1924). *Trans. Soc. trop. Med. Hyg.* **17**, 67.
Mayrink, W. (1961). "Antigenos Homologos em Calazar," Belo Horizonte, M. G., Brasil.
Mazza, S. (1926). *Prensa méd. Argent.* **13**, 139.
Mazza, S. (1935). *Reun. Soc. argent. Pat. reg. N.* **9**, 560.
Mazza, S. and Cornejo, A. (1926). *Bol. Inst. Clin. Quir. Buenos Aires* **2**, 140.
Medina, R. and Romero, J. (1959). *Arch. Venez. Pat. trop. Parasit. Med.* **3**, 298.
Mello, S. B. (1940). *Bras. Med.* **54**, 189.
Migone, L. (1913). *Bull. Soc. Pat. exot.* **6**, 210.
Minter, D. M., Wijers, D. J. B., Heisch, R. B. and Manson-Bahr, P. E. C. (1962). *Brit. med. J.* **2**, 835.
Mohiuddin, A. (1959). *E. Afr. med. J.* **36**, 171.
Montenegro (1923). *Bol. Soc. Med. Cir., Sao Paulo* **6**, 113.
Montenegro (1924). *Ann. J. trop. Med.* **4**, 331.
Moses, A. (1919). *Sem. med. Argent.* **19**, 144.
Moschkowsky, Ch. D. (1941). "Conférence interrépublicaine sur la leishmaniose cutanée et le problème de phlébotomes." U.S.S.R.
Muniz, J. (1953). *Hospital* (*Rio de J.*) **43**, 1.
Muniz, J. and Medina, J. (1948). *Hospital* (*Rio de J.*) **33**, 7.
Napier, L. E. (1946). "The Principles and Practice of Tropical Medicine." Macmillan, New York.
Napier, L. E. and Krishnan, K. V. (1931). *Ind. med. Gaz.* **66**, 603.
Napier, L. E., Krishnan, K. V. and Lal, C. (1933). *Ind. med. Gaz.* **58**, 272.

Neave, S. (1906). "2nd Report Wellcome trop. Res. Lab. Khartoum", p. 183.
Noguchi, H. (1924). "Proc. int. Congr. Hlth Prob. Trop. America." Boston, p. 455.
Noguchi, H. (1926). *J. exp. Med.* **44**, 327.
Nussenzweig, V. (1956). *Hospital* (*Rio de J.*) **2**, 217.
Paradiso, F. (1926). *Pediatria* **34**, 664.
Paraense, W. L. (1953). *Trans. roy. Soc. trop. Med. Hyg.* **47**, 555
Parrot, L. (1929). *C. R. Soc. Biol., Paris* **100**, 411.
Parrot, L. (1935). *Bull. Soc. Pat. exot.* **28**, 958.
Parrot, L., Donatien, A. and Lestoquard, F. (1932). *Algérie med.* **10**, extrait.
Patton, W. S. (1913). *Ind. J. med. Res.* **1**, 185.
Pavlovsky, E. N. and Petrischeva, P. A. (1958). "Proc. int. Congr. trop. Med. Malaria, Lisbon" Vol. 3, p. 811.
Pedroso, A. M. (1923). *Rev. Med. Sao Paulo* **23**, 42.
Pellegrino, J., Brener, Z. and Santos, M. M. (1958). *J. Parasit.* **44**, 645.
Penna, H. A. (1934). *Bras. Med.* **48**, 949.
Pessoa, S. B. (1961). *Arch. Hyg. Sao Paulo* **26**, 41.
Pessoa, S. B. and Arantes, S. (1944). Cited by Pessoa and Barreto (1944).
Pessoa, S. B. and Barreto, M. P. (1944). "Leishmaniose tegumentar." Americana Ministerio da Educacao e Saude, Rio de Janeiro.
Pessoa, S. B. and Cardoso, F. A. (1942). *Hospital* (*Rio de J.*) **21**, 187.
Pessoa, S. B. and Coutinho, J. O. (1941). *Hospital* (*Rio de J.*) **20**, 25.
Pifano, F. (1943). *Bol. Entomol. Venez.* **2**, 99.
Pifano, F. (1958). "Proc. int. Congr. trop. Med. Hyg., Lisbon", Vol. 3, p. 791.
Pifano, F. (1960). *Bull. Soc. Pat. exot.* **53**, 510.
Prata, A. (1957). Thesis Salvador Bahia.
Pringle, (1956). *Bull. end. Dis. Baghdad* **1**, 275.
Pulvertaft, R. J. V. and Hoyle. G. F. (1960). *Trans. roy. Soc. trop. Med. Hyg.* **54**, 391.
Pyne, C. K. (1958). *Exp. Cell Res.* **14**, 388.
Pyne, C. K. and Chakraborty, J. (1958). *J. Protozool.* **5**, 264.
Rés, J. F. (1957). *Anais Inst. Med. trop. Lisboa* **14**, 527.
Rodiakin, N. F. (1957). "Problems of Immunity and Specific Prophylaxis in Borowsky's Disease (Cutaneous Leishmaniasis)". Ministry of Health Turkmen S.S.R., Ashkabad.
Rodrigues, M. J. D. and Avilés, N. F. (1953). *Rev. Equatoriana Hig. Med. trop.* **10**, 35.
Rogers, L. (1904). *Lancet* **2**, 215.
Rossan, R. N. and Stauber, L. A. (1959). *J. Parasit.* **45**, Suppl. 50.
Sati, M. H. (1962). *Sudan med. J.* **1**, 198.
Sen, A. and Mukherjee, S. (1961). *Ann. Biochem. exp. Med.* **21**, 105.
Senekji, H. A. (1943). *Amer. J. trop. Med.* **28**, 53.
Senekji, H. A. and Beattie, C. P. (1941). *Trans. roy. Soc. trop. Med. Hyg.* **23**, 523.
Sen Gupta, P. C. (1943). *Ind. med. Gaz.* **78**, 336.
Sen Gupta, P. C. (1944). *Ind. med. Gaz.* **79**, 528.
Sen Gupta, P. C. (1945). *Ind. med. Gaz.* **80**, 396.
Sen Gupta, P. C. and Adhikari, S. L. (1952). *J. Ind. med. Ass.* **23**, 89.
Sen Gupta, P. C. and Mukherjee, A. M. (1962). *Ann. Biochem. exp. Med.* **22**, 63.
Sen Gupta, P. C. and Ray, A. N. (1954). *Proc. zool. Soc.* **7**, 113.
Sen Gupta, P. C., Das Gupta, N. N. and Bhattacharya, D. L. (1951). *Nature, Lond.* **167**, 1063.
Shattuck, G. C. (1938). *Publ. Carnegie Instn.* **499**, 113.
Sorouri, P. (1955). *J. Morphol.* **97**, 393.
Southgate, R. A. and Oriedo, B. V. E. (1962). *Trans. roy. Soc. trop. Med. Hyg.* **56**, 30.
Stephenson, R. W. (1940). *Ann. trop. Med. Parasit.* **34**, 175.
Strangways-Dixon, J. and Lainson, R. (1962). *Brit. med. J.* **1**, 297.

Strejan, G. (1963). *Bull. Res. Counc. Israel* **11** E, 21.
Swaminath, C. S., Shortt, H. E. and Anderson, L. A. P. (1942). *Ind. J. med. Res.* **30**, 473.
Symmers, W. St. C. (1960). *Lancet* **1**, 127.
Taj-el Deen and Aloasi, K. (1954). *J. Fac. Med. Baghdad* **1-2**, 15.
Taub, J. (1956). *Bull. Res. Counc. Israel* **6**E, 55.
Torres, C. M., Muniz, J., Cardoso, R. A. and Duarte, E. (1948). *Hospital* (*Rio de J.*) **33**, 405.
Trager, W. (1957). *J. Protozool.* **4**, 269.
Trejos, A. and Echandi, C. A. (1951). *Rev. med. Costa Rica* **10**, 91.
Velez, L. (1913). *Bull. Soc. Pat. exot.* **6**, 545.
Viera da Cunha, R., Xavier, F. S. A. and Alencar, J. E. de (1959). *Rev. bras. Malar.* **11**, 45.
Weiss, P. (1943). *Rev. Med. exp.* **2**, 209.
Weng, H., Chung, H., Hou, T. and Ho, L. (1953). *Chin. med. J.* **71**, 328.
Wenyon, C. M. (1911–1912). "Kala azar Bulletin Tropical Diseases Bureau", London.
Wenyon, C. M. (1914). *Tr. Trop. Med. Hyg.* **7**, 97.
Wertheimer, E. and Stein, L. (1944). *J. Lab. clin. Med.* **29**, 1082.
Yakimoff, W. L. (1915). *Bull. Soc. Pat. exot.* **8**, 474.
Young, C. W. and Hertig, M. (1926). *Proc. Soc. exp. Biol., N.Y.* **22**, 611.

Fascioliasis: the Invasive Stages of *Fasciola hepatica* in Mammalian Hosts

BEN DAWES and D. L. HUGHES*

Department of Zoology, King's College, University of London, and Allen and Hanbury's Ltd., Ware, Hertfordshire, England

I. Introduction

Most digenetic trematodes of more than one hundred families are parasitic in the intestines of vertebrate hosts. A few occur in the coelom or urinary system but the most notable exceptions are the liver flukes, lung flukes and blood flukes. Some liver flukes wander as juveniles through the opening of the bile duct from the intestines, for instance *Dicrocoelium* and *Clonorchis*. The adults of related genera such as *Hadwenius* and *Odhneriella* occur however in the intestines and the biliary systems respectively of their hosts and this is true of related species of the latter genus, *O. elongata* (Ozaki, 1935) and *O. rossica*

* Present address: Parasitology Section, Glaxo Laboratories Ltd., Greenford, Middlesex.

Skrjabin, 1915, which are Campulid parasites of marine mammals. *Metorchis xanthosomus* (Creplin, 1846), a parasite of various birds, occurs in both the intestines and the biliary system, although related species found in the cat, dog, fox and other mammals occur only in the biliary system. Little is known about the migrations of juvenile trematodes to internal locations in their hosts and *F. hepatica* is notable for the amount of controversy which its ill-understood wanderings has produced. This fluke is notorious because it is a ubiquitous parasite of man and domesticated animals and was found recently by Bezubik and Furmaga (1959) in monkeys *(Macaca mulatta)* from China. This fluke causes distress, disease and death as well as financial losses, and up to the present time it has been found impossible to control. The recent increase in our knowledge of the invasive phase has been due largely to the use of small mammals as experimental hosts. We are concerned with this invasive phase in normal infections but we shall turn to consider also attempts which have been made to influence its course.

It will profit us first of all to review briefly and chronologically some notable investigations during the period when the route of migration has been held in some doubt. This historical review is based mainly on the researches of about a dozen persons, including Lutz (1893), Sinitsin (1914), Shirai (1927), Montgomerie (1928 a, b, 1931), Susuki (1931), Shaw (1932), Vogel (1934), Schumacher (1938) and Krull and Jackson (1943), most of whom have been mindful of three possible ways of reaching the liver, which we can call the hepatic portal, the biliary and the abdominal routes. Their contributions to the controversial topic can be stated simply but deserve more detailed mention, because they have brought into use various methods of experimentation as well as a variety of hosts, and in one way or another have made creditably acute observations. Their findings may appear to be cursory and superficial but it is as well to remember that the technicalities of their pioneering efforts have been difficult and tedious. Some efforts to obtain numerical results have been unprofitable by comparison with what might have come to light by efforts to ascertain what young flukes are actually doing with the simple organs they possess at the time of migration. Only recently have we gained even elementary impressions of methods of feeding which prepare the way for physiological and biochemical studies. In our historical sketch, and also later, mention is made of forms such as *Dicrocoelium dendriticum*, which has adopted the biliary route of migration to the biliary system, and *Paragonimus westermani*, which closely parallels *F. hepatica* in its wanderings, as is clearly indicated here for the first time. It is the belief of the senior author at any rate that much effort could have been more profitably expended in research on *F. hepatica* during this invasive phase, had more notice been taken of the classical researches of Saduma Yokogawa on the wanderings of *P. westermani*, which has biological relevance to our topic.

II. Historical Review

After recounting the discoveries of Leuckart and of Thomas in respect of the life history of *F. hepatica*, Reinhard (1957) remarked that these classical researches "still left some dark corners unexplored". He was referring to the exact migratory route by which juvenile forms arrive at their final location in the bile ducts. Lutz (1893) first tried to fill this gap in our knowledge by infecting some guinea pigs, a rabbit, a brown rat and a goat by feeding cysts containing encysted metacercariae (which henceforth are referred to as young flukes), thus instituting a technique of great value in experimental fascioliasis. Alicata (1938) realized that Lutz was dealing with *F. gigantica* but it is unlikely that the routes taken differ in the two species of flukes. Lutz believed that the young forms which emerge from the cysts enter the wall of the intestines and are carried passively to the liver by way of the hepatic portal system, and he was supported in this belief by Railliet *et al.* (1913), Compes (1923), Nöller (1925), Nöller and Schmid (1928), Marek (1927), Bugge (1935) and other investigators. A step forward was registered by the rejection of Leuckart's opinion that young forms reach the biliary system via the bile duct.

Dimitry Sinitsin also considered the migration of young flukes by the biliary channels improbable. He fed cysts to rabbits and recovered flukes of twelve ages from 1–39 days, providing figures of 2, 7, 10, 14 and 24 day stages (Sinitsin, 1914, Fig. 2). By modern standards his figures were unsatisfactory but they indicated gradually increasing complexity with increasing age and made it possible to put the lengths of these young flukes at about 0·3, 0·5, 1·0, 1·5 and 4·6 mm. Sinitsin observed that the young flukes free themselves from their cysts 2–3 h after infection, pass through the wall of the intestines, creep about over the abdominal viscera (sometimes protracting their stay for 4–14 days) and penetrate the liver, each day drawing closer to the bile ducts. During the first 4 days of infection he never found any flukes in the liver and when subsequently he found flukes in the abdominal cavity even beyond 14 days from the time of infection, he could distinguish them from forms recovered at the same time from the liver by the well nourished state (Wohlgenährtheit) of the latter. He formed the opinion that young flukes feed on blood, mainly because the liver's delicate, vascular nature makes provision of such food possible, and this explained to his satisfaction why the young flukes aggregate progressively on the liver surface. He did not specify the intensity of his experimental infections but he noted the lesions made on the liver surface by older flukes leaving the organ and stated "bei einem Angriff seitens zahlreicher Distomen starker Bluterguss und in der Körperhöhle bedeutende Blutansammlung stattfindet...".

Mitsuji Shirai (1927) fed cysts to guinea pigs and recovered flukes measuring about 0·3 × 0·2 mm from the abdominal cavity and the surface of the liver 3–4 days after infection. On the intestinal serosa he noted haemorrhagic spots and in sections he found flukes and their burrows. Consequently, he rejected both the hepatic portal and the biliary routes as unlikely. In only one experiment (64th day infection of rabbit) were any flukes found in the bile duct, however, and this led Shirai to the erroneous conclusion that "in large animals such as the cow and the sheep, the worms appear to mature in the bile ducts", whereas "in small laboratory animals such as the rabbit, guinea pig and the mouse, the worms wander in the liver parenchyma, even if they have reached maturity". He also believed that after passing into the wall of the intestines some flukes "accidentally entering into the blood vessels, were carried away passively by the blood stream to the liver". Shirai also found in liver sections of a 10 day infection of the guinea pig "worms creeping out of the liver into the abdominal cavity" but he found it impossible "to make a distinction" between forms entering or leaving the liver and said it was "very hard to know which route is taken by these worms which were found in this locality (the abdominal cavity)". So he contrived an experiment in which "the abdominal cavities of two guinea pigs were united paratiotically (sic)"; one of the animals was then given cysts and at autopsy 3 days later "a few young worms measuring 0·29 mm in the body length and 0·15 mm in breadth" were found in the other.

Although engaged in researches on the life cycle of *F. hepatica*, Susuki (1931) freed young flukes from their cysts by means of artificial digestive juices and injected them into the abdominal cavity, portal veins and bile ducts of the goat, rabbit and guinea pig, killing the host "after a certain time". He claimed positive results for portal vein injections, negative results for the bile duct injections, and stated that forms which get into the portal vein "move on to the liver or some other organ, and develop into the adult fluke", with the proviso that "this, however, cannot be regarded as the ordinary route", because in the rabbit and guinea pig "the blood vessels distributed in the intestine are too small for the larvae to pass, and it is consequently only by accident that they reach the vena portae". When cysts are given through the mouth, Susuki added, the parasite "has to leave the intestine for the abdominal cavity, and from there enters some organ suitable for further development". About young flukes injected into the biliary system he stated "it may be assumed that the larvae, on emergence from the capsule, were not strong enough to withstand or overcome the pressure of the gall stream, but were carried down to the intestine or, maybe, they were killed by the chemical action of the gall".

Shaw (1932) found that cysts obtained from experimentally infected snails remained infective for 11 months after cold storage at –2·78 to 4·44 °C. One rabbit infection failed but two guinea pigs had burrows in the liver 6 days after infection and 1 fluke was recovered. Shaw also carried out transplantations of young flukes. He fed cysts to a guinea pig and 3 days later recovered

33 living flukes (one from the liver and the remainder from the abdominal cavity): 12 of these flukes were injected into the abdominal cavity of a second guinea pig immediately, and 17 days later at autopsy a single fluke was recovered from the liver. He repeated this experiment with lambs: he fed 361 cysts to a lamb, recovered 43 flukes from the abdominal cavity 3 days later, injected 15 of these flukes into a second lamb and recovered 3 flukes 16 days later. Some of the recovered flukes were injected into the jugular vein of a lamb, with negative results, and others were put into the pleural cavity of a goat and produced haemorrhagic areas in both lungs, but flukes were not found. In one of Shaw's experiments 246 flukes were recovered from the abdominal cavity of a guinea pig 3 days after infection, but none were found in the liver. Where a count was made of cysts at feeding, 8–30% of cysts were reported to have produced flukes.

In a paper dealing with the life cycle of *Opisthorchis felineus* Hans Vogel (1934) turned his attention to the problem of the "Invasionsweg" of *F. hepatica.* He mentioned the oft reported prenatal infection of the liver in newborn lambs and calves and the suggestion that this was proof of migration through the vascular system. Having found young flukes in mesenteric lymph glands and other organs, including the lungs, he profferred an alternative view that it is the "borehappy" juveniles in the abdominal cavity which enter the foetus through the wall of the gravid uterus. Vogel also described the results of some of his own experiments, carried out in 1931. After administering 200 cysts to a guinea pig and killing the host 24 h later, he recovered 11 young flukes from the abdominal cavity but failed to find any worms in the portal vessels, bile ducts or gall bladder, as well as in finely divided liver. A second guinea pig was given 4 500 cysts and killed 48 h later: in washings of the abdominal cavity Vogel found 110 flukes and after opening portal vessels and diluting the blood with sodium citrate he recovered from the liquid 11 flukes which, he carefully considered, might have been located in pockets of the abdominal cavity, or attached to the serosa. He opened the gall bladder, washed out the contents and found one more fluke; he cut off pieces of liver and in a search lasting 2 h he found 12 more flukes. A series of sections revealed 3 flukes only 1–3 "worm lengths" from the surface of the liver: they were situated at the end of a short "bore-canal" with a funnel-like opening, and this also contained cells with large nuclei and little cytoplasm, resembling monocytes, and also neutrophil polymorphs and some erythrocytes. Near the mouth of the fluke he witnessed the destruction of hepatic cells. In a third experiment 100 cysts were placed in fresh intestinal juice taken from a dog and maintained at 37° C. After $1\frac{1}{4}$ h 25 worms had excysted, after $2\frac{1}{4}$ h 65 worms, and he observed that "die Cystenhüllen waren nicht verdaut, sondern an ihrer Basis von den Egeln durchbrochen worden". The flukes were washed in saline and injected into the abdominal cavity of a rat, which was killed 41 days later and was found to have a very channelled and cirrhotic liver. In this remarkable digression from his main task Vogel displayed the master's touch and demonstrated the abdominal route taken by *F. hepatica* to the bile duct —in marked contrast with the results of his researches on *Opisthorchis*, which follows the biliary route.

Some investigators did not concern themselves with the route of migration. Montgomerie (1928a) was more concerned with the period which elapses between the time of ingestion of cysts and the time when eggs of the fluke appear in the faeces of the host, because such information would indicate when sheep could be treated with drugs to destroy the maturing flukes. He removed cysts collected on the sides of glass vessels, placed them on pieces of damp lettuce leaves, inserted these in hard gelatinous capsules, and administered these in turn to the sheep. In this manner he infected thirty-two sheep with 2,550 cysts and recovered a total of 954 flukes, which were seen to vary somewhat in size and degree of development at any of the various stages reached. The counts of eggs in the faeces of infected sheep also varied; in three sheep they were appearing in numbers large enough to be determined after 10 weeks, in sixteen sheep only after 11 weeks, in two sheep only after 12 weeks, in three sheep only after 13 weeks and in four sheep only after 14 weeks. His general conclusion was that sheep may commence to pass eggs 10 weeks after infection, more commonly about the 11th week, or not after even 14 weeks, and he attributed such variability to varying rates of development, influenced possibly by degrees of infection and other factors. Montgomerie (1931) concluded experimental infections of three rabbits, showing that flukes lived in these hosts for 1–3 years and that eggs appear in the faeces 8–9 weeks after infection and therefore 2–5 weeks sooner than he had reported for sheep.

The fecundity of the digenetic trematodes is a measure of the risks involved in continuing the life cycle and bringing it to completion. This topic was discussed by Dawes (1946:1956), who gave on the larval side of the picture a number of examples of daily rates of liberation of cercariae from snails and also cited the experiments of Krull (1941), who exposed twenty-one snails *(Pseudosuccinea columella)* individually to infection with single miracidia of *Fasciola hepatica*. Of these snails, eight became infected and began to emit cercariae 68–69 days after infection, continuing to emit them for 10–77 days. The total numbers of cercariae which developed from a single miracidium varied between 14 and 629, and Dawes remarked that under natural conditions the larger number might be exceeded. Realizing that one fluke egg may produce more than 600 potential flukes it is interesting to inquire how many eggs one adult fluke may produce. Montgomerie (1931) provided such information. He gave 40 cysts to a rabbit and found eggs in the faeces of this host 8 weeks later. The egg count was continued until this host died, which was 3 years and 1 month after infection; it rose to its highest level during the 18th week of infection at 766 eggs per g faeces, thereafter becoming stationary at about 200 for some time. One year after infection, however, the count was 533 and it remained at about this figure until the host died. This rabbit was producing about 190 g of faeces per week and the approximate mean egg count for 152 weeks during which eggs were passed was 450 per g of faeces. Montgomerie's estimation was therefore that this rabbit, from which 11 flukes were recovered, passed about 13 million eggs during the 3 years of infection, a yield of well over 1 million per fluke. Possibly this degree of fecundity is exceeded in instances where infection persists longer but it may be taken safely as one which is fairly representative in fascioliasis.

The wanderings of young flukes were very closely investigated by Schumacher (1938). His first series of experiments (1–9) involved guinea pigs, seven of which received 400 cysts and the remaining two 300 and 150. The hosts were killed serially at 30 min, 45 min, 1 h, 1¼ h, 1½ h, 2 h (twice), 3 h and 4 h after infection and the alimentary canal, body cavity, liver, main bile duct and gall bladder examined for flukes. In short-term experiments (up to 1½ h) the stomach, small and large intestines and caecum were scrutinized and unaltered cysts, inner cysts with or without young forms and free flukes were separately recovered and counted. The small intestine was divided into three or four measured parts for more precise delineation of results. Tables were prepared to give a clear picture of the distribution of cysts and flukes at specified times after infection. After 30 min unaltered cysts and complete inner cysts preponderate in the stomach, but 5 empty cysts and 5 free flukes were recovered from it, as well as 20 flukes from the first half of the intestines. After 45 min fewer unaltered cysts and inner cysts were seen in the stomach and 1 free fluke was recovered, but in various parts of the small intestines there were 36 inner cysts, 61 empty cysts and 26 young flukes. After 1 h 2 empty cysts and 1 fluke were found in the stomach and 14 empty cysts and 11 flukes in the intestines. After 1¼ h and 1½ h, here taken together, the small intestines yielded 34 complete inner cysts, 23 empty cysts and 31 flukes. Up to this time, nothing significant was found in large intestine and caecum, and no free flukes were seen in the abdominal cavity, liver, bile duct or gall bladder. After 2 h (two infections with 150 and 400 cysts respectively) 3 + 3 free flukes were found in the small intestines but 16 + 7 in the abdominal cavity. After 3 h and 4 h flukes were not found in the small intestines but 32 + 56 were recovered from the abdominal cavity. To summarize these results, during the period 30 min to 4 h after oral infection with cysts the stomach contained mainly unaltered cysts and cysts with the outer coat removed, empty cysts and free flukes being extremely rare. During the first 2 h the small intestines contained relatively few unaltered cysts but many inner cysts, some of which were empty, and many free flukes. After 2 h flukes began to enter the abdominal cavity from the small intestines and after 4 h their numbers increased, but the portal system and the bile duct remained devoid of free flukes, which were never found in the large intestines or caecum.

Schumacher's next series of experiments (10–23) involved fourteen guinea pigs which received 140–500 cysts each in infections of 8–48 h duration. The abdominal cavity, liver, portal veins (in which flukes were never found) and bile ducts were scrutinized for young flukes. Five other infections (2–400 cysts to each of five guinea pigs) were maintained for 3–6 days. After 48 h only 9 flukes had reached the liver (one of them after 24 h). After 12 h 130 flukes (from 400 cysts) were located in the abdominal cavity and in three 24 h experiments 104, 92 and 200 flukes were located therein (= 42%, 46% and 40% of cysts). In the 3–6 day infections 34 and 78 flukes (14% and 19·5% of cysts) were recovered from the abdominal cavity and 11 and 22 flukes (4% and 5·5% of cysts) from the liver. From 3 days onwards the numbers of flukes in the abdominal cavity diminished to 8, 4 and 7 at 4, 5 and 6 days (4%, 2% and 3% of cysts), while the numbers of flukes in the liver increased from 16 (= 8% of cysts) at 4 days to 18 and 57 (= 9% and 23% of cysts) at 5 and 6 days respectively. After 6 days nearly 25% of the cysts had produced flukes which had entered the liver, none by way of the portal veins or the biliary system. Summarizing these results, most young flukes have assembled in the abdominal cavity about 24 h after infection and begin to enter the liver after 48 h and have entered the liver in 4–6 days.

Other experiments of Schumacher were concerned with extended migration to the bile duct and gall bladder. Some (Expts. 29–38) were with guinea pigs and without numerical results. Guinea pigs were regarded as poor experimental hosts, because liver damage was serious and flukes wandered back into the body cavity after 23, 24, 34 and 35 days. Later experiments (39–56) were carried out with rabbits, but in this host also flukes emerged from the liver after 22, 49 and 66 days. However, some flukes had entered the bile ducts 45 days after infection, most of them after 55 days, proceeding along the smaller to the larger ducts. Exactly how long the young flukes remain in the hepatic parenchyma was not determined, however, and it was suggested that "dies könnte allein durch histologische Untersuchungen festgestellen werden". The final series of experiments concerned five sheep, each of which was given 2,000 cysts. The animals were killed 24 h, 48 h, 72 h, 35 days and 56 days after infection. In 1, 2 and 3 day infections 125, 186 and 115 flukes were recovered from the abdominal cavity but none from the hepatic portal system, main bile duct or gall bladder. After 35 days the abdominal cavity yielded 20 flukes ("in ihrer Grösse und Entwicklung stark zurückgebliebene Würmer") but about 500 were removed from the liver, many of these also being stunted forms, presumably as a result of the heavy infection. Even 56 days after infection 27 flukes were found in the abdominal cavity but these were not noticeably smaller than about 508 flukes recovered from the liver. In the bile duct of this 56 day infected sheep 8 other flukes were found, bringing the total number to 543.

During the course of researches primarily concerned with chemotherapy

Hughes (1959) produced notable results. In a group of rabbits given 15 cysts each, eggs of the flukes first appeared in the host's faeces 54 days after infection, the mean period for the group being 61 days. In a further experiment the minimum period was reduced to 51 days and all the rabbits had fluke eggs in the faeces by the 54th day. Other experiments with twenty-eight rabbits which received 15 cysts each produced a total of 267 flukes in groups of 2–15 per rabbit, giving a mean of 9·5 flukes per rabbit and a yield of 63·3% from cysts. In two instances the yield was 100% and in more than one-half of the rabbits (16/28) it was somewhere between two-thirds and the full number of cysts administered. One other set of experiments in which 225 rabbits received 15 cysts each, the mean number of flukes recovered was 7·6% (= approximately 50% of cysts), although some of these hosts had received drug treatment with inactive compounds. Experiments were varied by giving 50 cysts to each of ten rabbits, the yield of 242 flukes representing 5–36 per rabbit and 10–72% in terms of cysts administered. One interesting experiment in which some rabbits were given one cyst each proved that the fluke is self-fertilizing, because living flukes were recovered which were producing eggs 9 weeks after infection.

Hughes also experimented with mice, which were given 10 or 5 cysts. Mice did not die before the 19th day of infection in any of these experiments and immature flukes were recovered from abnormal locations. One mouse was dying on the 48th day and a fluke described as "quite large" was recovered. Two mice were killed on the 56th day and appeared to be healthy, although four flukes were recovered from one of them and one fluke from the other; fertile eggs of the flukes were found in the rectum, faeces and gall bladder. One mouse of a further experiment had eggs of the fluke in the faeces 44 days after infection and when killed on the 50th day produced six flukes. Mice used by Hughes survived infection 8 weeks and thus 2 weeks longer than Taylor and Parfitt (1957) believed possible and a prepatent period of 6 weeks was determined as against one of 11–14 weeks in sheep and 75 days in calves.

Krull and Jackson (1943) infected mice and guinea pigs orally with cysts, recovered flukes from the abdominal cavity and liver of these hosts and then transferred some of the flukes "in saline by means of a pipette" directly into either the abdominal cavity or the pleural cavity of guinea pigs, rabbits and sheep. Guinea pigs were regarded as unsatisfactory hosts because flukes did not mature in them, but twenty rabbits and three sheep were infected in this way and various results appear in their Table 1. In rabbits 66 transferred flukes were obtained from seventeen mouse infections of 7–30 days and 6 flukes from three guinea pig infections of 29–31 days. The numbers of flukes transferred varied, 2–11 from mice and 1–20 from guinea pigs. In addition, 41–48 flukes from 11–20 day infections of mice were transferred to two sheep. The total numbers of flukes recovered after transfer were 56 out of 72 from rabbits and (2 hosts) 10 + 17 = 27 out of 89 from the sheep. Eggs were found in the faeces of rabbits 34–68 days after transfer, giving ages of the flukes from the time of infection with cysts as 62–99 days. Breaking down these results according to the ages of flukes at the time of transfer to rabbits (i) six groups of flukes (Total = 17) were 7–9 days old, (ii) four groups (T = 27 flukes) were 16–20 days old, (iii) eight groups (T = 23 flukes) were 27–30 days old, and (iv) two groups (T = 5 flukes) were 31 days old. The four corresponding ranges of age at the time of egg laying by the flukes were (i) 70–74 days, (ii) 66–76 days, (iii) 62–76 days, and (iv) 99 days (one rabbit only). Excepting the last result, possibly indicating late maturation after transfer from a guinea pig, the mean age of transferred flukes at maturity (62–76 days) was approximately 70 days. One of the three sheep received 15 5-day-old flukes and these matured after a further 74 days (total age 79 days); the second sheep received 41 11-day-old flukes, which matured 90 days later (total age 101 days); and the third sheep received 48 20-day-old flukes, which matured 59 days later (total age 79 days). These experiments showed that flukes taken from mice 8–30 days after infection and transferred to the abdominal cavity of rabbits behave like young flukes which escape from their cysts, migrating through the liver to become mature in the bile ducts. The rate of maturation was not significantly affected by transfer, because three control rabbits given 11, 17 and 40 cysts respectively produced 1 + 6 + 6 flukes which matured 66, 69 and 81 days after infection, and one sheep which received 130 cysts produced 21 flukes which matured 75 days after infection. The experiments proved also that when flukes are transferred to the pleural cavities of rabbits they may enter the lungs or migrate through the diaphragm into the abdominal cavity, find and enter the liver.

It is known that some liver flukes do not reach their final location by the abdominal route, as does *F. hepatica*. Neuhaus (1938) suggested that *Dicrocoelium dendriticum* follows the hepatic portal route, but Krull (1956, 1958) proved that it takes the biliary route. In the earlier work he infected mice, rats, hamsters, guinea pigs and cats with this trematode and in the later work fed cysts to the host, carried out serial kills at appropriate times and removed the viscera. By teasing various organs in warmed saline he found that young worms may reach the biliary system by way of the opening of the bile duct. His data (Table I) show that in four infected hamsters 60 flukes were found in the anterior portion of the intestines and only 6 in the posterior portion, that 142 flukes were found in the common and cystic bile ducts, 3 in the gall bladder and 139 in the hepatic bile ducts. These figures are probably reliable but the exercise of recovering flukes took 48 h. Similar results were obtained with four mice which received 200 cysts each. After 6 h of infection 198 flukes occurred in the anterior part of the intestines, only 13 in the posterior part, but 70 flukes were found in the common and cystic bile ducts, 15 in the gall bladder and 21 in the hepatic bile ducts. Washings of the abdominal cavity did not yield any flukes, nor were any found on the surfaces of intact viscera. Under a binocular microscope flukes were seen inside various ducts in intact viscera and in three hamsters young flukes were seen to enter the common bile duct by its intestinal opening.

Although some young liver flukes do not follow the course taken by *F. hepatica*, lung flukes of the genus *Paragonimus* do follow this course and usually extend it. This is indicated when a comparison is made between findings already mentioned and results obtained for *P. westermani* by Saduma Yokogawa (1915a, b, c, 1916) and confirmed by Nakagawa (1915a, b). This early work, originally published in Japanese, has now been reviewed in English by Yokogawa *et al.* (1960). S. Yokogawa found that the young lung flukes excyst in the upper or middle part of the small intestines of puppies and kittens, penetrate the intestinal wall, wander about for two or more weeks in the abdominal cavity, sometimes entering but usually also leaving the liver, pierce the diaphragm, enter the pleural cavity and finally reach the lungs to become mature 60–70 days after infection. It may prove to be significant in this comparison that *Paragonimus* may leave the liver, almost invariably locating the lungs, while *Fasciola* often leaves the liver and may enter the pleural cavity and the lungs. Both flukes are found occasionally in ectopic situations, which is another indication of remarkable resemblance in their potentiality to wander and their capacity to enter many different kinds of tissue.

In the case of *P. westermani* the first young flukes were found in serous exudates in the abdominal cavity of puppies and kittens 5½ h after infection. Only five flukes were found in the earliest researches but it was clear that they had penetrated the wall of the intestines, because thirteen flukes were seen imbedded therein. Within the mucosa the flukes lay in the long axis of the intestines, those in muscles in the long axes of the fibres, the course of penetration seeming to be determined by the resistance offered by the tissue traversed. In the abdominal cavity young flukes appeared to wander at random, because they were found in various organs. The young flukes invaded the liver in large numbers (Yokogawa, 1919), and infections were more abundant on the upper and abdominal margin of the liver than on the lower margin. Most burrows ended blindly, possibly because flukes had turned about and vacated them. "Apparently, the liver tissue for some reason is unfavorable for the worms" (Yokogawa *et al.*, 1958). Subsequent penetration of the diaphragm "appears to be similar to that of penetrating the muscle layers of the intestine" and, once in the pleural cavity, young flukes enter the lungs through their outer surfaces. While some flukes penetrate the diaphragm within about 70 h of infection, others wander about for as long as 20 days before entering the thorax, growing meanwhile. Some young flukes stay in the abdominal cavity and penetrate different organs up to the time of maturity without ever reaching the lungs.

III. THE EXCYSTMENT OF METACERCARIAE

A. *In vitro* EXPERIMENTS WITH *F. hepatica*

A number of attempts have been made to induce the excystment of metacercariae *in vitro*, with varying success. Mention has been made of Susuki (1931), who placed cysts in artificial gastric juice (0·2% HCl + 1% pepsin) at a suitable temperature and then added artificial intestinal juice (0·2% $NaHCO_3$ + 1% pancreatin + 5–7% cows' gall). He realized that young forms are liberated only after reaching the intestines and stated "the larva must be active in the capsule and the pressure inside the capsule must be high enough if the larva is to penetrate the wall of the capsule". We have also seen that Vogel (1934) was aware that the young fluke breaks through the wall of the cyst. Schumacher (1938) infected a guinea pig with 400 cysts and killed the host 30 min later, recovering cysts which showed "an der Basis eine kreisförmige Öffnung, durch welche die Cercarie die Hülle verlassen hatte", adding later, "gesprengte innere Cystenhüllen waren nicht zu erkennen". Hughes (1959) also found that fresh intestinal contents induced excystment; 50 cysts were placed in intestinal juice from a rabbit and 43 flukes emerged 17 h later at a temperature of 37° C. He observed the active movement of young forms within their cysts and saw that emergence was anterior end foremost, but did not witness the entire process. The empty inner cyst had a circular hole on the more flattened part of its surface.

After pointing out that the "mouse test" for infectivity of juvenile flukes devised by Taylor and Parfitt (1957) requires a waiting period of about 3 weeks, Wikerhauser (1960) developed a method of testing viability by the use of digestion mixtures with cysts. He used both pepsin and trypsin solutions, but studied the action of trypsin alone, and also additions of 10%, 20% and whole ox bile. Without pretreatment of cysts with pepsin, a solution of trypsin either with or without additions of bile did not bring about any change in the appearance or behaviour of young encysted forms. Pepsin brought about partial digestion of the outer cyst and rare and weak movements of the young forms could be seen through the transparent inner cyst after 45 min in some instances but usually during the 2nd or 3rd h. Complete excystment never occurred in pepsin alone, even after 48 h. When trypsin was added to the pepsin solution containing cysts after 2–3 h, the young forms became active and excysted, and the highest rate of excystment (about 80%) was obtained with trypsin containing 20% ox bile. The rate was somewhat lower in 10% bile and very low (less than 1%) in trypsin without bile. In trypsin plus 20% bile most flukes emerged in the 2nd or 3rd h, but some excysted after 15 min, passing through a circular opening about 0·065 mm diameter, and moved about actively for 1–2 h but later died and disintegrated. Their empty cysts resisted proteolytic action for at least 48 h, confirming the *in vivo* findings of Schumacher (1938). When Tyrode's solution replaced digestive media after excystment, the young forms survived for periods up to 42 h.

Hughes (1963) reported his previous results and also noted that excystment took place only if cysts were treated with 1% acidified pepsin solution before being placed in an artificial intestinal juice made up of trypsin, pancreatin, Na taurocholate and cholesterol in 1% $NaHCO_3$ solution. The degree of excystment was then 50% after 3½ h, 55% after 4½ h and 60% after 5 h of final treatment at 37° C. In this instance pretreatment with pepsin lasted 1½ h but there was little difference within the period 1–2 h and 70% excystment was achieved.

In his experiments with cysts Shirai (1927) found that during the first few hours of encystment cysts are non-infective, although after 12 h a certain number reached an infective stage. Lengy (1960) treated cysts of *Paramphistomum microbothrium* with "extract of small intestine" and also observed differences in degree of excystment depending on the ages of cysts. With cysts 23–40 days old there was 58–54% excystment; with 11-day-old cysts the figure fell to 44% and with 8-day-old cysts to 8%; excystment did not occur from 2-day-old cysts. Drawing attention to these experiments, Dawes (1962a) suggested that the young forms of *F. hepatica* might not undergo excystment if cysts were less than 1 week old. Hughes (1963) found however that with 2-day-old cysts 63% excystment was achieved and with 1-month-old cysts 92% excystment. The two batches of cysts were not identical and it was difficult to know whether or not young flukes excyst more readily in older cysts. However, excystment does occur in cysts no more than 2 days old and the young forms are infective.

It is likely that the activation of young forms and excystment requires different sets of conditions in other trematodes, but in view of the paucity of information concerning the process in *F. hepatica* we might consider this matter further.

B. THE EXCYSTMENT OF *Paragonimus westermani*

According to Yamaguti (1943) the complete cyst of *Paragonimus westermani* is thick and rigid and measures 0·35–0·48 × 0·33–0·48 mm, and the inner cyst is only a little smaller and has a wall 0·01–0·02 mm thick. It resembles the cyst of *F. hepatica* but is much larger. Excystment was first investigated in this species by Yokogawa (1918, 1919), who found that in some instances young flukes emerged from their cysts when these were immersed in warm water. Inside cysts placed in artificial gastric juice at suitable temperatures the young forms began to move vigorously but within 2–3 h they became still and in 11–16 h they died without excysting. Fresh cysts were unaffected by treatment with dilute acidic or alkaline solutions, or water, but young forms in older cysts were sometimes able to free themselves in such media. Most fresh cysts were unaffected by artificial intestinal juice (0·5–1·0% pancreatin + 0·1% $NaHCO_3$) but the medium affected some young forms and they excysted in 30–90 min, and sometimes in 15–42 min, if they had been placed previously in artificial gastric juice. Note was taken that digestive juices do not break down the cysts but it was claimed that they make the cysts more fragile and more easily broken by the young flukes.

Khaw (1935) treated cysts measuring about 0·34–0·51 mm diameter in similar ways and confirmed Yokogawa's results, also noting that the young forms escaped through a triradiate opening in the otherwise intact cyst wall. Khaw extended his experiments to include treatment of cysts with artificial intestinal juice + ox bile, intestinal juice without bile and 12% bile only, obtaining excystment in each case and respectively after 105, 45–90 and 75 min. After making prolonged experiments with "dead cysts" and unaware of Yokogawa's work, Khaw realized that digestive juices are without action on the cyst wall.

Oshima (1956) followed up these investigations with a study of the effects of pH and temperature on excystment, which occurred in the absence of digestive enzymes and was clearly dependent on activation of the young encysted forms. In a series of experiments of 3 h duration carried out at 37° C excystment at pH 7·5, 8·0, 8·5 and 9·0 was (in round figures) about 10%, 20%, 40% and 30% respectively, corresponding figures for similar experiments at 40° C and pH 7·5, 8·5 and 9·0 being 5%, 70% and 60%. Temperature and pH were both significant factors in excystment. Oshima *et al.* (1958) studied excystment *in vitro* under various physical and chemical conditions. The most effectual temperature for excystment is 40° C, the optimum pH range 6·5–8·0 and the most favourable osmotic pressure $\Delta =$ –0·40° C. Excystment never occurred under conditions of high osmotic pressure ($\Delta =$ –1·0° C), or when acidity exceeded pH 6·0, or at temperatures lower than 30° C. Even under the most favourable conditions of temperature, pH and osmotic pressure it was 3 h before 50% excystment was achieved, but the addition of 0·5% bile salts (sodium cholate and sodium deoxycholate) to the medium brought about 50% excystment within 30 min. The importance of bile salts was thus clearly demonstrated, and it was found subsequently by Oshima and Kihata (1958) that excystment occurs within 10 min when bile salts are present, after pretreatment with artificial gastric juice. However, the effect was not due to either pepsin or degree of acidity but to low osmotic pressure and a temperature effect, and it was affirmed that proteolytic enzymes are not involved in the process of excystment.

C. *In vivo* EXPERIMENTS WITH *F. hepatica*

In the experiments of Dawes (1961b) mice were sometimes killed 2–3 h after cysts had been fed to mice in order to obtain young flukes, and excystment was witnessed. The activated young flukes within cysts removed from the intestines of the host moved about freely "extending and contracting the body in lively motion". Eventually, the young fluke was seen to attach itself to the broader part of the cyst by the oral sucker and to remain so attached for some time. At the point of attachment covered by the mouth a hole eventually appears and through this opening the fluke squeezed with considerable effort. The fibrils comprising the inner cyst were broken at slightly different levels, giving the opening a slightly ragged edge. An appended figure (Fig. 1, A) showed the body extremely constricted in the middle during emergence so as to have an hour-glass shape. In a later paper, Dawes (1963a) extended his observations by means of sections passing through excysting forms, indicating that the ventral sucker is able to grasp the fibrils of which the inner cyst is composed (Fig. 2, A) and pull the layers apart (Fig. 2, B), thus weakening the wall for the further action of the oral sucker, which is able to deal with the layers one by one in effecting the breakthrough. When this happens, the outer cyst may have been completely removed by enzyme action and churning peristaltic movements of the intestines. Otherwise, the outer cyst may be broken near the point of emergence or widely removed from it, and the young fluke has the task of pushing aside the remains of the outer cyst as it emerges by great extension of the forebody into the intestinal lumen (Fig. 5, A).

D. EXCYSTMENT IN UNUSUAL LOCATIONS

In addition to the methods of inducing excystment already mentioned it is known that young forms excyst when the cysts are put directly into the abdominal cavity (Hughes, 1959, 1963; Dawes, 1961a–c). Hughes (1959) infected a mouse by the intraperitoneal route, using cysts in distilled water; the mouse died 26 days later and one fluke was recovered from the thorax. Excystment probably occurred within the body cavity but it was possible that cysts could have been injected into the alimentary canal. In order to "test the possibility that activation of the metacercariae and excystment by perforation of the cyst are independent of the digestive processes in the alimentary canal" Dawes (1961b) gave 30 cysts to each of two mice, along with antibiotics, by intraperitoneal infection, care being taken that the cysts should not enter the viscera. These mice were killed 5 and 6 days later respectively and about one half of the cysts were recovered in each instance. Some cysts contained young flukes, others were empty and perforate, but all had been subjected to an intense foreign body reaction and were undergoing phagocytosis. Leucocytes had penetrated into some of the cysts and were attacking the young flukes, but in each of the two hosts three active young flukes were recovered from the abdominal cavity. The flukes were fixed, stained and mounted, then measuring 0·18–0·23 × 0·10–0·12 mm. It was stated that the experiment should be repeated but the result was categoric: the cysts may have been in contact with enzymes, but not enzymes of the alimentary system, and it was concluded that the digestive enzymes are not essential for either activation of young encysted flukes or excystment.

As young flukes will excyst in a wrong location within the host, it seemed possible that they might excyst in a "wrong" host. To test this possibility Dawes (1961b) orally infected six chickens about 1 week old with 50 cysts each. Two of the birds were killed 3 days later and the others at intervals during the period 7–34 days after the administration of cysts. From the abdominal cavities of the two birds killed after 3 days six flukes were recovered in one instance and two in the other. These flukes, when fixed and stained, measured 0·18–0·24 × 0·12–0·16 mm and were therefore somewhat smaller than flukes from similar infections in mice and recovered from the liver. One other fluke was found in the abdominal cavity in a 17-day infection of the chicken and this, when fixed and stained, measured about 0·47 × 0·35 mm. This fluke (Fig. 2, G) had grown to some extent but was clearly a stunted form by comparison with 17-day-old flukes from the liver of the mouse, which are more than 3 mm long. Evidently, young flukes can maintain themselves and grow a little by browsing on the serosa of the abdominal viscera, even when located in a "wrong" vertebrate host.

IV. Growth and Development of the Flukes

No sustained effort to investigate the growth to maturity of *F. hepatica* was ever made until Dawes (1962a) measured 356 flukes recovered from experimentally infected mice and arranged the data in nineteen age groups representing infections of 1–37 days, to which was added the measurements of one fluke from an infection of 20 weeks (Fig. 4). Some of these specimens were the basis of a demonstration (Dawes, 1962b) and in a later paper (Dawes, 1962c) an additional 60 flukes from thirty-four mice were used to supplement the original data from 25 days to and beyond the time of maturity of the flukes.

A. Size and Growth

Less than 24 h after mice had received cysts, Dawes (1962a) found young flukes 0·12–0·24 mm long creeping about over the abdominal viscera, and for a group of 45 flukes the mean length was 0·17 mm and the mean breadth 0·10 mm. On the 3rd day of infection and in a random sample of 20 flukes mean length had increased to 0·29 mm and mean breadth to 0·18 mm, indicating that nutrients must have been obtained by the flukes during their wanderings from the abdominal cavity to the liver. In the body of this paper it was noted that maturity is reached 37 days after infection, by which time length had increased to about 10·6 mm and breadth to about 4·2 mm, but in a footnote on p. 13 it was shown that maturity can be reached after 35 days. In the original grouping of data (Dawes, 1962a, Table I) size was related to the duration of infection, more than 350 flukes forming eighteen arbitrary stages prior to the attainment of maturity. This revealed very wide ranges of variability in size for infections of almost any age, some individuals of one and the same age group being more than twice as long and broad as others.

The regrouping of data in Table II as size groups irrespective of the duration of infection showed that flukes of a given size group could be referred to infections varying in duration by as much as 12 days. For instance, 35 flukes of Group 12 (mean length 3·23 mm) included specimens from infections maintained for 15, 20, 21, 22, 24 and 25 days. This effect might have been due to sustained variability of size which occurs in very young flukes, or related to the time taken to reach the liver and other inequalities of nutrition, but it was not due to significant differences in the intensity of infection. About 21 days after infection, however, growth diminished abruptly, and this was attributed to intense liver damage sustained by the selected hosts of the flukes during a critical period which arises in heavy infections but is rare in light infections. It may be significant that at this time and for a period of a week or more young flukes may come to lie in blood-filled cavities in the liver, in which case they show a definite tendency to migrate out of the liver and return to the abdominal cavity. Inhibition of growth at this time may accrue as a result of faulty nutrition, both within the liver and in the body cavity. In his later paper, Dawes (1962c) confirmed these results. Mice which showed any signs of distress were killed at once for autopsy, as before, so that those providing nutritional hazards of this kind for the flukes were automatically selected from the main infected stock. Opinion was strengthened, therefore, that when the young flukes must feed upon blood in a badly damaged liver they do not grow at normal rate and, indeed, may not grow at all. An important note must be made, however, that many mice either do not experience critical liver damage or else recover quickly, with the result that normal growth is continued or resumed. In six periods of infection (25–32 days) the range of length was 2·7–5·9 mm and the means 4·0–4·9 mm. In older infections there was some improvement of growth just prior to the attainment of maturity at 35–36 days, and in the combined results of both series of experiments 33 flukes in groups of 1–6 in seventeen mice were laying eggs for periods increasing to a maximum of 14 weeks. The complete ranges of sizes for these flukes were 8·1–20·9 mm (length at 36–71 days) and 4·2–9·6 mm (breadth at 37–140 days). The largest flukes were recovered 71 days after infection, when the ranges of length and breadth were 19·3–22·5 × 8·3–9·9 mm. In most instances when adult flukes were recovered liver damage was negligible (see Fig. 10), so that not only can well-grown flukes be nurtured in mice (Figs. 2–4) but serious liver damage is not inevitable. Growth curves of length and breadth are shown in Fig. 1.

Other investigators have found that flukes grow well in mice. In one instance, Lagrange and Gutmann (1961) obtained a fluke measuring about 16 × 8 mm in a healthy mouse 91 days after infection. This host had also sustained for 49 days an experimental infection of *Schistosoma mansoni*, so that the liver fluke must have been mature at the time of infection with blood flukes, because these investigators found mature *F. hepatica* in mice 33 days after infection. In their infections adult flukes were 9–10 mm long at 36 days, 11 mm long at 53 days, and 11·5–16 mm long at 108–150 days, the largest specimens in the shortest of these long infections.

Striking features of differential growth were revealed by measurements of the suckers, the anterior cone and other parts of flukes by Dawes (1962a). In the ultimate analysis, the anterior cone (which in the living fluke is the highly

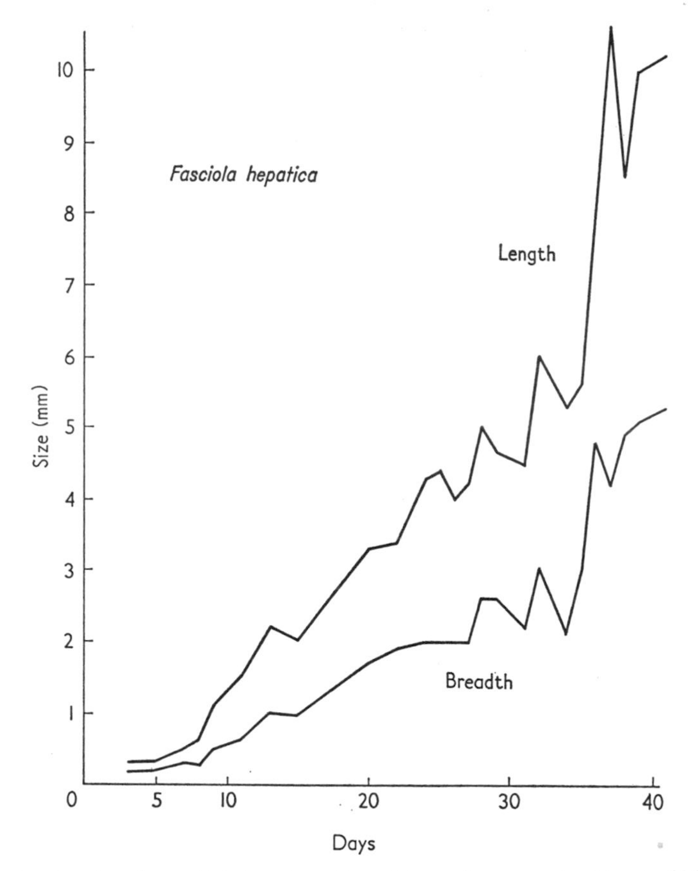

FIG. 1. Graphs of growth in length and breadth (mm) to maturity (Dawes, 1962c).

extensile and contractile region of the body in front of the ventral sucker) takes up only 15–16% of the length of the body in flukes 10–15 mm long but develops from almost the entire anterior half of the body of newly excysted flukes. The oral and ventral suckers diminish in relative size during growth, ranging from 24% to 5% and 27% to 6% of body length respectively. We might imply in consequence that during feeding operations the oral sucker will inflict relatively less damage to liver tissue in proportion to

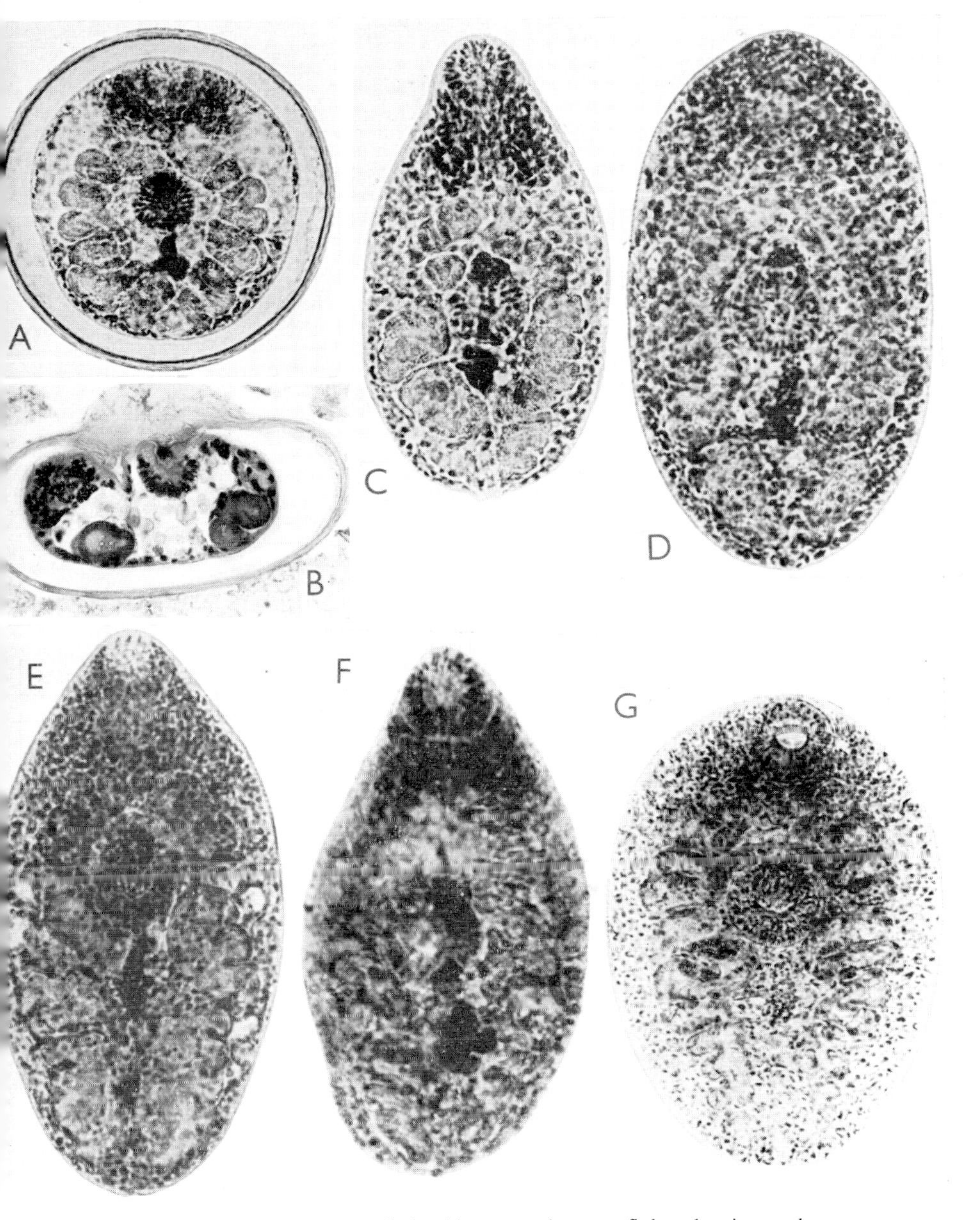

Fig. 2. A, Inner cyst (0·19 mm dia.) with encysted young fluke, showing suckers, caeca and genital rudiment. B, L.S. cyst and fluke; ventral sucker grasping fibrils, weakening cyst for action of oral sucker. C-F, flukes from mice; ages of infection and actual sizes specified. C, 1 day (0·21 × 0·12 mm); note genital rudiment of hour-glass shape. D, 3 days (0·28 × 0·16 mm). E, 5 days (0·39 × 0·21 mm). F, 6 days (0·30 × 0·17 mm). G, Fluke recovered from abdominal cavity of chicken 17 days after oral infection with cysts (0·47 × 0·35 mm).
(B, Dawes, 1963a; G, Dawes, unpublished; remainder, Dawes, 1962a.)

body size with increasing age and development. Smaller increments are added to breadth than to length of the body during growth, so that the proportions of the fluke are changing continually as size increases. In the smallest flukes the centre of the ventral sucker is situated just in the posterior half of the body, whereas in flukes which are attaining maturity it is situated in the anterior quarter of the body. The more intensive growth of the posterior region of the fluke's body thus creates the illusion that the ventral sucker is moving anteriorly throughout the period of growth to maturity and, in fact, to some extent throughout life. During a period of 20 weeks, that part of the body which lies in front of the ventral sucker increases in length by less than 3 mm; the posterior region by more than 12 mm. The changing form of the body does not mask the identity of this fluke, but we are reminded that such features of allometric growth may place under suspicion specific taxonomic allocations which are based on differing proportions of flukes of various sizes in some circumstances.

B. MATURATION

Dawes (1962a) also noted the general features of maturation in *F. hepatica*. Contrary to previous statements that the genital rudiments appear during juvenile life, he showed (1962b) that they are present in the cercariae. In young juveniles this rudiment is a condensation of embryonic cells of hourglass shape and about 0·1 mm long (Fig. 2, C). The anterior club-like portion is destined to give rise to the cirrus, cirrus pouch and the terminal part of the uterus. The similarly enlarged posterior portion forms the gonads, the shell gland complex and related ducts, and the intervening portion also contributes to the genital ducts, uterus and vas deferens. The hindmost part of the genital rudiment is U-shaped, and from this portion both testes develop; they are connected at the 3 day stage (see Fig. 2, F at 6 days and C at 1 day also) but completely separated by the 8th day of infection (Fig. 3, A). Slightly anterior to this testis rudiment there is a single, spherical cell cluster which henceforth stains intensely with basic dyes; this ultimately gives rise to the ovary and the ootype, with its accompaniment of shell glands. The ovary originates as a small single lateral outgrowth at about the 8th day of infection and during the next two or three days it becomes finger-like and at about the 11th day shows a terminal bifurcation (Fig. 3 C); subsequently, these outgrowths undergo further branching. By this time the rudiments of the testes have passed through stages in which the surfaces are crenate and then lobed (Fig. 3, B), and the intestinal crura have produced lateral sacculations (Fig. 3, A) and then long, finger-like caeca which are undergoing primary and secondary branching, with tertiary branches appearing (Fig. 3, B). The body of the fluke is ovoid but becoming elongate during this period, with as yet no indication of the characteristic shape of the adult fluke. The only indication of the anterior cone is a slight difference in the thickness of the cuticle ante-

FIG. 3. Flukes from mice showing features of maturation; ages of infection and actual sizes specified. A, 8 days (0·81 × 0·43 mm). B, 11 days (small specimen; 1·17 × 0·55 mm). C, 13 days (2·31 × 1·16 mm). D, 23 days (3·55 × 0·55 mm). E, 24 days (5·0 × 2·7 mm). F, mature fluke of 37 days (10·6 × 4·2 mm).

(D, Dawes, unpublished; remainder, Dawes, 1962a.)

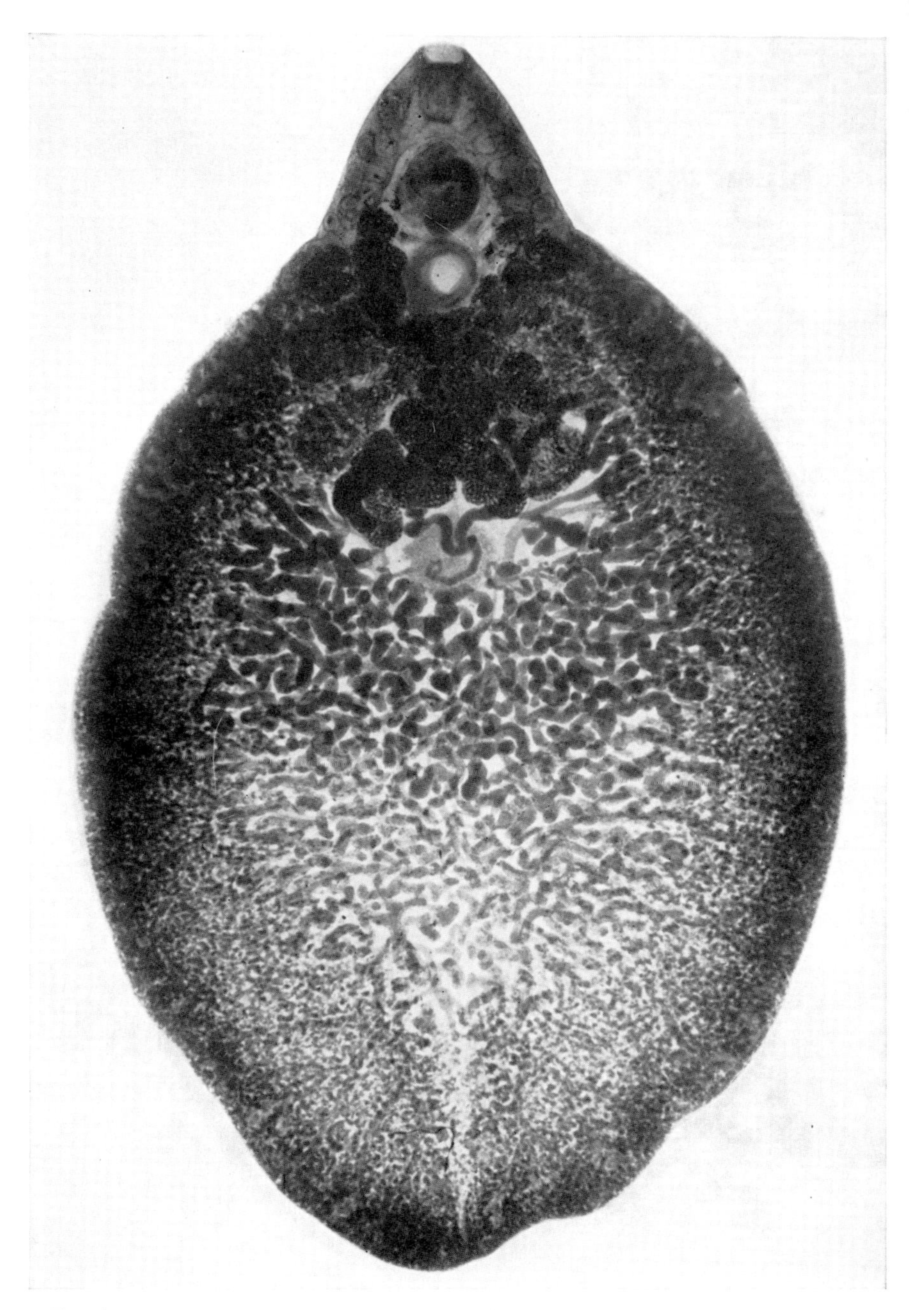

Fig. 4. Mature fluke from mouse 140 days after infection. Note masses of normal eggs *in utero* and masses of sperms in seminal vesicle. (Dawes, 1962a.)

riorly and the more elongate and sharply pointed nature of the spines there. By the 13th day of infection (Fig. 3, C), however, the anterior region is becoming more or less sharply defined in front of the ventral sucker. In the broadening posterior region the tertiary outgrowths of the caeca are approaching the margin of the body more closely, the testes have long, branching outgrowths and the ovary is also showing early branches. By the 21st to the 24th days of infection the hindbody is much broader and the gonads and caeca considerably branched. The cirrus and its pouch, which were evident in the previous stage, are also more advanced and the folds of uterus extend across the intestinal crura and into the lateral fields of the body (Fig. 3, D, E). The young fluke is now clearly recognizable as a young *Fasciola* and during the 4th week of infection this appearance has become even clearer, with the development of follicular testes, branched ovary (still connected with the ootype and oviducal rudiment), and a cirrus with an associated seminal vesicle. Both the uterine folds and the caeca are now even more extended and the young fluke is about 1 week from maturity. The rudiments of the vitellaria are evident, particularly in sections, but the follicles are still unripe and must still accumulate their secretion. During the final few days before maturity is attained this secretion forms rapidly so that quite suddenly the branching caeca are to some extent masked by the vitellaria (Fig. 3, F). About 37 days after infection the uterine folds are filled with eggs, which first appear on the 35th day and are then often malformed and infertile, which is not the case on the 37th day. It is likely that drugs directed towards interference with egg production might at this early stage of maturity condemn the maturing fluke, because in guinea pigs blockage of the genital ducts by shell-forming materials derived from the vitellaria was seen in several instances to result in the complete degeneration of the vitellaria. Perhaps, further study of infections of the guinea pig in a quest for factors which inhibit maturation could be very profitable.

V. The Invasion of the Host's Organs and Tissues

A short account of the burrowing and feeding activities of young forms of *F. hepatica* in mice (Dawes, 1961a) was based on a more detailed study (Dawes, 1961b) and led to a demonstration (1961c) which proved that the flukes eat their way through the liver, feeding mainly on hepatic cells. At that time sustained efforts were being made to track the juvenile flukes in the intestinal wall of the host. When this arduous task was completed a further paper (1963a) confirmed the tissue feeding habits of the young flukes not only in the wall of the intestines but also in the abdominal cavity, in which it was claimed (Dawes, 1961a) browsing occurs. Features of bile duct histology in relation to the nutrition of both maturing juveniles and adult flukes also came into the general picture of tissue feeding (Dawes, 1963b) and further observations were made concerning modification of feeding intensity during early and later stages of liver penetration (Dawes, 1963c). These findings also supported previously sustained series of observations relating to the diet of the flukes, because it was shown that the damage caused to the liver in late stages of penetration is augmented by the processes of the local tissue reaction of the host. In this way, a picture has been built up steadily to indicate that *F. hepatica* is never a blood feeder in the true sense of the term, as is and has long been commonly supposed. The observations made by Dawes are descriptive and mainly histological, but they have paved the way for future research of a physiological and biochemical nature by their direct approach to the problem of what the flukes are actually doing with their organs of feeding and locomotion during juvenile and preadult stages of migration and during adult life in the bile duct. In all this work there is no denial of the fact that under some circumstances of infection the flukes ingest blood and it is realized that burrowing in such a vascular organ as the liver and feeding by a combination of suctorial and pumping methods the fluke cannot fail to ingest blood. The observations of Dawes have shown also what is happening to the tissues of the host through which the parasite is passing, or in which adults must live for some years, so that his results form a basis of work which is of value to experimental pathologists.

Investigating the chemotherapy of fascioliasis in sheep, Kendall and Parfitt (1962) confirmed the observations of Montgomerie (1926, 1928a, b) that juvenile flukes are comparatively unaffected by carbon tetrachloride, which affects only flukes which are at least 8–10 weeks old. This relative insusceptibility of young flukes to a drug was related to the fact that during its development the fluke occupies different locations in the host's body, and also to the likelihood that at different stages there is a differential physiological response, perhaps depending on the degree of maturity attained. They accordingly tabulated the stages of migration of young flukes from the intestines of the sheep to the bile duct after dosage with cysts as 18 h spent in the abdominal cavity, 90 h to reach the liver and 40 days to reach the main bile ducts. Flukes were considered to mature in the bile ducts (eggs in the fluke) after 43 days and (eggs in the faeces) after 55 days. They considered that "there seems good reason to suppose" that flukes in the abdominal cavity may be relatively insusceptible to therapy, and that the period is "so brief that there can hardly be any significant effect on the efficacy of chemotherapy". Movement into the liver could expose young flukes to drugs which attain a therapeutic concentration in the blood and tissue liquids. They quoted Ashworth and Sanders (1960) to the effect that the hepatic cells are bathed continually in a liquid which has all the constituents of blood, excepting the formed elements, and that the biliary canaliculi communicate directly with the intercellular spaces". Kendall and Parfitt believe "there is no sound basis for differentiating the situation of a young fluke in the hepatic parenchyma from that in the bile ducts; yet experiment shows that during these first five weeks of life the flukes are resistant to the drug". In this paper they also showed that hexachlorophene in a relatively high dosage can have a marked effect on flukes 3–4 weeks old.

A. FLUKES IN THE INTESTINAL WALL

The problem of following the fluke's movements in the wall of the intestines is formidable because of the short period concerned, no more and perhaps much less than 24 h, and the great amount of tissue which must be searched. Dawes (1963a) sometimes used operative techniques on anaesthetized hosts, using a ligatured section of the duodenum, but turned to a method of administering as many as 500 cysts to a mouse in two or more doses so as to obtain heavily infected intestines. The sampling of selected regions was also discarded in favour of the laborious but forthright method of sectioning the entire intestine after careful removal of stomach and rectum. The dry dissection of a mouse can be completed in a few minutes to the stage at which the entire intestines and mesenteries are safely in fixative. This operation, coupled with serial sectioning of the entire intestine, was carried out six times, with adequately rewarding results. Many stages of penetration were found, some of them in plentiful amounts, along with excysting flukes, free flukes in the intestinal lumen and flukes lying in the folds of the villi. Flukes were also caught in their movements over the mesenteries and within lymph nodes and adipose tissue, clearly supporting statements about the fluke's wanderings and browsings in the abdominal cavity which Dawes had made previously.

Free flukes contract and extend the anterior region of the body in what can be called exploratory movement and on an artificial substratum the young flukes move rapidly by alternate use of the oral and ventral suckers. The amount of movement depends on the amount of extension which the anterior region can command, as well as on the speed with which the suckers can be engaged and disengaged. Even before excystment is completed the large and scanty cells of the caecal epithelium show an eosinophilic reaction and secretion droplets can be seen in the distal cytoplasm. The young fluke (Fig. 5, B) is thus quite ready to commence feeding. For some weeks, or even months, it has been wrapped in its cyst and denied access to sources of nutriment, so that its nutritional resources must stand at a low level. The young fluke soon begins to move over some substratum, also having an evident urge to feed. Newly excysted flukes are soon creeping about over the mucosal surface, coming to lie in folds between the villi, and their browsings lead to the formation of lesions which are minute and difficult to observe. Sites of penetration are inconspicuous but many instances were seen of flukes already located within the villi (Fig. 5, C), where the fluke excavates part of the core and compresses the remainder against the overlying epithelium. Flukes were also found in the submucosa, sometimes checking and turning as if to avoid some obstacle. Many more stages of penetration were found in the

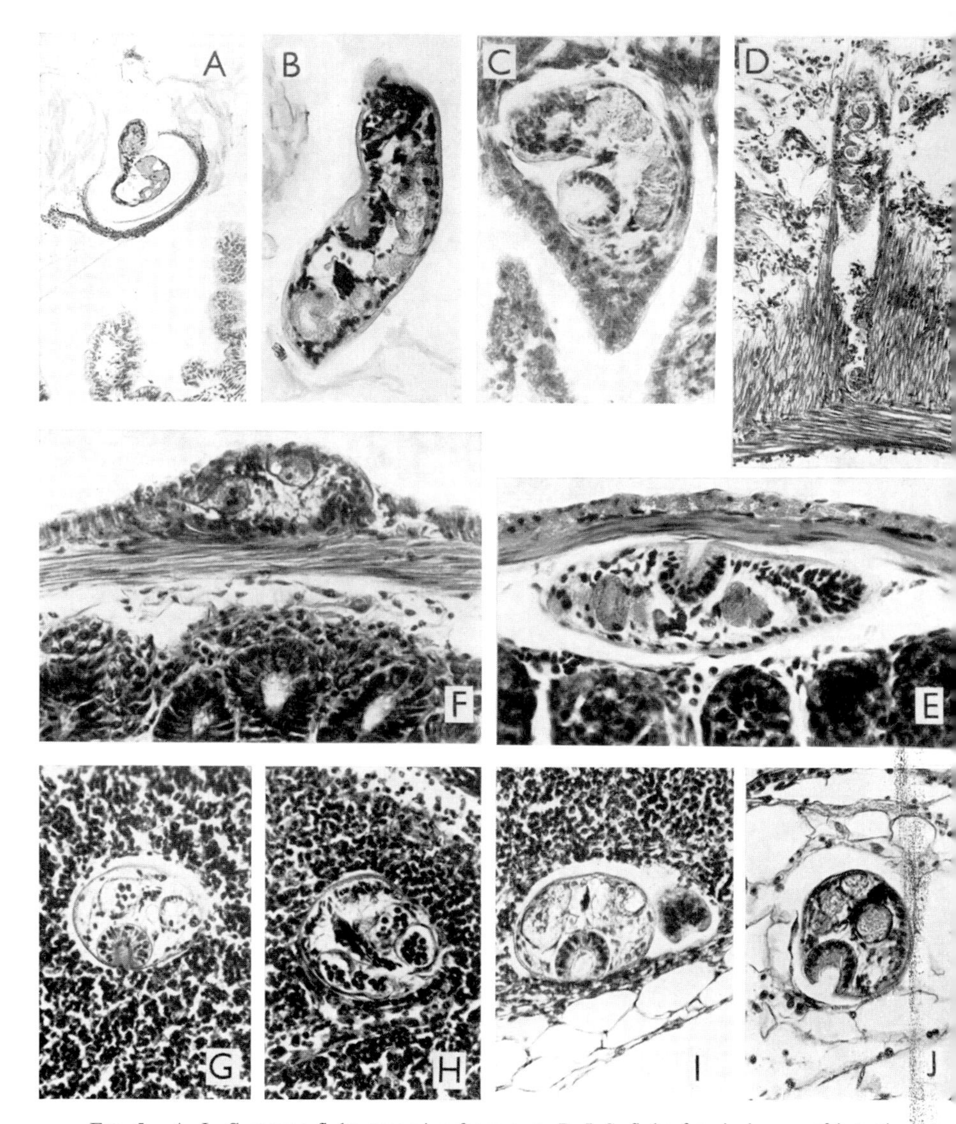

FIG. 5. A, L, S. young fluke emerging from cyst. B, L.S. fluke free in lumen of intestines. C, L.S. fluke in villus showing feeding posture. D, Section of fluke in formed burrow within circular muscles of intestinal wall; note (below) debris in wake of fluke. E, L.S. fluke about to enter circular muscles (from left to right). F, L.S. fluke within longitudinal muscles of intestinal wall, enclosed only by serosa, muscle cell debris in caeca. G and H, T.S. fluke within lymph node; note lymphoidal cells in caeca. I, L.S. fluke emerging from lymph node and about to enter adipose tissue. J, Fluke in adipose tissue, caeca containing connective tissue debris. (Dawes, 1963a.)

muscle layers of the intestines than in the mucosa and submucosa and this was not entirely due to time factors but more probably due to the greater difficulty which the fluke has in penetrating this kind of tissue. Flukes soon aggregate beneath the muscles (Fig. 5, E) and it is evident they find these layers more resistant to their browsing activities than those already encountered.

Numerous examples were seen of damaged unstriped muscle cells and recognizable remains of fragments of such cells were often found in the caeca of many flukes (Fig. 5, D, F). Torn fibres showed curled ends suggestive of abrupt recoil at the time of breakage. The elongate nuclei of the fibres become more rounded in broken fibres and clearly of the same type as nuclei in the caecal contents. In eating its way through the circular muscle layer the fluke makes a characteristic burrow, the wall of which contains broken and compacted cells (Fig. 5, D). Within this, pyknotic nuclei and other cellular debris remain in the wake of the fluke, which squeezes its way through the muscles, travelling along their main axes but breaking some of them in its slightly diagonal progression, which is probably determined by the subterminal position of the mouth and oral sucker. Within the longitudinal muscle layer the translatory movements of the fluke cut through the cells transversely but progression is still diagonal, so that in many instances only the thin serosa forms the roof of the burrow (Fig. 5, F). Damage and debris are more conspicuous throughout this layer and Dawes gave several photomicrographical examples of it. Throughout this part of their migration, the flukes travel by devious routes, sometimes adopting the line of least resistance in following muscle fibres but at other times cutting across these. With the final breakdown of the serosa, the flukes emerge into the abdominal cavity, there to wander into various unpredictable situations such as lymph nodes (Fig. 5, G–I) and adipose tissue (Fig. 5, J), filling their caeca with recognizable debris of distinctive kinds, lymphoidal cells in the one case and connective tissue fragments but not lipoidal materials, as might be supposed, in the other. During the first 24 h of infection, therefore, the young flukes display a readiness to feed on several different kinds of tissue—epithelial cells, connective tissue, unstriped muscle cells and glandular tissue—and although some vascular damage must be caused in some parts of the migration erythrocytes are rarely seen in their caeca. The young flukes at this stage are indubitably tissue feeders.

B. FLUKES IN THE ABDOMINAL CAVITY

The recovery of young forms of *F. hepatica* from the abdominal cavities of various hosts has been a common feature of research in this field since Sinitsin (1914) found them creeping over the viscera and then entering the liver. After tracking young flukes through the intestinal wall and over mesenteries we get a more vivid impression of what Vogel (1934) referred to as "bore-happy" flukes attempting to invade organs other than the liver, and even getting through the diaphragm into the pleural cavity and the lungs. The flukes often remain in the abdominal cavity much longer than is necessary to reach the liver, as many investigators would testify, and Dawes (1961b) found them in all mouse infections of 3–8 days, long after others had entered the liver. Because of pretreatment of hosts by means of intraperitoneal injections of trypan blue, Dawes (1962a) was able to recover from the abdominal cavity 4 days after infection clear and colourless living flukes having caeca filled with blue-stained tissue debris. This was interpreted as the result of the browsing activities of the young flukes. Just as newly excysted forms browse on the mucosa of the intestine, so the fluke which reaches the abdominal cavity must also travel along the only available substratum, the serosa, there to find nutriment by browsing, sometimes entering this organ or that before arriving at the liver. Flukes which remain in the abdominal cavity for unusually long periods become stunted forms; their liveliness indicates that maintenance needs may have been satisfied but retarded growth results from a failure to satisfy total needs. Browsing is inadequate to satisfy fully the nutritional requirements of the flukes, which seek an adequate diet. Arrival at the liver is not inevitable, and there is no numerical assessment which shows how many young flukes are lost on their way to this organ, although there have been reports of flukes lodging in many ectopic locations. To gain access to the liver from the surfaces of abdominal viscera entails but a small step, because of sharpness of its edge, and once the liver has been reached browsing becomes a successful exercise in the sense that the fluke is able quickly to bury itself in tissue which is well suited to its needs, and occurs in practically unlimited amount. The fluke must spend very little time wandering over the surface of the liver, as a rule, because early infections are generally found near the margins of the liver lobes. Up to this point, *F. hepatica* seems to behave in much the same manner as *Paragonimus westermani* and in leaving this organ during the later stages of development extends the comparison. There appears to be good reason to believe that in the matter of digestive equipment the two flukes would bear very close comparison, because both species are equally capable of penetrating into and feeding upon an equally great variety of tissues.

C. FLUKES IN THE LIVER

Mention has been made already of many instances in which young flukes have been found in the livers of natural and experimental hosts. Most of the findings are by investigators concerned with the route of migration, some with numerical result of infection with cysts. Little or no effort has been made to determine what flukes are doing during their brief stay in the liver. Schumacher (1956) gave a guinea pig 1,000 cysts, killed the animal 64 h later and prepared sections of the infected liver. The brief analysis of this material, including five photomicrographs, provided "only a slight and somewhat tantalizing picture of early migration through the liver " (Dawes, 1961b). However, his Fig. 1 shows in section a young fluke entering the liver through its capsule, giving a clear indication of the size and nature of the burrow.

In a preliminary account of liver invasion, Dawes (1961a) noted the assembly of flukes in the abdominal cavity of mice 24 h after infection and the entry of some flukes into the liver during the next two days. Flukes at this stage are about 0·3 mm long but nutritional needs are well satisfied in the liver and during little more than 1 week the young fluke quadruples its length. The early burrow is slightly greater than the breadth of the fluke and equal to about six hepatic cells, so that by the time the fluke is just buried in the liver several hundred cells have been obliterated. As the fluke grows the burrow enlarges and even in moderate infections the liver may be riddled with burrows and the toll of hepatic cells enormous. In a later paper Dawes (1961b) gave a fuller account of this research. Burrowing follows an arc-like course, which brings some flukes quickly to the surface, occasionally to leave the liver. Each burrow comprises a consecutive series of ovoid chambers connected by narrower regions (two are seen in Fig. 6, C), and the course which it takes is determined by periodical reorientations of the fluke's body. The oral sucker and pharynx are used systematically to burst and break down hepatic cells and the size of each chamber is an indication of the extension which the mobile anterior forebody is able to make. Within the burrow a kind of cytoplasmic homogenate flows about the body, and this is aspirated by the pharynx and pumped into the caeca. It contains largely the liquid cytoplasm of hepatic cells but also the pyknotic nuclei of these and Kupfer cells, band-type neutrophil leucocytes and some erythrocytes. The walls of the burrow comprise masses of neutrophils and broken and compressed hepatic cells, and the superficial part stains deeply with basic dyes of all kinds in the same manner as cells of the caecal epithelium of the fluke. This probably indicates that materials lost from the caeca during defaecation are responsible for some alteration of their cytoplasm, probably enzymes. Many examples have been seen of flukes biting into masses of hepatic cells as in Fig. 6, D. At first entry into the liver the oral sucker can grasp an individual hepatic cell and burst it by physical action, but with continued growth of the fluke an enlarged oral sucker takes greater toll of the hepatic cells, so that haemorrhage from sinusoids which inevitable occurs is also somewhat enhanced. Instances were given in which flukes approached close to portal blood vessels, sometimes partially breaking down their walls, but flukes which could have been accommodated therein turned away from them and continued to move directly into apparently normal hepatic parenchyma (Fig. 6, C). The mouth of the fluke is almost invariably found pressing into the hepatic tissue in this manner, clearly indicating that in the liver there is a continuance of the tissue feeding habit which is so well displayed during penetration of the intestinal wall.

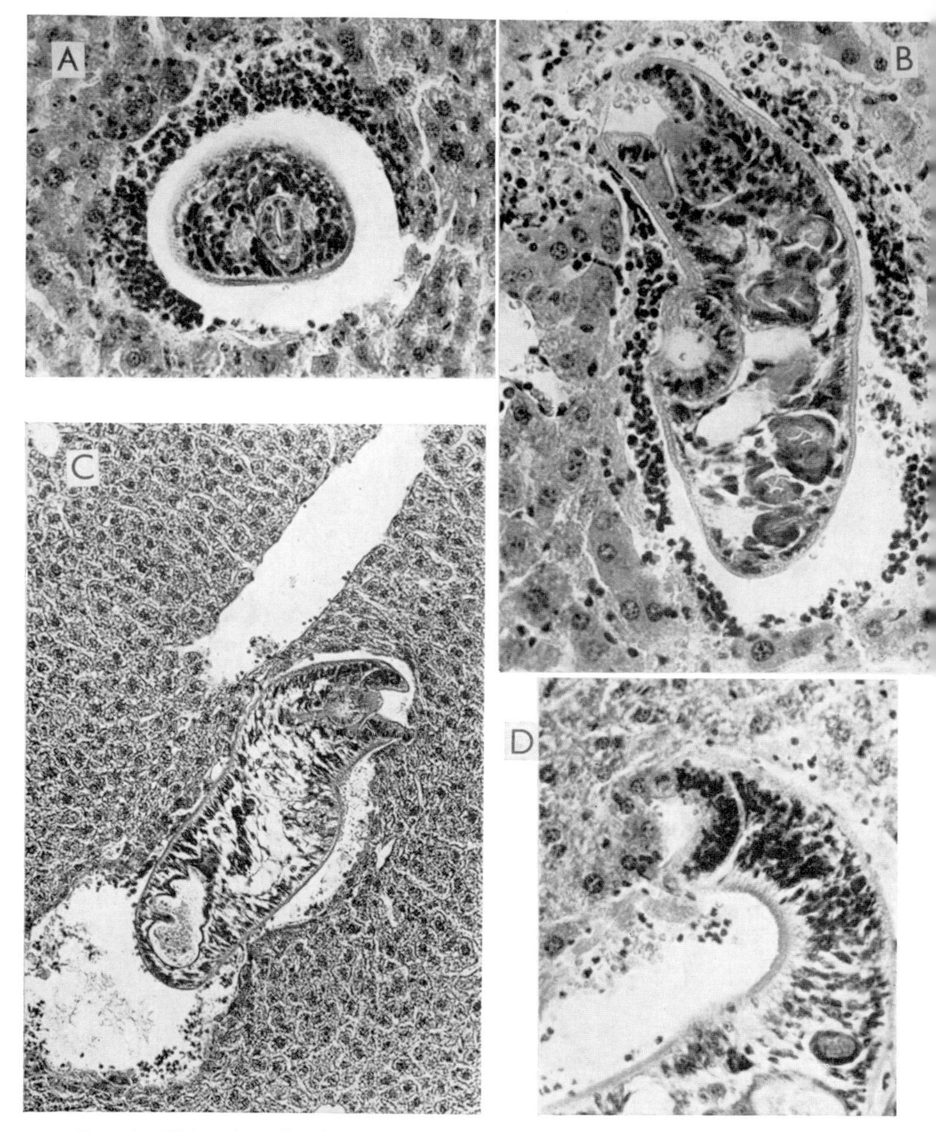

FIG. 6. Flukes invading liver. A, T.S. fluke (pharyngeal region); 3-day infection; note leucocytic infiltrations compressed into broken hepatic cells in wall of burrow. B, L.S. fluke in burrow; 3-day infection. Note masses of leucocytes and some erythrocytes in burrow. C, L.S. fluke in burrow; 6-day infection, showing "avoidance" of a portal vessel large enough to accommodate the fluke, after partial rupture of its wall. D, L.S. anterior end of fluke, in 6-day infection, showing characteristic "bite" into the hepatic parenchyma, the anterior end flexed in the typical feeding posture. (Dawes, 1961b.)

In a later paper (Dawes, 1963c) consideration was given to the critical conditions which arise in the liver about the 4th week of infection in mice. Many hosts then die as a result of lesions which arise and pathological conditions which ensue. During early infection liver damage consists mainly of typical excavations of hepatic cells without complications, although leucocytic aggregates occur around the portal vessels and bile ductules and at indeterminate points elsewhere. The local tissue reaction of the host is concerned primarily with repair of the burrows and does not add significantly to the damage caused by flukes in such early infections. In 6-day infections flukes are 0·4–0·5 mm long and the oral sucker about 0·03 mm diameter. By about the 17th day of infection, however, the fluke is almost 3 mm long and the diameter of its oral sucker has increased to about 0·25 mm, that of the mouth (through which the pharynx partially protrudes during feeding) to about 0·12 mm. The internal organs of the fluke are well advanced and appetite must be increasing in proportion to size. The burrow is now much larger than formerly. The smaller and older parts of a burrow contain an organization of leucocytes of various types superimposed on a fibrin network previously laid down, and reticulin fibres are seen in these tissue formations. Macrophages and multinucleate giant cells are also seen in peripheral regions of the damaged areas. The damage here is much less than that seen in the larger and more recent portions of the burrow and the processes of repair are proceeding with characteristic success.

In the terminal regions of the burrow in 17-day or subsequent infections during hepatic invasion it is sometimes difficult to see how repair can be effected, because these are not cylindrical spaces or tubes but spreading pools containing much broken hepatic tissue as well as blood. In 25-day infections such damage may spread more widely and sometimes involves the liver capsule, with the result that serious haemorrhage occurs into the abdominal cavity. This damage, however, is much too considerable to be attributable to the direct feeding activities of the flukes. Allowing that the fluke is now 4 mm long and that increasingly more copious amounts of enzymatic digestive secretions are being evacuated into the burrow than was formerly the case, the lesions are too large. What is happening is widespread breakdown of hepatic tissue around the original burrow as a result of inordinately large leucocytic infiltration (Fig. 7, A, B). This is producing a tremendous amount of debris, including very many entire hepatic cells at some distance in front of and at the sides of the mouth, and also around the attachment of the ventral sucker. The hepatic parenchyma outside the actual burrow is being shattered by the outpouring leucocytes, and a point may be reached at which the damage becomes irreparable. That this does not suit the requirements of the flukes is evinced by their migration into the abdominal cavity, and that the damage is not inevitable even in such a small host as the mouse is shown by the fact that in some much older infections liver damage caused by flukes

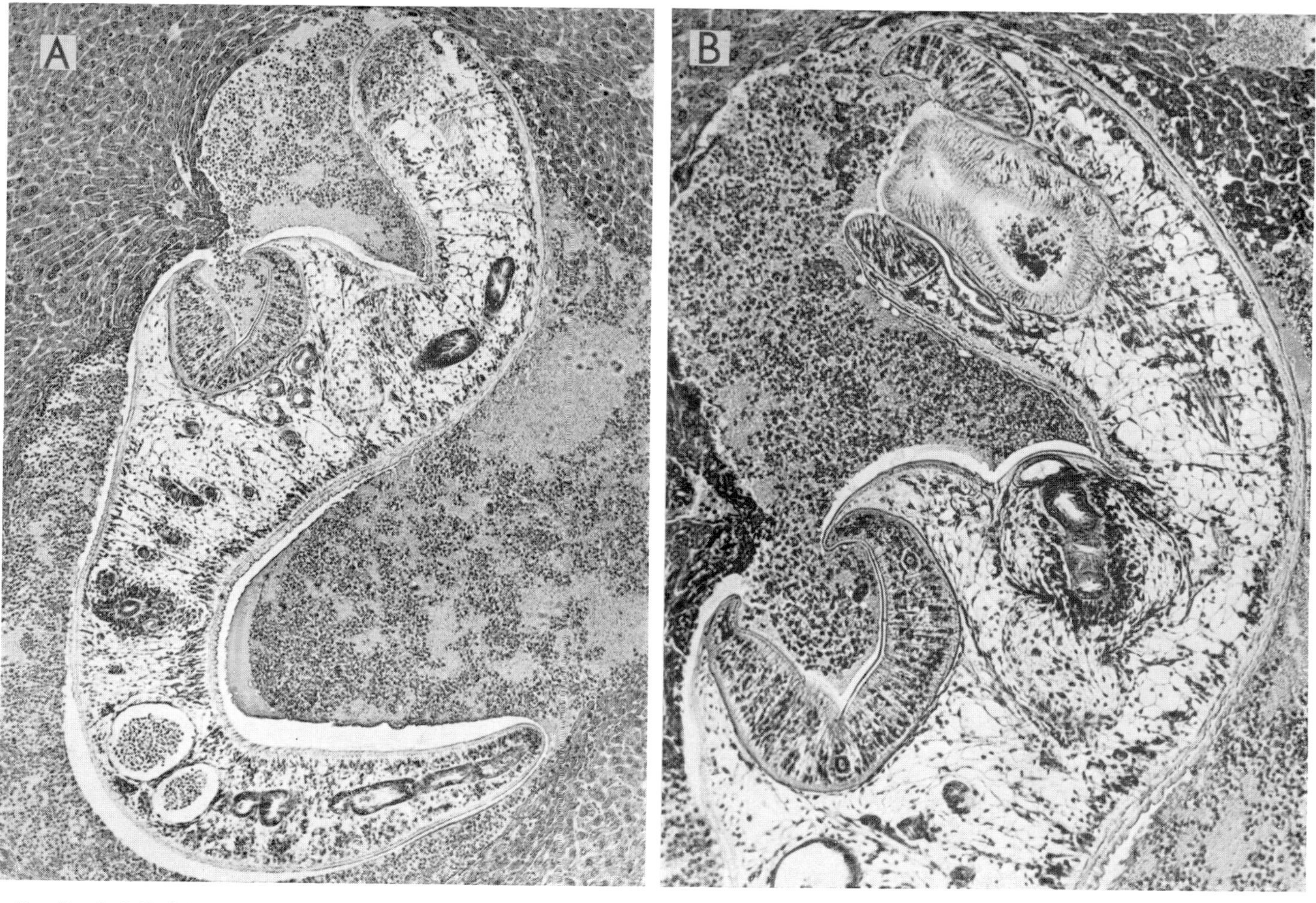

FIG. 7. A, L.S. fluke *in situ* in liver of mouse; 17-day infection. Note damage spreading around fluke in consequence of local tissue reaction of host, converting burrow into a vascular pool containing numerous infiltrating leucocytes and masses of hepatic cells derived from the shattered walls of

B, L.S. anterior end of fluke seen in A, showing in greater detail the devastating effect of infiltrating leucocytes on the hepatic parenchyma near the mouth and suckers; note broken tissue in lumen of pharynx and damage to tissue held by ventral sucker. (Dawes, 1963c.)

which have come to maturity in the biliary system has been either avoided, or effectually repaired. We cannot ascribe the critical situation which sometimes arises to the direct feeding activities of the flukes. Dawes (1963c) suggested that possibly flukes which enter the bile ducts at a comparatively early stage produce damage which can be repaired by a usual and controlled inflammatory process, while flukes which remain for an inordinately long period in the hepatic parenchyma and have reached a much larger size produce greater damage which evokes a less restrained and even unrestrained inflammatory reaction and that it is this which increases liver damage to the point at which it becomes irreparable. Early entry into the bile ducts obviates the critical effects, late entry facilitates them.

D. ENTRY OF FLUKES INTO THE BILIARY SYSTEM

The precise time of entry of flukes into the bile duct will remain a matter for speculation until the manner of entry has been observed. Shirai (1927) considered it "reasonable" to find flukes in the hepatic parenchyma but he was surprised not to find flukes in the bile duct, except in one of his rabbit infections (64 days). His results are not now surprising, because none of the other infections extended beyond 38 days. What he had on his mind was "if these worms would really persist and grow in the bile ducts they should make their way there while the size of the body is still small and favourable for entering". The advantage of early entry can hardly be denied. Clunies Ross and McKay (1929) also made a point of noting that flukes had not apparently entered the bile ducts of rabbits in infections of up to 24 days, or those of guinea pigs up to 39 days. Schumacher (1938) noted first entry into the bile ducts after 45 days in rabbits, but found that most flukes did not reach this location for another 10 days. He stated also that in guinea pigs a period of about 7 weeks was required and that for sheep more than 8 weeks. Urquhart (1955) confirmed this result for rabbits: individual flukes had entered the bile ducts as early as 5 weeks when about 3 mm long but it was usually 8 weeks before they were all in the duct. In mice, flukes mature much more quickly than in other hosts and entry into the bile ducts may take place after 24 days but in some instances only after 29–32 days (Dawes, 1961b, 1962a). What appears to be an atypical mode of entry by way of broken ducts in a damaged liver was not accepted as the normal mode. Lagrange and Gutmann (1961) stated that flukes may attain maturity in the bile ducts of mice after 33 days. They found two mature flukes in the hepatic parenchyma but in all other infections maturity was not attained until flukes were in the bile duct. Lagrange and Kemele (1962) tried to ascertain why young flukes enter the bile duct and when they do so. The first question could not be answered and in respect of the second they affirmed that entry does not usually occur in mice before the 28th day of infection, though in some instances it occurred after 25 days. They also gave 33 days as the time of attainment of maturity but in a postscript noted that in two mice signs of maturity were evident 30 and 32 days after infection. They accepted the period of 24 days suggested by Dawes (1961b, 1962a) as minimal. In the latter of these two papers by Dawes, it was stated "what is remarkable about the invasive stage of *Fasciola hepatica* is the shortness of the period during which the fluke lives and feeds in the hepatic parenchyma, by comparison with the whole possible duration of life, which is several years. From the point of view of host survival or weakening of the host it would seem that the latter part of this phase just before the bile duct is reached, is the most precarious. Liver damage may mount to very serious proportions from the end of the third week onwards, so that the most vital stage concerned is just prior to entry into the biliary system". It is possible that the failure of flukes to enter the bile ducts at a suitably early stage is linked in some way with the serious liver damage which ensues. There is also the further possibility that in such a small host as the mouse flukes which do not enter the bile ducts at an appropriate age and size may find entry impossible after even a short further period spent in the hepatic parenchyma on account of their increased size and the small size of the ducts. Dawes (1963d) concluded "that flukes enter the biliary system when continued feeding on hepatic cells would greatly endanger the life of the host". It seems that the failure of flukes to enter the bile ducts at a suitably small size possibly implies lethal consequences for the host. Many instances of long sustained infections in mice could be explained in terms of timely entry into the biliary system.

E. THE "WASTAGE" OF POTENTIAL FLUKES

A discrepancy commonly occurs in experimental infections between the number of cysts administered to a host and the number of flukes later recovered from it. This "wastage" of potential flukes occurs in all the hosts of *F. hepatica* and its amount will depend in the first instance on the degree of skill exercised during manipulation of the cysts and on the care taken in the process of recovering the flukes. The disparity of the numbers of cysts and flukes will also tend to vary according to the size of the host and the duration of infection. It is comparatively easy to recover flukes from the bile ducts of their hosts and accuracy will be greater because it will be difficult to overlook any flukes. To recover flukes at an early stage of infection from a location such as the abdominal cavity is technically very difficult, if accuracy is required, and much more difficult in a larger than in a smaller host. Even the most careful worker may overlook flukes which have wandered into unusual locations. In the following discussion of this topic we shall refer to the percentage of cysts which yielded flukes in parentheses, thus (x %).

Montgomerie (1928a) reported four sets of "observations" on his infections of sheep. In the first set, seven sheep were given 100 cysts each and the numbers of mature flukes recovered were 25, 17, 52, 27, 8, 13 and 43 (= 8–

52%). A total of 185 flukes were thus recovered after administration of 700 cysts (= 26·5%). In the second set, one sheep received 50 cysts and 12 flukes were recovered (= 24%) and four sheep were given 25 cysts each and 12 + 14 + 19 + 10 = 55 flukes were recovered (= 55%). In the third set, ten sheep received 100 cysts each and the total yield of flukes was 403 (= 40·3%). In the fourth set, ten sheep were given 70 cysts each, the numbers of flukes per sheep were 9–41 and the total number 299 (= 42·7%). Montgomerie's conclusion that the degree of infection which arises from the administration of a known number of cysts "varies within quite wide limits" was well justified, although his suggestion that a larger proportion of cysts may develop when the number given is small cannot carry much weight.

Schumacher (1938) carried out five experiments with sheep, each of which received 2,000 cysts. The first three were infections running for 1, 2 and 3 days respectively, and the yields of young flukes were 125 (= 6%), 186 (= 9%) and 115 (= 6%). The remaining two experiments were 35-day and 56-day infections which yielded in one instance 20 flukes in the abdominal cavity and about 500 flukes in the liver (= 26%), and in the other 27 flukes in the abdominal cavity, 108 in saline washings of the liver, 400 in the liver itself and 8 flukes in the gall bladder (= 27%). This was a remarkable effort but percentage yields of flukes were much lower than in Montgomerie's experiments, probably because of the much greater technical difficulties involved in the recovery of younger flukes.

Sinclair (1962) gave four sheep (not treated with drugs) 600 cysts each, two sheep in his Group A receiving the cysts in one dose and two in his Group B in four doses with 3 intervening weeks. One sheep in each group died and the remaining two were killed. In the A group the sheep that died yielded 221 mature flukes (= 36·7%) and the slaughtered sheep yielded 199 flukes (four of them immature) (= 33·2%). In the B group the corresponding results were 188 flukes (= 31·3%) and 173 flukes (one mature) (= 28·8%). For the two sheep of the A group, therefore, the total yield was 420 flukes (= 35%) and for the two of the B group 361 flukes (= 30%). Sinclair overlooked the paper of Montgomerie but he remarked that about the "usual" level of infection was obtained, with a close approximation to Schumacher's figures.

To sum up the results of these three sets of results, Montgomerie recovered 954 flukes from thirty-two sheep which were given a total of 2,550 cysts (= 37·4%): Schumacher's two sheep of the 35-day and 56-day infections received a total of 4,000 cysts and yielded a total of 1,063 flukes (= 26·6%): and Sinclair recovered 781 mature flukes from four sheep which received a total of 2,400 cysts (= 32·5%). To these we might add the results of Hughes (1963), who gave 200 cysts to each of ten sheep, recovering 49–119 flukes per sheep (= 24·5–59·5%) and a total of 782 flukes from the ten sheep given a total of 2,000 cysts (= 39·1%).

Using the rabbit as hosts, Kerr and Petkovich (1935) gave three hosts 13 cysts each and recovered 5, 9 and 11 flukes after 111, 116 and 124 days (= about 38 %, 69 % and 85 %). Urquhart (1954) infected thirty-two rabbits with a total of 1,600 cysts, 37·2 % of which yielded flukes. He studied the effects on yields of various methods of administering cysts. After feeding 50 cysts in water by means of a pipette in 63-day infections the mean number of flukes recovered was 8·7 (= 17·4 %). By feeding cysts on small cellophane sheets inserted in small pieces of cabbage stalks in thirty-two experiments of the same duration he obtained a mean yield of 18·6 (= 37·2 %). In four groups of rabbits given 50 cysts each but from different snails there was only slight variability of yield, 16·3–24·3 flukes per rabbit (= 33–49 % approximately). The mean number of flukes per rabbit for the four groups was 19·4 (= 38·8 %). No significant difference of yield accrued in respect of sex; ten castrated males and ten females gave yields of 36·6 % and 37·0 %. These results forced Urquhart to the conclusion that the most important factors determining the numbers of flukes which develop from a given number of cysts administered to rabbits were the technique of infection and the natural resistance of the hosts.

Using the oral method of administering cysts to rabbits, Hughes (1963) gave twenty-two of his control animals 15 cysts each and obtained a total of 100 flukes (= 30·3 %). In six sets of experiments with three to five rabbits he obtained yields of 17·8–46·7 %. In his experiments with mice, twenty-four hosts (of Exp. 4) received 20 cysts each and in 8–26 day infections the yield was 15·8 %. In another instance (Exp. 3), eight mice given 10 cysts each produced in infections of 22–29 days 34 flukes (= 30 %). This was his Group B: his Group A (challenge infection following the administration of X-irradiated cysts) comprised eleven mice which were given 10 cysts each and in 15–28 day infections 39 flukes were recovered (= 35·5 %). Dawes (1964) gave 5 or 10 cysts to each of fourteen mice, which yielded 62 flukes in groups of 1–8 per mouse, giving means of 4·4 flukes per mouse (= 55·4 %).

In hosts of very different sizes, therefore, the number of flukes which result from infection with a given number of cysts varies considerably and the "wastage" of potential flukes could hardly be greater in small experimental hosts such as the mouse than it has been shown to be in sheep. Apart from considerations already made, the principal sources of loss are to be sought in the early stages of infection, when technical difficulties of recovering flukes are greatest, making reliable results difficult to obtain. Schumacher (1938) concerned himself with the period 8–48 h after infection with his experiments 10–23 with guinea pigs. After 8 h a yield of 58 flukes from 400 cysts gave the value 14·5 %; after 12–20 h the range of variability was 28–46 % (with 28 % at 30 h), and after 48 h 15–27 %. For the entire period of 8–48 h the yields were 14·5–46 %, during the period 3–6 days (when flukes were entering the liver) 11–26 %. The fact that only about one-quarter of the cysts administered yielded flukes in the abdominal cavity was explained by Schumacher as due to the failure of some forms to excyst, but partly to a belief that some remained behind in the intestines. He observed how firmly young flukes adhere to the sides of glass vessels and he fancied that some flukes fasten themselves to the mucosa in this way and fail to penetrate it. If we look at other of his records, however, we find that after 2 h unaltered cysts as well as empty inner cysts were found in the large intestine and caecum of guinea pigs. This was true also of 3–4 h infections. Beyond 4 h there were no numerical results but mention was made of finding unaltered cysts and inner cysts with contained flukes in both the large intestine and the caecum. Considering the timing, this is evidence in favour of the view that in many experiments unaltered cysts (and potential flukes) are swept out of the alimentary canal with the faeces. A simple experiment with the help of drugs which reduce or abolish peristalsis in the host's intestines could be used to test this hypothesis but has not yet been carried out.

Higashi (1960a) studied the effect of intestinal micro-organisms on the intensity of infections resulting from the feeding of cysts. A suspension of *Bacillus subtilis* was given orally and cysts later. In the absence of bacteria 75% infection was claimed, in their presence 100%. Pathological observations indicated changes in peritoneal inflammation and especially abdominal oedema and hepatic degeneration, which was followed by an enhanced death rate of the hosts. In a later paper (1960b) the suggestion was made that the penetration of the Gram-positive bacilli into the tissues of the liver supported and reinforced the infection with *F. hepatica*.

VI. The Nutrition of the Flukes

Little is known about the physiological processes of trematodes and the study of nutrition has been much neglected. According to von Brand (1952) and Bueding (1948) carbohydrate is the major if not the only source of energy for many parasitic helminths, although the end products of carbohydrate metabolism vary between species. Such helminths utilize glucose and glycogen anaerobically, with the production of volatile fatty acids (Bueding, 1948). According to Weinland and von Brand (1926) *F. hepatica* produces *in vitro* higher fatty acids, possibly butyric acid. Mansour (1959), using a Gulf Coast strain of fluke in place of the European variety, found that propionic acid and acetic acid (in the ratio of 3/1) are the main end products of carbohydrate metabolism in this liver fluke. It was shown that *F. hepatica* utilizes carbohydrate anaerobically at a high rate; with glucose excluded from the external medium the rate of glycogen utilization was 84–97 μmoles of glucose units/g wet wt/6h, and glucose utilization from the outside medium was 110–180 μmoles/g wet wt/6 h. Adult flukes in the bile duct live in "a predominantly anaerobic environment" and in the presence of atmospheric oxygen there was only a slight sparing effect on glucose utilization and only slight reduction in the amount of propionic acid, which suggested that

F. hepatica "may require a small portion of its oxidative metabolism". Oxygen is used when it is available and the R.Q. was given as 1·56–2·2. One other point can be mentioned because it is very relevant here. Although there is much evidence that *F. hepatica* ingests large amounts of food, both solid and liquid, the very high rate of glucose uptake from the external medium raised the question as to whether this substance is absorbed after entry into the alimentary canal of the fluke or can be absorbed through the cuticle. By making determinations after ligaturing the body between the suckers the conclusion was reached that neither glucose uptake nor the excretion of metabolic products are carried out through the caeca.

A. JUVENILES

After tracking young flukes through the hepatic parenchyma of mice by means of serial sections of many infected livers, Dawes (1961a, b, c) provided the first clear impression of the feeding activities of young flukes in the burrows which they excavate in the hepatic parenchyma. In later papers (Dawes, 1963c, d) it was suggested that newly excysted flukes have oral and ventral suckers the prime functions of which are adhesion to and locomotion over some substratum. The young flukes may have been confined within their cysts for some weeks or months and a low state of nutrition probably impels them to seek a substratum over which to creep and to browse. They settle on the lining of the intestines and browsing soon produces local breakdown of epithelial cells, enabling them to enter the mucosa as they continue to feed, an event which is exceedingly rare in the digenetic trematodes, and which must be correlated with the potency of the fluke's digestive secretions. In the wall of the intestines, as in the liver, flukes make tortuous burrows, leaving behind them recognizable debris which is clearly of the same nature as food seen in the caeca. Much photomicrographical evidence has been produced to prove that young flukes can break down epithelia, connective tissues, unstriped muscle fibres and other tissues. That these materials are utilized by the fluke is clearly indicated by the rapid rate at which it grows. Unless the wandering fluke reaches the liver in a few days, its needs are unsatisfied with the result that growth is reduced in a stunted body. Dawes suggested that it is probably this lowered nutritional state which impels the young fluke to continue its wanderings in the abdominal cavity, either until it reaches the liver or until it is lost in some ectopic location. Movement to the liver seems to be produced by a kinetic effect—flukes checking, turning and moving off in new directions in quest of food—and is neither predestined nor inevitable, but once this organ has been found the fluke is able to burrow into it with ease, again to demonstrate its tissue feeding propensities. It is very unlikely that structural differences in the suckers or pharynx determine the ability of some trematodes to penetrate tissues and organs, which is likely to depend on the characteristics of digestive enzymes. This is a subject about which nothing is known. A branched intestine is not a prerequisite of life in the biliary system of some host, but it is remarkable that the development of dendritic caecal branchings in *F. hepatica* occurs mainly during the second week of juvenile life in the hepatic parenchyma. During a few days the caeca ramify to a much greater extent than occurs during the entire life span of all other digenetic trematodes, with the exception of *Fascioloides magna*.

Within the liver the excavations which produce burrows indicate clearly that hepatic cells play a large part in the nutrition of *F. hepatica* (Dawes, 1961b). The fluke consistently and continually presses its mouth into normal parenchyma. That haemorrhage occurs from sinusoids is inevitable when hepatic cells are ruptured and the use of a suctorial pharynx implies that some blood will be included in the diet. It is contended that, unless liver damage be excessive, the amount of blood ingested (though not the amount lost to the host) is probably insignificant in amount by comparison with the amount of cytoplasmic homogenate and cellular debris which is ingested. When serious liver damage occurs the true manner of feeding is masked, but there is some evidence that young flukes leave the liver when serious damage occurs. This return to the abdominal cavity is seen not only in mice but occurs in all other hosts, experimental and natural, and it may be taken to indicate that the fluke is not satisfied with a vascular pool as its environment. Moreover, such damage does not favour the survival of either parasite or host.

B. ADULT FLUKES

The problematical mode of nutrition of adult flukes located in the bile duct has received some attention. The earlier researches were reviewed by Müller (1923) who from meagre but careful discerning observations of his own concluded that the adult fluke "ernährt sich nich von Blut, sondern der zähflüssige, eiweisshaltige, schleimige Inhalt der Gallengänge, die abgestossenen Gallengangs-epithelien und emigrierte Leucocyten bilden ihre einzige Nahrung". Later investigators tended to dismiss this conclusion too readily, although it was based on methods more direct than their own. Flury and Leeb (1926) detected haemoglobin and related pigments in saline washings of flukes, Hsü (1939) found erythrocytes and leucocytes in the caecal contents and Stephenson (1947), using a combination of simple spectrometrical and imperfect histological methods, concluded that flukes "feed mainly, if not exclusively, upon blood". The most categoric statements about blood feeding have been made. Clunies Ross (1928) stated: "owing to the fact that the fluke sucks blood, there is a continual drain on the body, so that the tissues become pale and anaemic". It is odd, however, that none of the investigators who have postulated or supported this hypothesis have ever shown how adult flukes can secure blood.

Van Grembergen (1950) briefly reviewed earlier work, mentioning the finding of Sommer (in Leuckart, 1886–1901) that the caecal epithelial cells bear what have been called pseudopodial processes. According to Van Grembergen the only proteinase found in secretions of the caeca is a cathepsin, an intracellular proteinase, so that some support is gained in favour of the hypothesis of phagocytic digestion. He regarded Müller's views about the digestive and absorptive functions of the caecal epithelium as diametrically opposed to prevalent opinion, and he supported the view that the adult fluke is haematophagous. However, he also stated "mais il semble que le sang ne soit pas la seule substance enlevée par le parasite à son hôte", finding in his experiments concerning arginase reason for belief "que la Grande Douve ingêre des cellules hépatiques en même temps que le sang". He ended his paper with a remark about our familiarity with many parasites which have a red colour when alive and that this colour is too often accepted without further proof as an indication of haematophagy, adding, "que cette assertion est erronée est maintenant suffisament clair; chaque cas exige un examen sous le double point de vue de pigment autochtone et de sang préléve par le parasite à son hôte."

Jennings *et al.* (1956) also supported this belief that *F. hepatica* feeds on blood. They used ^{32}P to label erythrocytes and ^{131}I to label serum albumin in an attempt to measure the quantity of blood lost in the anaemia suffered by infected rabbits. They estimated that 0·2 ml of blood per day per fluke was "a measure of the quantity of blood consumed by the flukes". Dawes (1963b) objected to the misleading expression "consumed by the flukes" in reference to an enigmatical anaemic condition with other possible contributory causes. The possibility had also been noted that the radioactivity of flukes was due to uptake of radioactive phosphorus from the bile and not the blood, deciding in favour of blood because of the large quantitative difference in radioactivity between the flukes and bile.

Pearson (1963) also used radioactivity in experiments with two guinea pigs, five naturally infected sheep and five sheep experimentally infected. He also attempted to estimate blood loss but by the use of ^{51}Cr in sterile solution as sodium chromate in isotonic saline, a method employed by Gray and Sterling (1950) for tagging erythrocytes and by Roche *et al.* (1957) for measuring intestinal blood loss associated with hookworm infection. His main conclusion was that flukes from the main bile duct had ingested more blood than those from smaller ducts. It was claimed that washing in saline quickly lowered radioactivity, mainly because flukes expel their caecal contents. Mean figures for a 2 h period in the main duct were 0·027 ml of blood per washed fluke and 0·058 ml per unwashed fluke, comparable figures for the hepatic bile ducts being 0·014 ml and 0·052 ml. Lower figures were obtained for guinea pig experiments. It was noted also that bile from the gall bladder showed very low activity but bile in the vicinity of flukes in the main bile duct was relatively active. Because of the rhythmic feeding pattern of the fluke, it was added, the method of estimation used over a short period could not give an accurate figure for the volume of blood ingested, but efforts were being made to obtain a daily or weekly figure. In fairness, we quote "flukes were removed carefully from the bile ducts so that they did not come into contact with blood, and most were then washed in physiological saline".

Pearson paid some attention to certain findings of Weinland and von Brand (1926), who examined 12,000 flukes recovered from various locations in the liver and biliary system. In some flukes the caeca were empty, in others full, but out of considerations relating to mean weight, and to nitrogen, glycogen and fat content of the flukes they concluded that flukes feed on hepatic tissues and blood in the smaller bile ducts and then migrate to the main bile ducts to process food and perform other activities. Food is absorbed in the caeca, residues being ejected through the mouth, and hunger then causes the flukes to migrate into the smaller ducts and resume feeding operations. Pearson considered it unnecessary to postulate such a feeding migration, stating, "relatively stable fluke populations with slightly smaller flukes could explain their results equally satisfactorily, and different rates of metabolic activity such as required (sic) for oogenesis might account for the different levels of ingestion of blood". It is difficult to visualize "stable fluke populations" engaged in sucking blood, and there seems to be no chance of basing any reliable conclusion on the possibility that the flukes remain in one place. We must note also that in Pearson's results, bile from the gall bladder showed low radioactivity in seven sheep (0–0·01 ml blood/ml bile) but a small sample from the main duct of one sheep showed relatively very high activity (0·26 ml blood/ml bile) after tagged cells had circulated for 2 h, and bile from another sheep had an activity equivalent to 0·17 ml blood/ml bile after 80 min.

Urquhart (1956) infected rabbits with 50 cysts each and reported on infections of 2 weeks, 3–7 weeks and 8 | weeks. In the oldest of these infections bile ducts large enough to contain adult flukes showed great increase in diameter and in the thickness of their walls, and the changes were most striking in the common bile duct, in which most of the flukes were found free in the lumen but sometimes grasping the mucosa with their suckers. The sharp spines of the flukes were often seen to be imbedded in the epithelium, causing extensive damage due to the movements of the flukes. The state of the epithelium varied according to the presence and activities of the flukes, but was often hyperplastic. In long standing infections the wall was thickened and fibrotic because of proliferative changes of subacute and chronic inflammation. Dawes (1963b) remarked that in these observations there are several indications of the availability to the flukes of nutrient materials but no suggestion that these might serve to nourish the flukes. Dawes agreed with the conclusions of Lagrange and Gutmann (1961) that flukes in the bile ducts are nourished "soit du contenu de canal qui passe à leur portée soit en se fixant a la paroi distendue du canal", size preventing migration towards the interior of the liver. "Cette partie terminale des voies biliaires est en somme la seule qui se prête à héberger des parasites de dimension".

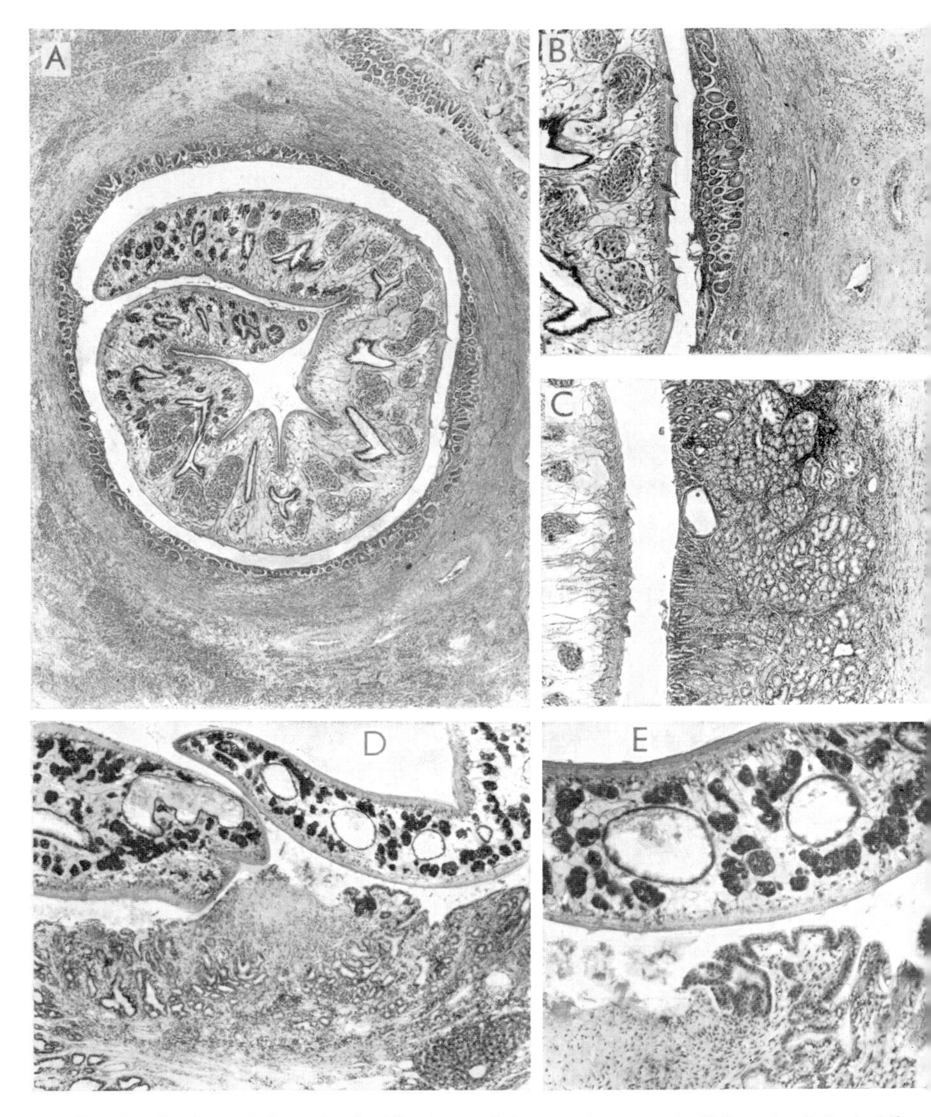

FIG. 8. Sections of flukes in the bile ducts of sheep and mouse. A, T.S. entire fluke in bile duct of sheep, showing hyperplasia of the ductal epithelium and fibrosis in an intrahepatic duct. B, much enlarged portion of A seen on right, indicating denudation of superficial part of epithelium by spines of fluke. C, Parts of bile duct and fluke seen in sections in a 46 days infection of mouse. Epithelium of duct adenomatous, superficial denudation complete. D and E, sections of flukes in bile duct of mouse; 30-day infection, showing patchy denudation of hyperplastic epithelium. (Dawes, 1963b.)

The development of the hyperplasia of the bile duct was observed by Dawes (1963b) in the mouse, rat, sheep and other hosts. At a time when young flukes are burrowing through the hepatic parenchyma and not later than the third week of infection, minute cytoplasmic blebs appear on the free surfaces of the epithelial cells of the bile ducts over extensive areas, until the entire epi-

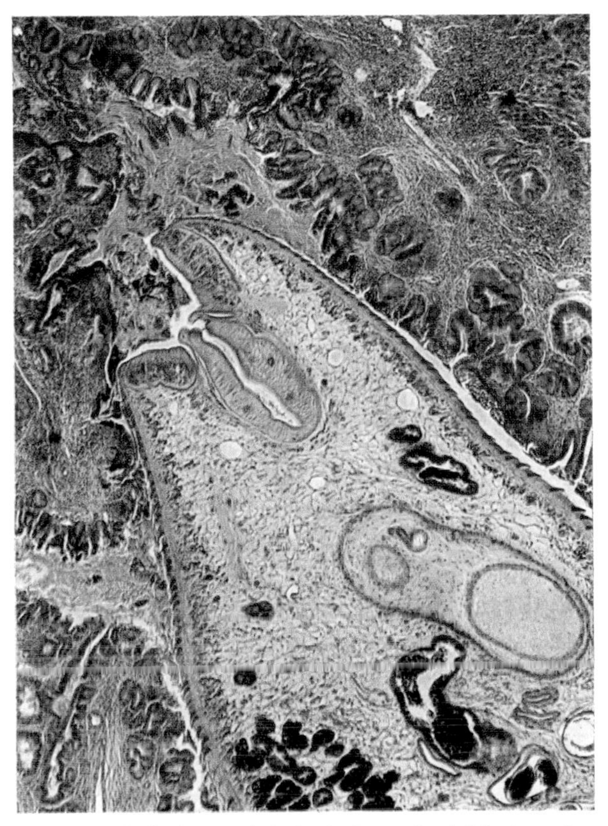

FIG. 9. L.S. anterior end of a feeding fluke in the hyperplastic bile duct of a mouse, showing the secretory activity of epithelial cells and also masses of secretion and cellular debris available to the fluke. Note the dark pigmented debris of non-vascular origin in the caeca. (Dawes, unpublished.) (Haematoxylin, acid fuchsine, alcian blue, tartrazine.)

thelium of some regions is studded with them, one to each cell. This transitory effect occurs at a time when the intrahepatic bile ducts are becoming surrounded by dense aggregates of band type neutrophil leucocytes. In some regions the epithelium is slightly thickened by deepening of the cells and mitotic figures indicate that these cells are multiplying. This becomes more evident with the production of minute invaginations of the epithelium, possibly the earliest indications of enormous overdevelopment and folding which ensues. At a later stage and when flukes have entered the bile duct in the mouse, the epithelium has become folded into numerous crypt-like formations with crowded nuclei. Some few invaginations are flask-shaped and may extend for some distance underneath the superficial cells, which are covered with a slimy secretion. In many situations small amounts of damage are caused to this superficial epithelium by the spines of the flukes and the subepithelial tissue is fibrous and densely crowded with nuclei of cells derived from the local tissue reaction of the host. Blood vessels are not very evident in this tissue, unless crushed and occluded. The bile duct is similar in sheep. The crypt-like invaginations extend deeply beneath the surface and there is an intense fibrotic condition of the subepithelial tissues, so that the flukes fit snugly into the duct, widely removed from vascular or hepatic tissues. Spines, eggs and the suckers of the flukes clearly inflict much damage to the hyperplastic epithelium, sometimes completely denuding it locally. This damage may be severe or it may show signs of gradual attrition, as when the entire superficial epithelium is removed, leaving invaginations intact. Superficial debris evidently due to abrasion caused by spines can be seen in the bile ducts and in the vicinity of flukes drifts of it may arise. It is evident that regenerative processes are at work, for as cells are broken fresh cells are increasing the depth of the epithelium until in some instances remarkable adenomatous thickening in great depth occurs. In such formations there is a copious secretion of a mucus-like substance from the free surfaces of the cells and the simple epithelium of the bile duct has been transformed into a thick, glandular formation. Over this, flukes move about, scraping and browsing, here and there completely removing a piece of the now compound epithelium. Conceivably, the feeding fluke breaking the surface in this way causes some haemorrhage, but this was never seen. The illustrations (Fig. 8, A-E) have been selected from a much wider range and give only an incomplete impression of the states of the bile duct actually seen. At the time of writing the section (taken from a series) photographed in colour (Fig. 9) more fairly indicates as well perhaps as any one figure can, the nature of the hyperplasia, with the glandular cells, the mucoid secretion and the denudation of the superficial portions of the epithelium, coupled with a longitudinal section through the anterior end of a feeding fluke in the characteristic posture often seen elsewhere during liver invasion. There can be no doubt; this fluke is actively engaged in feeding on hyperplastic epithelium and there is no trace of blood anywhere in the picture, although dark, pigmented materials are seen in the caeca of this fluke in copious amount. Whatever the nature of this pigment may be, it can hardly have been derived from a diet of blood.

As a result of such observations, Dawes (1963d) concluded that "flukes enter the biliary system at a time when continued feeding on hepatic cells would greatly endanger the life of the host, and that as a result of the inflammatory reaction the flukes are provided there with a "pasture" of tissue on

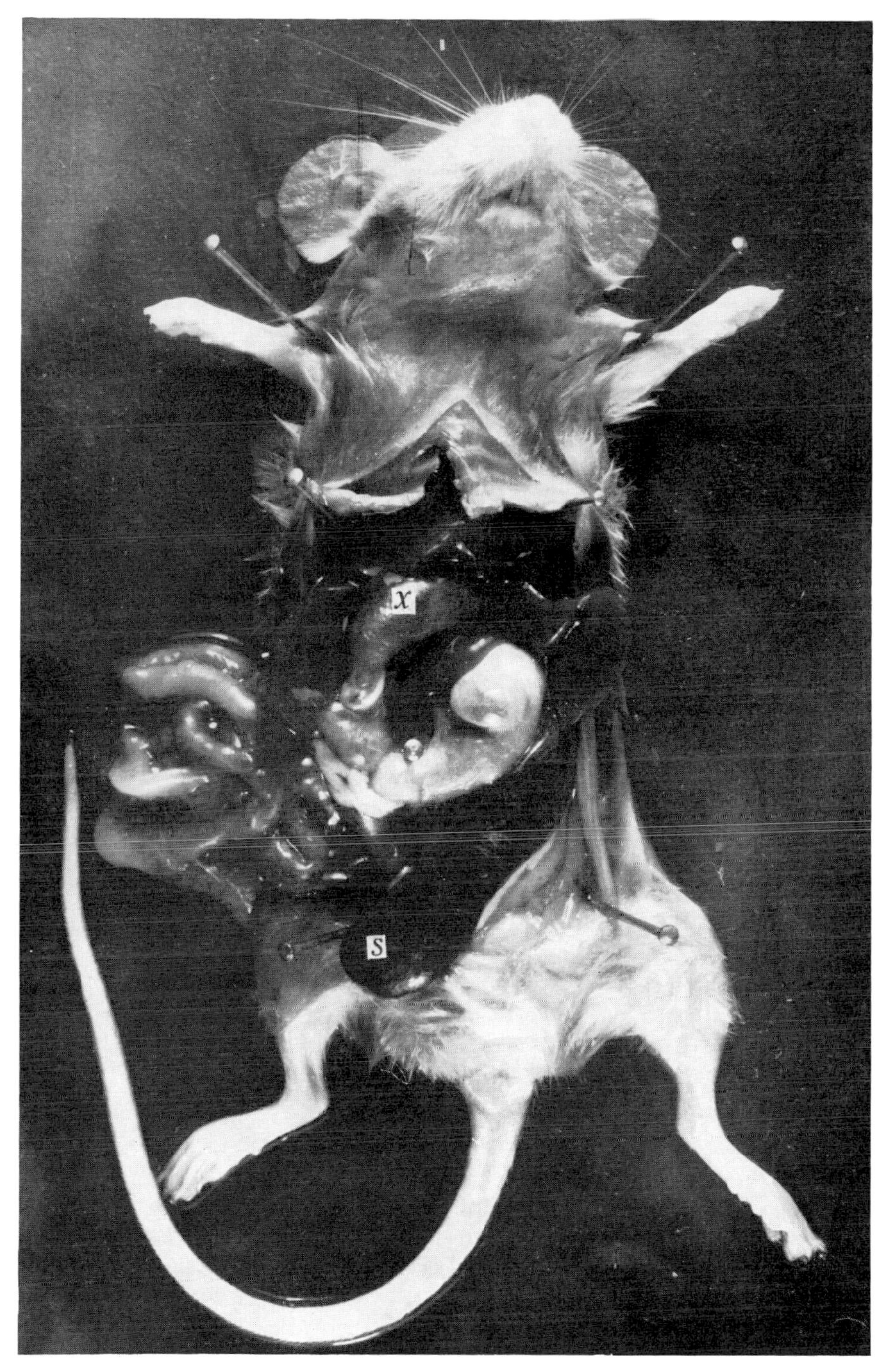

FIG. 10. Mouse which received five normal cysts and was killed 17 weeks after infection. Abdominal cavity opened to reveal the viscera. Note the greatly enlarged and distended bile duct (from which three adult flukes were recovered) and the enlarged spleen: liver damage very slight.

which to feed. In moderate infections, these conditions tend to favour the maintenance of a well-balanced host–parasite relationship, whereas a blood-feeding habit could scarcely be expected to sustain either the parasite or its host". Previously, Dawes (1963b) had emphasized that the ubiquity of *F. hepatica* testifies to its fitness to participate in such a relationship, and pointed out that throughout a 3–4 year "static period of mature worms" (Belding, 1958, p. 282) the hyperplasia of the biliary epithelium is upheld, along with cystic dilatation of the biliary passages and periductal deposition of connective tissue and fibrous encapsulation around the ducts. This condition may be upheld much longer because, according to Durbin (1952), adults may continue to live in the bile ducts for as long as 11 years. The precarious hepatic stage of migration occupies the fluke for only a few weeks; but the biliary stage is extended for as many years, often with security for host and parasite alike.

C. THE CAECAL EPITHELIUM

Several investigators have studied the caecal epithelium of *F. hepatica* in its connexion with the digestion and absorption of nutrients. Dawes (1962d) reviewed the relevant literature and upheld the view of Müller (1923) who observed that granules and droplets appear in the caecal epithelium and that these cells develop a fringe at the free surface, and in other ways indicated the occurrence of a secretory cycle. In the opinion of Stephenson (1947) evidence of form change in these cells "is based on the assumption that a short cell is necessarily a small one" and that shallow cells are produced by a stretching of the epithelium as the lumen distends. Gresson and Threadgold (1959) used both the light and the electron microscopes in a study of this epithelium. They confirmed earlier observations that shallow cells line caeca which contain much food, while taller cells occur in those containing little or no food. They concluded that short cells are absorptive and that fine cytoplasmic processes seen at their free ends play some part in this process, giving these the name absorptive processes. The electron microscope revealed, however, that these processes often appear as loops connected at each end with the same cell and sometimes series of three or four loops one within another. They also recognized another type of cell as glandular but did not observe any transformation of secretory cells into flat cells and suggested that the presence of food in an anterior region of the intestine acts as a stimulus resulting in the discharge of secretion from glands situated farther back in the body. It is difficult for a reader to determine what their views really were, but these were summarized by Pantelouris and Gresson (1960) who confirmed "that both tall and short cells are absorptive in nature", that "both tall and short cells are transformed into either absorptive or glandular cells" and that "the epithelium passes through absorptive and secretory cycles". They also indicate that this is in general agreement with the conclusions of Müller

(1923). They pointed out that gland cells were not shown in the electron micrographs of Gresson and Threadgold and reiterated the conclusion that gland cells "are not present in parts of the alimentary canal containing food". They were concerned with the possible differentiation of so-called absorptive cells and secretory cells and used the method of autoradiography with ^{59}Fe partly to determine this. In their experiments the protoplasmic processes of the caecal epithelium did not "appear to be labelled" but they found a high degree of activity in the caecal epithelium and the cuticle of the flukes, putting forward the view that this layer (cuticle) "is either excretory or secretory in function, serving for the elimination of excess metabolites, or alternatively for the secretion of mucus on to the surface of the body". Their final tentative conclusion was that such substances are eliminated from the cuticle in the mucus. Dawes (1962d) criticized the opinions of Gresson and Threadgold but did not mention the later paper of Pantelouris and Gresson because it contributed nothing to the topic under consideration and it is mentioned here in part because it gives the consolidated opinions of Gresson and Threadgold. Dawes reached different conclusions, as the following abstract from his summary indicates. The basal portions of tall cells of the caecal epithelium contain mitochondria-like bodies which become numerous and arranged in vertical arrays which extend into the middle parts of their cells. In the terminal parts of these cells vacuole-like formations appear, usually one to each cell, until the lumen of the caecum becomes occluded by a frothy mass. When food enters the caecum in small amount the vacuoles are broken down and some secretion is liberated and mixed with the incoming food. Meanwhile, vacuoles encroach more deeply into the cell and eventually residual secretion is liberated. New cell boundaries are established near the nuclear zone and after complete discharge of secretion the distal parts of the cells comprise aggregates of long cytoplasmic filaments, which form a fringe on the free borders of now much shortened cells. The delicate fringes are abraded, probably by the physical action of the food streams, so that in typical short cells of the caeca they are much shorter, or may even disappear. Dawes examined the epithelium in juveniles as well as adults and in the young fluke denudation of cytoplasm during the formation of secretion appears to be more rapid. His conclusion was that in young and mature flukes alike the secretory cycle is of the apocrine type (not merocrine, as erroneously reported by Vickerman, 1963) and in a comparison with stages seen in the cycle of the mammary gland fits very closely in a number of particulars. The association of an apocrine secretory cycle with suctorial feeding is appropriate in both instances, but in the trematode the pharyngeal mechanism is responsible for the stream of food which enters the caeca and liberates secretion at what must be the appropriate moment. As in the case of the mammary gland the regeneration of cytoplasm in the shortened caecal epithelial cells is probably rapid and difficult to observe. In the young, growing fluke new cells must come into existence at a rapid rate, particularly during the first 4 weeks of infection when the ramifications of the caeca are in process of development.

VII. The Pathology of Experimental Fascioliasis

A very clear, elementary account of liver fluke disease was given by Lapage (1956) and by Lucker and Foster (1957). Lapage wrote about the haemorrhage which young flukes may cause during their migration to the liver and within this organ, about the irritation caused by the fluke's spines and resultant inflammatory processes with consequent fibrosis and about toxic substances secreted by the flukes which, some experts believe, are the chief cause of the anaemia that results. Flukes may obstruct the flow of bile and distended ducts may become calcified and "pipey". Obstructions may cause jaundice and restricted biliary flow may disturb the digestion of fats. If inflammation is severe, abscesses may form round the flukes in the liver and during healing processes much hepatic tissue may be replaced by scar tissue. If the resistance of the host is reduced bacteria such as *Clostridium oedematiens* may multiply, produce toxins and contribute to the condition known as "black disease". The less common acute form of the disease is commoner in sheep, which may die within a few days while apparently in good condition. According to Taylor (1951) 10,000 cysts must be given at one time to produce this acute disease in sheep and about 200 in a rabbit, though this would depend on the nutritional state of the host. This form of fascioliasis usually occurs towards the end of the summer, whereas the chronic form arises from June or July and gets worse as cysts are repeatedly ingested during autumn and winter, so that sheep weaken steadily and die from exhaustion.

Urquhart (1956) described the pathology of experimental fascioliasis in the rabbit, dealing with livers representing 2 week, 3–7 week and 8+ week infections. The early stages of penetration were not described, possibly not seen. In 2 week infections cream-coloured or pink lesions on the liver surface formed spots 1–2 mm diameter or streaks up to 10 mm long and 2 mm wide. Extrahepatic bile ducts were of normal size and there was no gross peritonitis on the liver surface. Microscopically, the lesions were resolved (somewhat ineptly) as "necrotic tracts" caused by the passage of young flukes through the parenchyma and "the space left in the wake of the fluke is filled with cell debris, neutrophil leucocytes, lymphocytes and red cells". Liver cells adjacent to the "tract" are atrophied, their nuclei pyknotic and sinusoids in this area are distended with large mononuclear cells, lymphocytes and granulocytes. At a later stage two zones are seen: an inner composed of dead liver cells, erythrocytes and masses of granulocytes, mostly eosinophils, and an outer consisting of mononuclear cells, with a few fibroblasts near the margin. Where a zone of acute inflammation diminishes in size, fibroblasts are more numerous and small capillaries appear at the outer edges of the tract, which is filled eventually with loose connective tissue containing large mononuclear cells and lymphocytes.

The livers of rabbits which had sustained 3–7 week infection had a superficial covering of firmly adherent fibrin patches, and there were adhesions between liver and stomach, intestines, omentum and diaphragm, as well as between individual lobes of the liver. There were many tract lesions, the livers in some instances containing thirty or more flukes. The tracts were as seen in the previous stage but larger and more numerous "owing to increase in size of the parasites and their continued wanderings". In the centre there was "cell debris, polymorphonuclears, lymphocytes and large mononuclears; erythrocytes may or may not be present in large numbers". Amongst the granulocytes eosinophils predominate, and often appeared to undergo degeneration by loss of cytoplasmic granularity, Recently formed tracts are surrounded each with a ring of necrotic liver cells and distended sinusoids filled with mononuclears and granulocytes. Older tracts had a reduced zone of necrosis and the space filled with large mononuclear cells with foamy cytoplasm, plasma cells, lymphocytes and eosinophils. At the edge of the lesion concentric arrangements of fibroblasts occurred but not in great numbers. Tracts are formed until about the 8th week of infection and "in heavy infections the liver may be honeycombed with them", although by this time the flukes are all in the bile ducts.

In the three livers of rabbits infected for 8 weeks Urquhart found adhesions between the lobes of the liver and between the liver and other viscera, and the liver was unusually pale. The gall bladder was often twice its normal size and filled with black or brown bile. The walls of extrahepatic bile ducts were thickened and the lumen several times its normal size. These ducts contained flukes. Sections showed the liver to be fibrotic and intrahepatic bile ducts were distended and had thickened walls. Sometimes, flukes were found in these also. Fluke tracts were undergoing advanced stages of repair; granulation tissue was more heavily collagenized, granulocytes and mononuclears were scarce and the tissue was relatively non-vascular. In some regions portal tracts remained unaltered; in others with many parasites they are enlarged, very fibrous and have many small bile ducts. Fluke eggs were commonly found in smaller bile ducts, one of them filling the lumen of a small duct and five or more completely obstructing larger ducts. In bile ducts large enough to lodge flukes the most striking lesion is the increase in their diameter and thickness of their wall. Most flukes were found in the common bile duct, their bodies folded one over another; many were free but others grasped the mucosa with their suckers. Sharp cuticular spines are often embedded in the epithelium of the duct, which is extensively damaged by them and the movements of the flukes. Eggs occur in the duct, which may contain red and white blood cells and cellular debris. The duct is often hyperplastic and in some ducts the epithelium was lacking, only fibrous tissue remaining in the wall.

We have drawn liberally on Urquhart's report because pathological effects depend on the host's reaction to its flukes and should be considered in the whole biological effect. The description is diagrammatic, misleading in some particulars and probably over-simplified in others, and some passages are not clearly expressed. It is suggested that the gross fibrinous peritonitis seen in later infections could not be ascribed to penetration of the liver capsule by young flukes migrating from the abdominal cavity, but it is associated with their intrahepatic migrations. More probably, we believe, it is brought about by older flukes which leave the liver and re-enter the abdominal cavity. The statement "dedifferentiation (sic) of liver cells into bile duct epithelium was often observed" is difficult to accept.

In a study of the clinical pathology of fascioliasis in sheep, Sinclair (1962) made counts of eggs in the faeces and carried out tests involving blood samples, blood cell examination, estimations of serum calcium, magnesium and also serum proteins. There was no reason to believe that infection followed an unusual course. After noting that migrating flukes have often been said to cause much haemorrhage in the liver, Sinclair stated; "the migration of young flukes did not cause sufficient haemorrhage to affect the number of erythrocytes and the counts did not begin to fall until 56 days after infection, at a time when adult parasites were arriving in the bile ducts". He gave support on this account to the statement of Dawes (1961a, b, c) that young flukes are tissue feeders but misquoted this writer as stating that "only adults ingest blood". He also recognized that loss of condition and oedema in the host is accompanied by anaemia and mentioned some differences of opinion regarding its cause—the production of a toxin which damages erythrocytes (Cameron, 1951) and the belief of Urquhart (1955) that the anaemia is similar to that brought about by repeated bleedings, which would produce the usual response, increased production of blood corpuscles in the bone marrow, with immature forms appearing in the circulation. Sinclair's results, however, showed the anaemia to be normocytic and normochromic; reduced erythrocytes were not observed, except when sheep had received carbon tetrachloride. Interference with bone marrow function did not involve leucopoiesis, and there was no evidence of an iron-deficiency induced anaemia. Liver damage was expected to produce changes in serum protein. At first there was an increase in globulin, particularly gamma-globulin, and when this neared its peak a progressive hypoalbuminaemia developed which could hardly be due to simple blood loss or haemodilution and was likely to be due to impaired synthesis or abnormal loss. Reduced synthesis of albumin might occur after liver damage, but albumin loss could be attributed to exudation from damaged bile ducts. A fall in serum calcium when flukes were maturing was believed to be due to hypoalbuminaemia, and a fall in serum magnesium occurred during early infection but persistent low levels of magnesium involved adults also.

Variation of white cell counts are shown in Sinclair's Fig. 6. A progressive leucocytosis began soon after administration of cysts, rising to a peak about day 63, its main cause (as shown in his Fig. 7) an absolute eosinophilia. Eosinophils in all sheep began to increase within 14 days of infection and (in Group A) the counts rose to a peak of 5,700 cells/mm^3 on day 63. Individual counts to 53% of circulating white cells were recorded. According to the figures cited considerable increase in mean leucocyte counts had occurred, in regular fashion from the time of infection, after 4–6 weeks (Fig. 6) and after 4+ days of infection the eosinophil count rose steeply to its maximum, presumably on day 35 (Fig. 7).

Ibrovic and Gall-Palla (1959) described the clinical and pathological effects of acute massive infections of sheep with liver flukes, and made comparisons of the blood of five sheep having sub-acute fascioliasis and three sheep having sub-chronic disease. In comparison with data from healthy sheep, their results indicated marked decrease of the iron level, significant decrease in the phosphorus level and slight decrease in the levels of magnesium and potassium. The calcium level was said to be somewhat higher, and the value of total serum proteins increased, likewise the value of globulins, particularly gamma and beta fractions, but the levels of albumins decreased.

Gresham and Jennings (1962) briefly considered fascioliasis (pp. 179–82), referring to early stages in which the migrating immature flukes produce "haemorrhagic tracts through the liver surface and focal areas of fibrinous peritonitis", evidently unmindful that flukes cause such lesions when emerging from the liver long after original entry. They also affirmed that such flukes "may cause excessive liver damage which results in a fatal haemorrhage, but this latter is rare". They noted that spines of adult flukes cause "severe irritation of the bile ducts so that a chronic cholangitis results with occlusions of the smaller ducts". Infected bile ducts are said to contain inspissated bile and "cholesterol gallstones are sometimes to be found in the bile ducts and these contribute to the biliary obstructions". How the liver fluke produces disease other than in a mechanical fashion is described as still obscure and "it is probable that other mechanisms are concerned". They are categoric that "the cause of the severe anaemia is unknown" but after referring to the fluke as "a blood sucking parasite" which produces a haemolysin, they remark that it also liberates metabolic products which are harmful to the host and "all these factors may be concerned in the causation of the anaemia".

It is clear from this and other recent works that the pathology of fascioliasis is also a rather neglected subject about which much more information will be required before a reliable assessment can be made of the parts played by parasite and host respectively. The nature of the injuries caused by young flukes in the hepatic parenchyma by direct action during feeding has been delineated, but to what extent the inflammatory responses of the hosts are involved in the damage caused has yet to be fully ascertained but is apparently significant. This will be a task for the experimental pathologist, who will need to consider the early invasive stages in the liver as well as later stages in the biliary system. Dawes (1963c) has suggested that serious liver damage, which may result from the protracted stay of flukes in the hepatic parenchyma, may call forth "an unrestrained inflammatory reaction which enhances liver damage to the point at which it becomes irreparable, the host's attempts to repair the damage leading to pathological effects bordering on malignancy", which in itself should be an encouragement to further research.

VIII. Human Fascioliasis

During the past 60 years at least sixteen reports have been made of human infection with *F. hepatica* in Great Britain, mostly isolated cases. Facey (1959) noted some cases of sickness which suggested a helminthiasis—pain in the right hypochondrium, irregular pyrexia and eosinophilia—in a district where fascioliasis was prevalent in farm animals. Facey and Marsden (1960) described six cases, in five of which eggs appeared in the faeces, and the patients (a doctor, a factory worker and four housewives) were living during the rainy summer of 1958 in the Ringwood district of the Avon Valley, and all were fond of eating watercress. There was a high eosinophilia in all cases and one leucocytosis, but anaemia was not a prominent feature of the sickness, although in one patient the haemoglobin value diminished almost 3 g in 1 month. Other cases in Britain were recorded by Ward (1911), who mentioned four earlier cases, Owen (1928), Patterson (1928), Biggart (1937), Manson-Bahr and Walton (1941), Shaw and Clyne (1941), O'Donnell (1949), Murphy and Pascall (1950), Ramage (1951) and Catchpole and Snow (1952).

Human fascioliasis is much more common in some other countries, past records mainly concerning Cuba (Kouri and Arenas y Martorell, 1931a, b, 1932a, b; Kouri, 1948; Arenas y Martorell *et al.*, 1948), Uruguay (Rial *et al.*, 1951), Argentina (Roderiguez, 1952), France (Lavier and Marchall, 1942; Lavier and Deschiens, 1956), Germany (Bürgi, 1936) and Russia (Levinam, 1950). Facey and Marsden (1960) have compiled a more exhaustive list of references than can be given here and in some instances more than one hundred cases are considered in a single paper, 186 cases in Germany by Bürgi (1936). In some of the cases the symptoms of the invasive phase of fascioliasis are described. After a short period of dyspepsia there may be mild or high fever and stabbing or dragging abdominal pain which is likened to cramp and is aggravated sometimes by movement or coughing. These symptoms, together with an enlarged liver and an eosinophilia are said to be almost diagnostic of the condition. Some recovery may occur as a result of the loss of parasites by evacuation or calcification and in later stages a latent period may occur, symptoms subsiding abruptly so that for some months or even years the patience suffers nothing worse than indigestion, although relapse does sometimes occur. Biggart (1937) discussed the case of a 36-year-old woman in Scotland who died in 1934. About "a dozen parasites in all" were recovered and measured about 25 × 13 mm. The bile ducts were dilated and thickened and in many of them "a curious hyperplasia" was seen, epithelium transformed into long tubular glands surrounded by a stroma containing eosinophils and plasma cells. In his Fig. 1 some denudation was seen. This patient had lived all her life in a sheep-rearing district of Fifeshire, and showed characteristic lesions of fascioliasis.

IX. Attempts to Influence the Invasive Stages

At the present time there is no method available to prevent the establishment of *F. hepatica* in animals placed on infected pastures; nor is there any treatment which will kill this parasite before the liver is badly damaged by its harmful wanderings and feeding excursions.

The early work of Montgomerie (1928b) in sheep, using a dose of 10 ml of carbon tetrachloride per animal, showed that some of the parasites were killed when only 4 weeks old and that almost all the flukes were dead in 5, 6, 7 or 8 week infections. This finding has been supported by subsequent research in this field by a number of authors. The fact that individual flukes from the same batch of cysts do not develop at the same rate will affect their susceptibility to drugs and the interpretation of results must be made with this in mind.

More recently, hexachlorophene [2,2′-methylene bis (3, 4, 6-trichlorophenol)]* has been shown by Kendall and Parfitt (1962) to be capable of killing flukes 3–4 weeks old, but at a dose (40 mg/kg) which is not without toxic hazards to the host. Williams (1963) has also reported on the activity in rabbits of 4,4′-diaminodiphenylmethane against the very early stages (1–4 weeks old), but unfortunately she found this compound to be unsafe for use in cattle. Further details of the chemotherapy of fascioliasis can be found in the literature (Lämmler, 1955, 1956, 1959; Lämmler and Loewe, 1962a, b; Hughes, 1959; Kendall and Parfitt, *loc. cit.*). In the present volume Gibson (1964) also deals with this subject.

The ideal drug, which has yet to be found, would be one which is very well tolerated by the host and which has a long-term prophylactic effect on the early migratory stages of the fluke in the intestines, abdominal cavity and hepatic parenchyma of the host. Alternatively, vaccination would provide the requisite answer, although at the present time little or no success has been achieved by such methods.

Attempts to produce resistance to helminth infections without having to resort to infection of the host with the pathogenic parasite initially fall into four main groups: (a) passive immunization by means of sera; (b) vaccination by means of injections of dead parasitic materials or the secretions and excretions of living parasites; (c) vaccination by means of exposure to a non-pathogenic strain of the parasite; and (d) vaccination with the living parasite

* This form is used in the British Pharmaceutical Codex, although some chemists prefer the equally correct form given on p. 241.

which has been weakened either by ionizing radiations or chemicals. Many examples of these methods are to be found in the literature and the reviews of Taliaferro (1929), Culbertson (1941), Soulsby (1961) and Hughes (1963) and will not be described here. It is perhaps relevant to mention, however, that the only vaccine commercially available against a species of helminth is the X-irradiated vaccine against *Dictyocaulus viviparus*, the nematode which causes parasitic bronchitis in cattle. This vaccine, based on the researches of Jarrett *et al.* (1957) and subsequently developed in the commercial field by Poynter *et al.* (1960), has contributed in no small way to the control of this harmful disease of cattle. For an account of this work see Poynter (1963).

A. NATURALLY ACQUIRED RESISTANCE TO *F. hepatica*

There is little evidence to be gleaned from the literature of any significant naturally acquired resistance to *F. hepatica* in the field. Leiper (1938) using goats as hosts and Taylor (1949) working with sheep found no evidence of the development of resistance after infection. Montgomerie (1931) and Taylor (*loc. cit.*) found that this infection persisted in experimentally infected rabbits. Durbin (1952) determined that infection can persist for at least 11 years in experimentally infected sheep when care has been taken to avoid any possibility of reinfection. Montgomerie did point out, however, that cattle may possibly differ somewhat from sheep in their reaction to this parasite, stating, "*F. hepatica* does not continue to infect cattle for a long period. The livers of 2½ years old cattle very commonly show evidence of severe infestation without the presence of a single fluke". Lederman (1958), reporting on *F. hepatica* infections in cattle in the Walungu area of South Kivu in the Belgian Congo, believed that stock develop considerable tolerance to this trematode and indicated that if they are removed from the source of infection they will free themselves of infection after one or two years. Sinclair (1962) found little evidence that several doses of cysts induced an immunity, although he considered that there was possibly some evidence of a delayed onset of egg production and a reduction of the numbers of eggs in later maturing adults. In human facioliasis, Facey and Marsden (1960) state that "recovery may occur by the parasite being evacuated through the intestine or they may become shut off and calcified (visible radiographically). If live adult flukes are retained in the bile passages, the disease passes into the next stage". It is interesting to note that of the six cases reported on, which were treated with chloroquine or hydroxychloroquine sulphate, at least four were still infected as judged by eggs in the faeces at the time of their discharge from hospital. They were, however, free from the acute symptoms. Taylor (1961) held the opinion that many cases of *Fasciola* infection in man undergo spontaneous cure.

B. ARTIFICIALLY ACQUIRED RESISTANCE TO *F. hepatica*

The methods used to produce immunity to helminths referred to above have all, with the exception of (c), been tried with *F. hepatica.* The reason why method (c) has not been used is because as far as is known there is not a strain of *F. hepatica* which is non-pathogenic. The attempts made by various investigators to induce artificially acquired resistance in the hosts of *F. hepatica* we must now briefly review.

1. *Passive Immunization*

It is well known that young flukes can be induced to excyst *in vitro* (Susuki, 1931; Vogel, 1934; Hughes, 1959, 1963; Wikerhauser, 1960). Wikerhauser (1961a) also showed that if the young artificially excysted flukes were placed in"immune" sera of rabbit or bovine origin, microscopically visible precipitates were formed, particularly at the mouth and the excretory pore of the fluke. If guinea pigs were injected with this "immune" serum, either by the subcutaneous or the intraperitoneal route, there was no noticable inhibitory effect after subsequent oral infection of the host with cysts. Hughes (1963) also injected sera from rabbits which had been previously infected, or in which previous attempts at vaccination had been made, by the intraperitoneal route into mice. These mice and also appropriate control groups were then infected orally with five normal cysts but no difference was found either in the degree of infection or in the time of death of the mice in the various groups. Hughes concluded that the sera from rabbits with previous experience of liver fluke failed to influence the course of an oral infection in mice, results which support the findings of Wickerhauser. This latter author found, however, that if the early forms which had excysted *in vitro* were placed in the abdominal cavity of guinea pigs a somewhat lighter infection resulted when the young flukes had been either immersed in "immune" serum or when this serum was administered by intraperitoneal injection at the time of infection. It was pointed out that these conclusions are based on a limited number of inoculated guinea pigs and that the results have only a relative significance.

2. *Vaccination with Materials from Dead Flukes or the Secretions and Excretions of Living Flukes*

Kerr and Petkovich (1935) stated the serological studies led them to believe that it might be possible to induce active immunity to liver fluke in rabbits, without stating what these studies were. They gave intraperitoneal injections to seven rabbits of a 1% suspension of dried fluke materials which had been prepared by drying flukes at 55° C. They then infected the seven animals together with three rabbit controls 5 weeks after the last of nine of the intraperitoneal injections with 13 cysts each and, from the results obtained, indicated that it was possible to establish this immunity. The seven rabbits which had received injections were found at post-mortem examinations 92–118 days after infection (challenge) to have a total of 15 flukes (1+2+2+3+2+2+3). In two of these rabbits a total of three flukes were found to be calcified. It was also noted that eggs of the flukes did not appear in the faeces of infected hosts. By contrast, the three rabbit controls became patent and yielded at post-mortem examination 11+5+9 flukes, none of which were calcified. It is difficult to know how significant their result was considering that only three controls were used and in view of the great variability which exists in the yields of flukes from cysts.

Urquhart *et al.* (1954), using an essentially protein antigen, claimed to have produced an artificial immunity to *F. hepatica* in rabbits. Immunity was shown, they state, by an inhibition of development in the flukes when the hosts were killed 63 days after infection. The flukes from the injected rabbits were smaller than those from the control hosts, and total nitrogen estimations of the flukes in the two groups confirmed this finding. However, they did not find any evidence of a significant reduction in the numbers of flukes in the injected hosts. In contrast with their result is that of Healy (1955), who found no evidence of resistance to infection when rabbits had been given potential immunizing extracts of worm tissue or with pooled regurgitated caecal contents of *F. hepatica*. Other efforts along these lines were made by Shibanai *et al.* (1956), who injected seven rabbits with an antigen which was administered by various routes; the rabbits, and also three control rabbits, were infected orally with cysts 70 days later, and after a further 56 days the rabbits were killed and their livers examined macroscopically and histologically. It was claimed that the lesions found in the inoculated rabbits were least severe by comparison with the controls in hosts which had received subcutaneous and intradermal injections, whereas lesions of intermediate severity were found in those hosts which had received intraperitoneal or oral injections. Unfortunately, these investigators did not give full details of their experiments and as they used only a small number of rabbits it is difficult to interpret their results.

Ershov (1959) prepared an antigen which was considered to be a polysaccharide albumin complex and claimed that this produced an immunity in sheep, 10–25% of these animals being found to be immune after infection. If there was partial immunity, however, it was short lived because sheep infected 45 days after immunization procedures were carried out did not show any immunity. Although Ershov stated that the toxicity and immunogenicity of this antigen was examined first of all in laboratory animals, he did not give any details about immunization procedures.

Hughes (1962b, 1963) attempted to immunize rabbits with antigens prepared from lyophilized adult flukes. One experiment, in which the immunization schedule consisted of nine intraperitoneal injections and in which the experimental design was similar to that of Kerr and Petkovich (1935), showed a degree of protection that was statistically significant. Nine injected rabbits submitted to challenge with 13 orally administered cysts were found at post-mortem examination 14 weeks later to have a mean of 1·7 flukes per rabbit as compared with a mean of 5·6 flukes in the six rabbit controls. The experiment was repeated, with the addition of other experimental groups, but this time and with twenty-six rabbits in all Hughes failed to demonstrate any degree of immunity in the vaccinated hosts. In the experiment with sheep, ten hosts (Group A) were given a series of nine intraperitoneal injections with adult fluke antigen over a period of 3 weeks. The vaccinated sheep, together

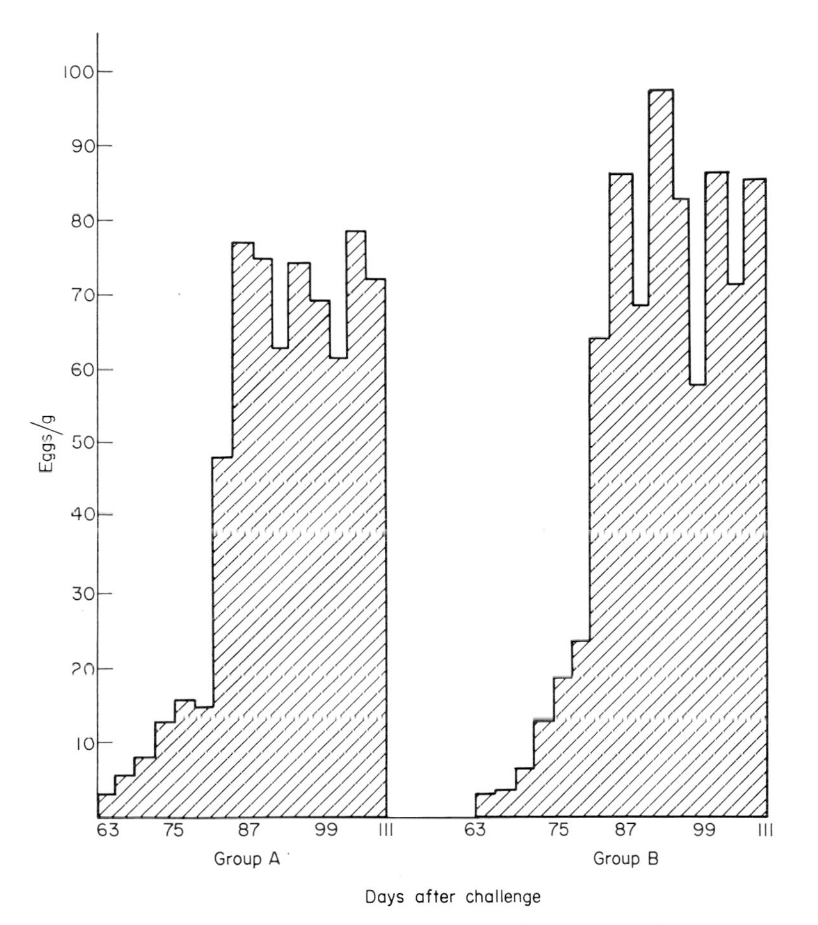

FIG. 11. Histograms of egg counts in sheep. (Expt. 12 of Hughes, 1963.) Group A received nine intraperitoneal injections of 10 ml of 1% adult fluke antigen per sheep every 2 or 3 days over a period of 3 weeks and were then challenged with 200 normal cysts per sheep. Group B received only 200 normal cysts. Mean peak egg counts: 80 eggs per g in Group A, 97 eggs per g in Group B. (Hughes, 1963.)

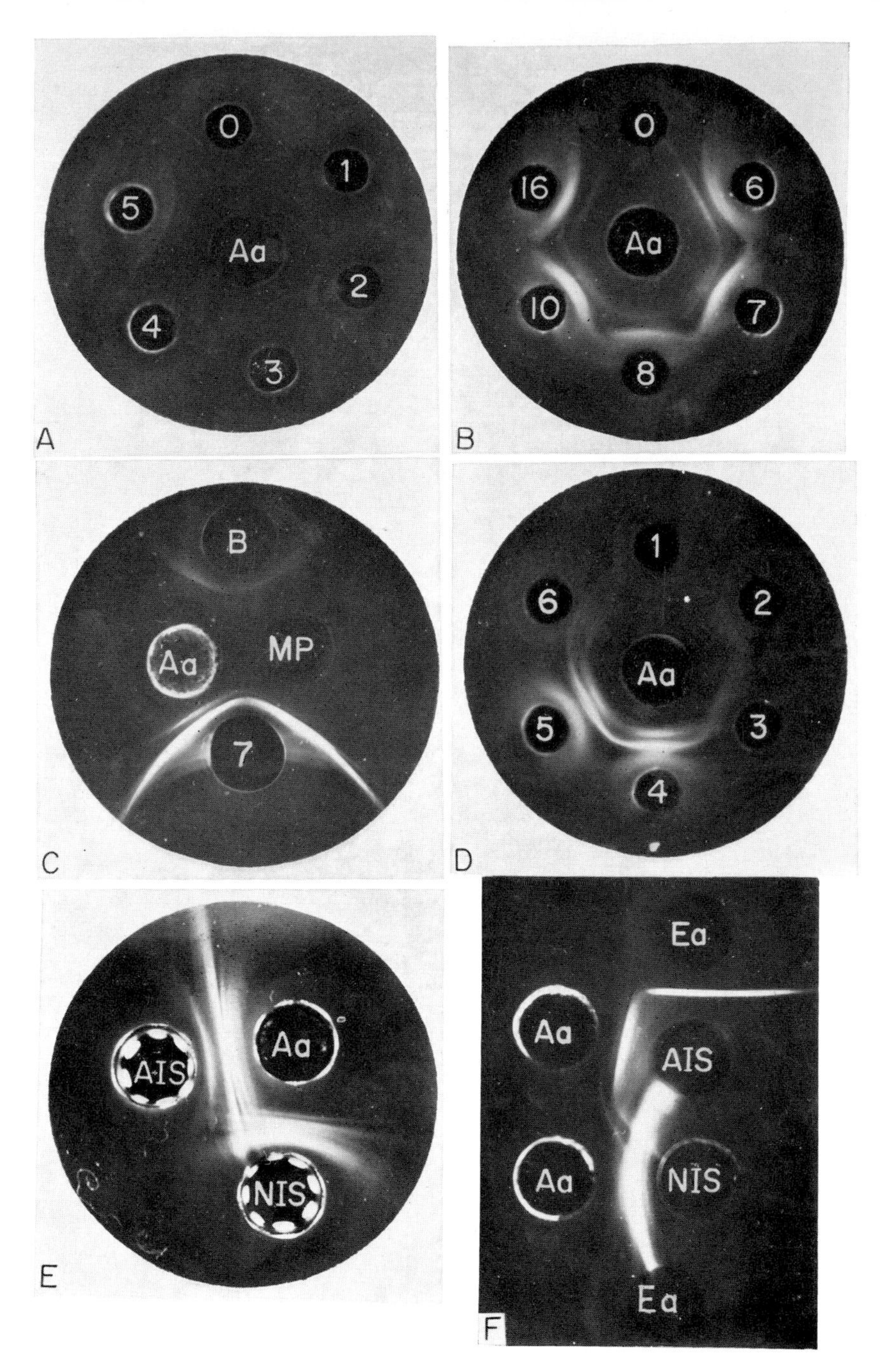

FIG. 12. Serological reactions. Antigen abbreviations: Aa, 4% (w/v) suspension of ground lyophilized adult flukes in saline; Mp, Antigen obtained from Earle's solution after adult flukes had been maintained in it for 48 h at 37°C; Ea, suspension of approx. 50,000 ground lyophilized fluke eggs per ml saline.

In A and B, appearance of lines in the sera of rabbits infected with 100 normal cysts. Numbers in outer wells refer to weeks after oral infection. C, Aa and Mp antigens set up against the sera from rabbits with normal infections (rabbit B and rabbit 7); at least three lines show reactions of identity. D, Response in a rabbit given two subcutaneous injections of adult antigen mixed with Freund's complete adjuvant, 3 weeks apart. Numbers in outer wells refer to weeks after first injection. Note the marked secondary response at well 4. E, Reaction of non-identity between sera from a rabbit with a normal infection (N.I.S.) and that from a rabbit which had been injected with adult antigen (A.I.S.) when set up against Aa. F, Showing the marked line produced with Ea and A.I.S. and its absence in the N.I.S./Ea system. (Hughes, 1963.)

with ten sheep controls (Group B), were challenged 5 weeks later with 200 orally administered normal cysts and there was no statistically significant difference in the time of patency of the two groups of sheep (Fig. 11) or in the mean numbers of flukes per sheep. At post-mortem examination 16 weeks after challenge a mean of 59·6 flukes per sheep was obtained for the treated sheep as compared with a mean of 78·2 for the control group. Hughes also found that an antigen prepared as described by Ershov (1959) failed to produce immunity in rabbits.

Hughes (1963) carried out more serological work than can be reported here, examining the sera of rabbits which had been infected with cysts and also sera of rabbits which had been injected with antigens of fluke origin. These serological investigations were concerned mainly with the reactions of antigens and antibodies in agar gel. Hughes stated his original hope, that these antigens might produce resistance in experimental animals, but also that it soon became evident that such hopes are unduly optimistic. Some of these serological reactions are shown in Fig. 12, A–E. In A and B the gradual appearance of lines of precipitation is shown after a rabbit had been infected with normal cysts. Lines against the adult antigen (Aa) first appeared about 4 weeks after infection and, as the infection progressed, the precipitation lines increased in number and in intensity. All the infected rabbits showed this characteristic pattern, except for a few minor variations in the exact position of the lines. The pattern persisted and did not alter appreciably in infections which had been sustained for more than 6 weeks. Sera which showed this full complement of lines also reacted with a "metabolic products" antigen (Mp) and in this instance at least three precipitation lines showed a reaction of identity (Fig. 12, C). In rabbits injected with Aa antigen mixed with Freund's complete adjuvant a marked secondary response was seen when the second injection was given 3 weeks after the first (Fig. 12, D). When sera from a rabbit with a normal infection was set up together with sera from an artificially immunized rabbit against the adult antigen Aa (Fig. 12, E), the majority of the lines produced appeared to show reactions of non-identity. The sera of these artificially immunized rabbits showed a single line of precipitation with the fluke antigen Ea and this line was identical with the strongest line obtained against Aa (Fig. 12, F). No lines were seen when Ea was set up against the sera of rabbits with normal infections. Hughes also showed that although the use of Freund's complete adjuvant resulted in a marked antibody response when mixed with fluke antigen in rabbits, as shown by agar gel diffusion, this response failed to produce any significant effect on a challenge infection when 15 cysts per rabbit were administered orally.

3. *Vaccination with X-irradiated Cysts*

a. The reduction of the pathenogenicity of flukes in X-irradiated cysts. It has been shown clearly that the pathenogenicity of young *F. hepatica* in rabbits can be reduced by X-irradiation of the cysts before administration (Wikerhauser, 1961b) and Hughes (1962a, 1963) confirmed this for rabbits and extended the findings to include sheep and mice as well. Dawes (1964) confirmed the observations for mice. In all three hosts used by Hughes there was a marked host response to flukes from X-irradiated cysts as judged by serological and histological methods. Both Wikerhauser and Hughes found that marked reduction of pathogenicity occurs with flukes from cysts at irradiation levels greater than about 3,000 r. Hughes (1963) pointed out that flukes from X-irradiated cysts (which henceforth will be referred to as X-irradiated flukes) appear to behave normally; they excyst *in vitro* as well as *in vivo* in about the same proportion of cysts as untreated forms, migrate through the intestinal wall and abdominal cavity and penetrate into the liver, where they feed and grow. After 8–10 days approximately they cease to grow normally (Fig. 13) and the developing intestine does not differentiate in the normal complex manner, so that the appearance of the intestine in X-irradiated flukes 20–30 days old (Fig. 14, A–E) is essentially similar to the normal appearance in 8–10-day-old flukes from untreated cysts (Fig. 2, A). The study of sections indicated that the X-irradiated flukes appear to burrow more slowly than normal flukes in the liver and that, eventually, they die (Hughes, 1962a, b, 1963). "The reaction of the host to the presence of X-irradiated flukes is essentially similar to that provoked by normal flukes; but as the X-irradiated flukes slow down, so the host reaction appears to catch up with the parasite and eventually surrounds it" (Hughes, 1963). Few forms X-irradiated at the level of 3 Kr reached maturity, in one experiment with sheep 6/2,000, and this is in agreement with the findings of Wikerhauser (1961b), who experimented with rabbits. The few flukes which Wikerhauser obtained by the use of cysts X-irradiated at either 3 Kr or 8 Kr produced eggs, but these were largely infertile. Moreover, eggs which developed took 36 days as compared with about 12 days for the controls. Hughes' findings on this point differed; eggs of flukes X-irradiated during the encysted stage at 3 Kr developed normally in 12–14 days. Jarrett *et al.* (1959) stated that the level of 5 Kr is sufficient to inactivate young *F. gigantica* within their cysts, without disclosing whether or not the effects of lower dosages were investigated. The findings of Hughes (1963) do not lend support to the statement of Pearson (1960) (in a general discussion of a paper by Mulligan, 1960) that "100 r would produce inactivation of cercariae", although Pearson did not state whether this applies to *F. hepatica* or *F. gigantica*. It is interesting also to note that 3 Kr is within the dosage range which is required to inactivate the cercariae of *Schistosoma* spp. (Villella *et al.*, 1961; Smithers, 1962; Hsü *et al.*, 1962). This range is very different from that required to attentuate the infective stages of nematode parasites, which is 40–60 Kr. We have seen that X-irradiation does not interfere with excystment in *F. hepatica*, although Wikerhauser (1961b) found that gradually increasing doses of irradiation resulted in gradually decreasing survival of young forms as judged by their ability to excyst *in vitro*. In respect of his finding that the effects of X-irradiation do not become apparent until the young flukes are about 8 days old Hughes (1963) stated: "it may be of course that the X-rays have damaged some essential enzyme system of the fluke which only comes into operation at about 8 or 10 days. It is interesting to note that no drug will kill flukes before this stage. A good deal more

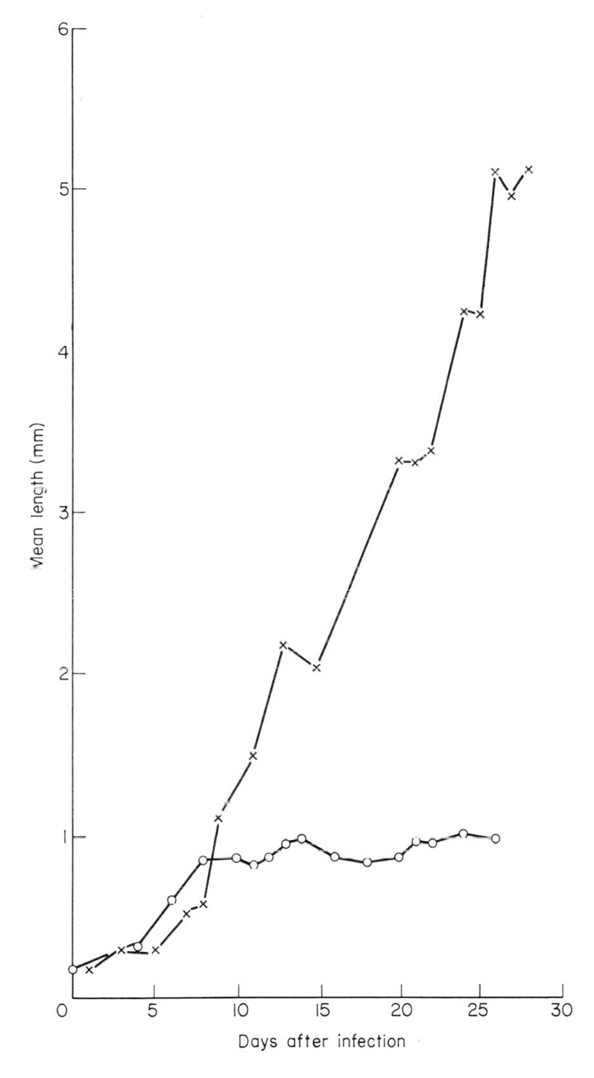

FIG. 13. Growth curves of mean length for mouse infection with normal cysts (x—x) (Dawes, 1962a) and cysts X-irradiated at 4 Kr (o—o) (Hughes, 1963).

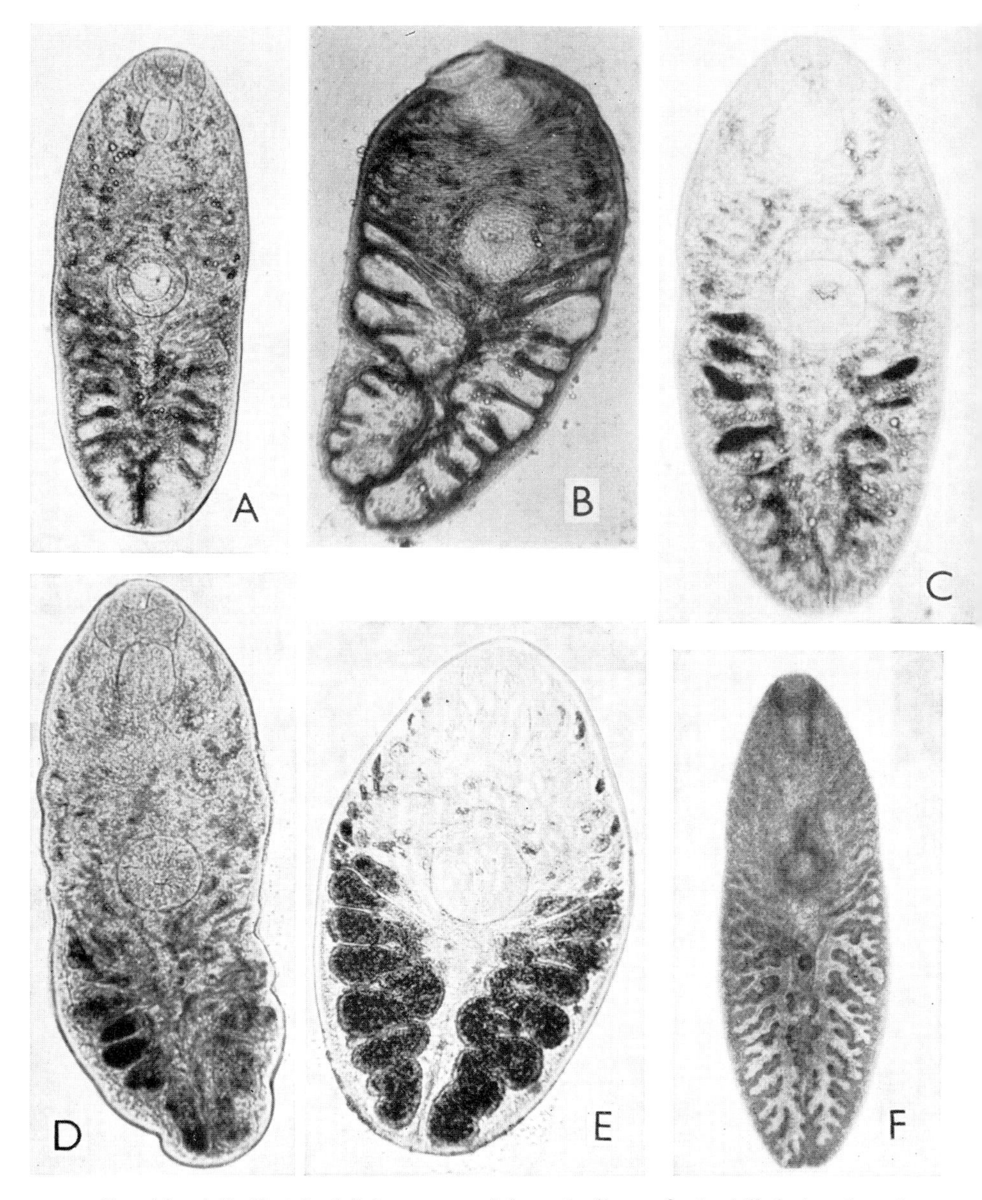

FIG. 14. A-E, Unstained flukes recovered from the livers of mice killed at various times after infection with cysts X-irradiated at 4 Kr. A, 8 days; B, 11 days; C, 14 days; D, 20 days; E, 32 days. All of these flukes were approximately 1 mm long. F, Stained fluke recovered from the liver of a mouse which received 40 X-irradiated cysts (3 Kr) followed by challenge with 10 normal cysts 22 days later and was killed 11 days after challenge infection. This fluke measured 1·6 × 0·5 mm and shows a degree of caeca branching not evinced in any of the X-irradiated flukes. A-E, Hughes (1963); F, Dawes (1964).

information on the biochemistry of *F. hepatica* will be required before this idea can be either accepted or rejected".

b. The death of young flukes weakened by X-irradiation. Dawes (1964) confirmed the results of Hughes (1963) regarding the weakening of X-irradiated flukes in mice. The intestine of the stunted flukes does not develop beyond the stage at which ten to fourteen short sacculations arise from the crura of the two sides of the body, and with this feature is correlated unusual slenderness of the fluke's body. The small flukes produce relatively small burrows, because their translatory movements are less extensive and because movement is along a tortuous course and not a wide arc. Consequently, the young flukes are beset by dense aggregates of band type neutrophil leucocytes (polymorphs) which infiltrate into the burrow and, prevented in many instances from contacting hepatic cells, the feeding fluke engorges itself with these leucocytes (Fig. 15, A, C, D, E). Many of the leucocytes within the burrow adhere to the body of the fluke and some of these penetrate the epicuticle and form extensive arrays within the core of the cuticle (Fig. 15, A). Entry seems to be effected at only a few points, suggesting that the epicuticle offers some resistance to leucocytic action, and the basal portion of the cuticle (lamina vitrea) is also resistant, as is indicated by comparative rarity of invasion into the parenchyma of the fluke at this stage.

Leucocytes continue to pass into the core of the cuticle until they form an extensive layer four to six cells deep, raising the epicuticle in blister-like formations (Fig. 15 C, D). Subsequently, the broken epicuticle undergoes fragmentation, presumably by the action of enzymes of leucocytic origin, and both the fragments and the invading leucocytes are thrown into the burrow outside the body of the fluke. The fluke is now a dying fluke and Dawes (1963c) drew attention to the fact that this phenomenon of cuticular invasion with lethal consequences had never before been witnessed in *F. hepatica* and probably provides a unique example of what happens to a weakened trematode when it is overtaken by events which do not affect it adversely in the normal state. In normal infections the young fluke burrowing in the liver is constantly moving out of the range of action of infiltrating leucocytes into what appears to be normal hepatic parenchyma. In such infections leucocytic infiltrations are more concerned with the repair of damage caused by young flukes than with the dispatch of the flukes themselves, which behave as if nothing untoward were happening. The young X-irradiated flukes on the other hand are trapped, injured, killed and finally disintegrated by the agency of the inflammatory reaction. Dawes (1963e)—summarizing results and conclusions already in the press (Dawes, 1964)—suggested that we must abjure the widespread opinion that the local tissue reaction of the host is directed against the trematode parasite in order to bring about its extermination. The evidence indicates that this reaction is really directed towards the repair of injury caused by flukes burrowing into the liver and may be without effect on

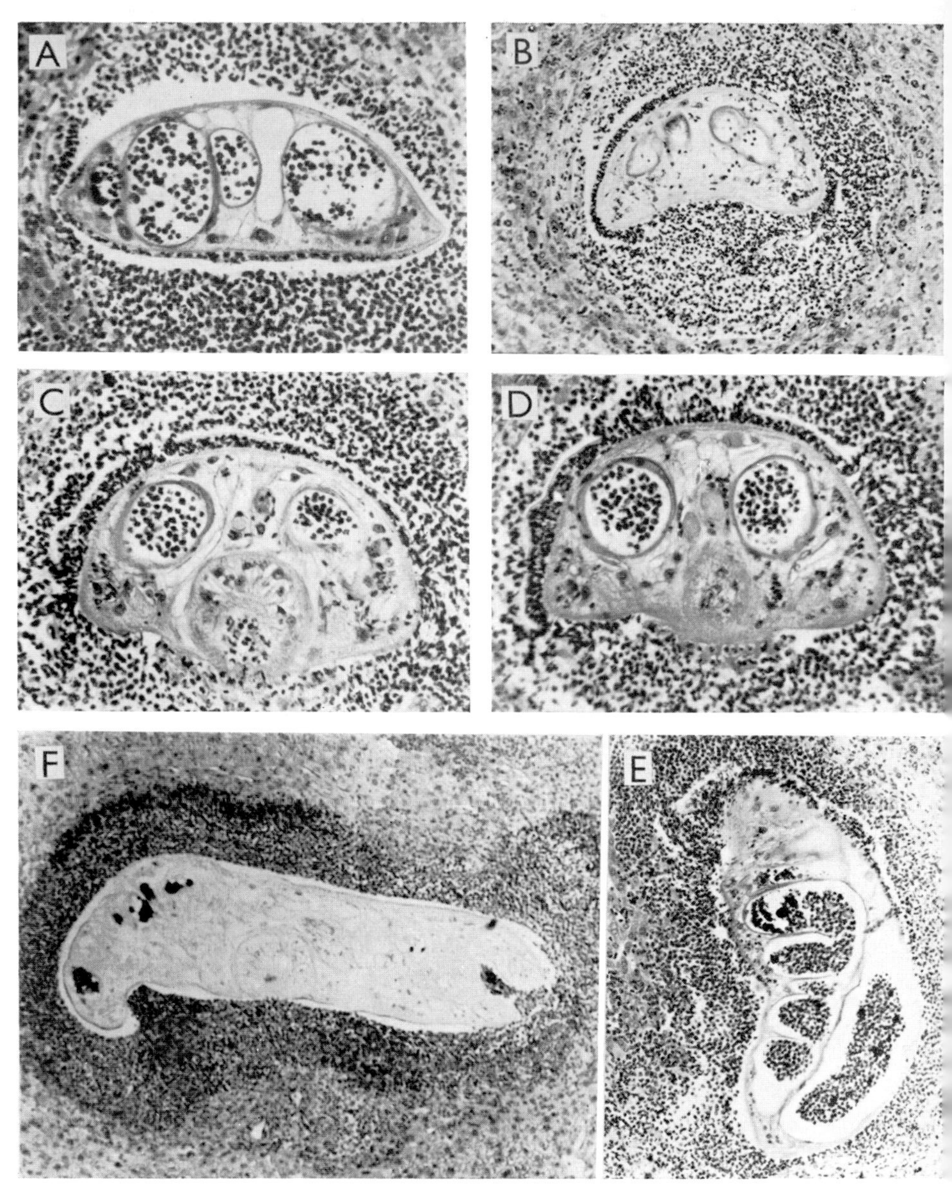

FIG. 15. Sections of flukes in the livers of mice infected for 12 days with cysts X-irradiated at 3 Kr, illustrating invasion of the cuticle by leucocytes and subsequent death of the fluke. A, T.S. fluke showing caeca and excretory canal. Leucocytes have entered the cuticle, forming mainly a single array; many leucocytes have also been ingested. B, T.S. fluke and burrow, showing more massive leucocytic infiltrations into cuticle on the left and shattered epicuticle on the right. C and D, Sections near the ventral sucker showing blister-like elevations of epicuticle due to the infiltrating leucocytes; on lower side, some stripping of epicuticle. Many ingested leucocytes seen in caeca. E, Oblique L.S. posterior end of fluke with damaged cuticle. Note intense leucocytic infiltrations into burrow and masses of leucocytes in the caeca and their sacculations. F, Dead fluke seen in L.S. showing ghostly appearance, heavily pigmented debris in caeca and mouth filled with leucocytes of final meal. Excessive infiltrations of leucocytes surround the remains of the fluke. (Dawes, 1964.)

the parasites, unless these are injured in some manner—in this instance by X-irradiation—and thereby brought into the sphere of influence of the reaction.

Hughes (1963) found a number of dead X-irradiated flukes and prepared sections of livers infected with both living (Fig. 14, A–E) and dead flukes, in order to demonstrate the reaction seen around a dead fluke in liver tissue. He figured a number of sections of both living and dead flukes and noted in dead ones that "the structure of the fluke differs from the normal fluke in that the cuticle is no longer clearly visible and it is also difficult to make out the usual internal structure of the fluke". He did not observe the stripping of the epicuticle as a result of leucocytic invasion, which is an indication that the entire process occurs quickly. Dawes (1963e) noted that in the livers of mice which are supporting flukes X-irradiated (in the encysted stage) at 3 Kr weakening of the fluke results in the death of some but not all of them during the period under investigation. Up to the 22nd day of infection living flukes outnumbered dead flukes in the ratio of 101/3 but in older infections were outnumbered by dead flukes in the ratio of 10/13. The dead flukes seen in sections have a ghostly appearance which is due to the loss by their tissues of affinity for stains and the body shows little infiltration by leucocytes as a rule but is surrounded by very dense aggregates of leucocytes which represent secondary infiltrations from the hepatic parenchyma (Fig. 15, F). The young X-irradiated flukes, then, are attacked by the local tissue reaction as living entities and not merely as "foreign bodies" which they later become and which are dealt with by further action and gradually disintegrated and dispelled.

c. The effect of infection with X-irradiated cysts on the course of subsequent infection with normal cysts. At first sight, the partial inactivation of flukes by X-irradiation would appear to be promising for the purposes of "vaccination". We have seen that X-irradiated flukes have only low pathogenicity and yet develop sufficiently to induce a host reaction which can be identified by both histological and serological methods (Hughes, 1962a, b, 1963). The picture is comparable with that of X-irradiated larvae of the nematode *Dictyocaulus viviparus*, which are now used to "vaccinate" calves in the field. In this instance, however, there is a marked resistance to reinfection in the field once the host has experienced and recovered from the mild disease, whereas in *F. hepatica* any significant degree of resistance has yet to be demonstrated.

Hughes (1963) attempted to immunize twenty mice with two doses of 20 X-irradiated cysts (4 Kr) given with an interval of 3 weeks; 3 weeks after the second dose each mouse was infected orally with 10 normal cysts, and twenty untreated mice received 10 normal cysts and served as a group of controls. Two other groups of five mice which received only X-irradiated cysts served as controls of the "attenuation" procedure. After challenge, all the mice which had received treatment with two doses of X-irradiated cysts and then normal cysts were found to die at the same time as the infected control group, 22 days after infection. Furthermore, no difference was found in the mean number of flukes recovered (2·1 flukes per mouse) in these two groups. The conclusion drawn was that this experiment had not provided any evidence of immunity. The mice which had received only X-irradiated cysts were killed 13 weeks after this treatment; their livers appeared normal and no flukes were to be found. In a subsequent experiment the level of X-irradiation was reduced to 3 Kr and the number of X-irradiated cysts given was increased to 40 per mouse at each infection. As before, the mice received two lots of X-irradiated cysts separated by an interval of 3 weeks, but this time the mice were not challenged with normal cysts until 7 weeks had elapsed from the time of the second dose. After challenge, mice in both experimental and control groups died at approximately the same time and there was no significant difference in the mean numbers of flukes recovered in the two groups (3·5 per mouse as compared with 4·0 per mouse). Dawes (1963e) repeated the experiments of Hughes with mice. The fourteen mice of his Series 3 experiments each received 40 cysts X-irradiated at 3 Kr and this infection was then maintained for 22 days, after which nine of the mice were challenged with 10 normal cysts and killed serially 8–25 days afterwards, while the remaining five mice were challenged with 5 normal cysts and killed serially 25–53 days afterwards. The controls of Hughes (1963) were regarded as controls in this experiment also, because both cysts and mice were from the same stocks. The total number of flukes recovered was 62 (in groups of 1–8 per mouse), providing a mean of 4·4 flukes per mouse. In terms of percentage of cysts yielding flukes 55·4% as against 35·3% in Hughes' experimental animals and 42·5% for the control hosts, in which a mean of 4·25 flukes per mouse was obtained. In the results of neither series of experiments with mice, therefore, was there any evidence of immunity in terms of intensity of infection (fluke burden). In the experiments of Dawes, three almost mature flukes were recovered after 32 days of challenge infection, and at 39 days one mature fluke had shed many eggs into the bile duct and gall bladder, indicating that there was no immunity in terms of delayed maturation either. It was noted by Dawes, however, that there is an apparent diminution of growth intensity which occurs from about 3–4 weeks after challenge infection and which requires further investigation.

Attempts by Hughes (1963) to immunize rabbits with two doses of cysts X-irradiated at 4 Kr also failed. The four experimental animals received two doses of 500 X-irradiated cysts each with an interval of 3 weeks between doses and they were then challenged with 15 normal cysts each 4 weeks later. At post-mortem examination 10 weeks after challenge the mean number of flukes per host recovered from treated animals was 5·7 as compared with 5 in the control group. Further attempts by Hughes (1963) to immunize ten sheep with two doses of cysts X-irradiated at 3 Kr with an interval of 3 weeks between doses were also indicative that immunity does not arise. These sheep were challenged with 200 normal cysts 7 weeks after the second dosage with X-irradiated cysts. Two sheep (Group 1) which received X-irradiated cysts only became patent, although only a few eggs of the flukes were found in their faeces. After challenge, the "vaccinated" sheep (Group A) and the controls (Group B) became patent at about the same time (see the histograms of egg counts in Fig. 16). When the sheep were slaughtered 16 weeks after

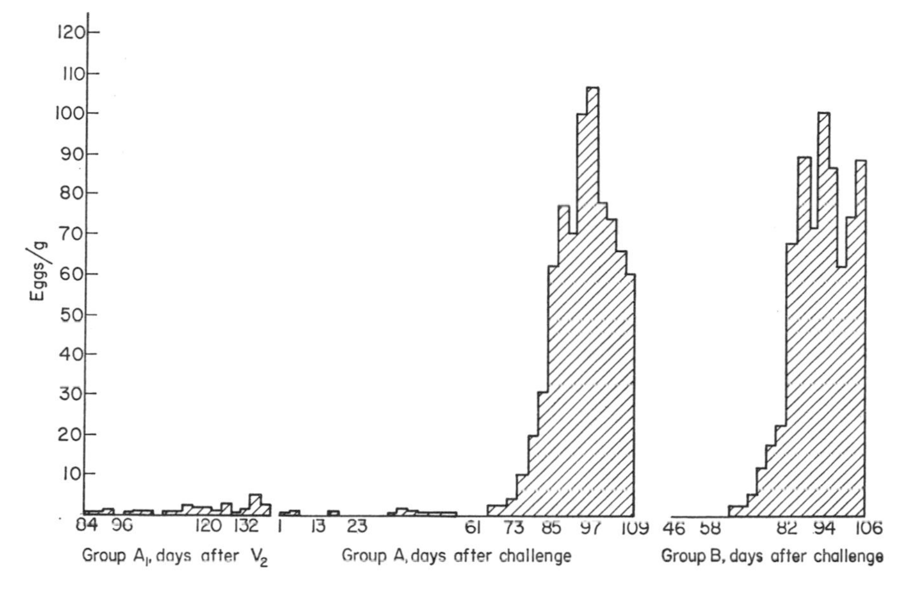

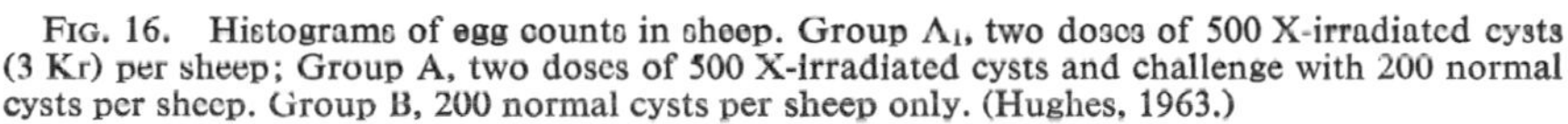

FIG. 16. Histograms of egg counts in sheep. Group A$_I$, two doses of 500 X-irradiated cysts (3 Kr) per sheep; Group A, two doses of 500 X-irradiated cysts and challenge with 200 normal cysts per sheep. Group B, 200 normal cysts per sheep only. (Hughes, 1963.)

challenge, there was no significant difference between the mean fluke burden in Group A (which was 66·6 flukes per sheep) and in that in Group B (which was 78·2 flukes per sheep) and statistical analysis also showed that there was no significant difference in respect of mean wet weights of flukes and weights of livers, all of which showed characteristic damage. From the two sheep in Group A_1 which had received 1,000 X-irradiated cysts each 4+2 flukes were recovered at post-mortem.

These findings in respect of mice, rabbits and sheep are in marked contrast to the claims of Thorpe and Broome (1962) who experimented with rats. These hosts received a single dose of cysts irradiated at various levels in the range 1–10 Kr, and were challenged either 7 weeks or 11 weeks later. The claims were "following oral vaccination with metacercariae irradiated at 1,000 roentgens there is evidence of immunity to challenge" and "at a level of 2,500 roentgens there is also evidence of a significant immunity with a reduction of approximately 50% in the challenge take at both 7 and 11 weeks after immunisation". These conclusions are open to several criticisms. Firstly, Thorpe and Broom did not give complete details of the fluke burdens of individual hosts and secondly, they measured the degree of "immunity" obtained by referring their results to a theoretical "control group fluke burden". Thus, in rats which were given 40 cysts irradiated at 1 Kr they recovered a mean of 7·0 flukes, and in rats which were given 20 normal cysts a mean of 4·5 flukes. Their suggestion was that hosts which had been given 40 cysts irradiated at 1 Kr and then challenged with 20 normal cysts should have a mean fluke burden of 7·0+4·5 if there were no immunity. As their "vaccinated" and "challenged" rats had a mean fluke burden of 7·7, they claimed that this "reduction" from a theoretical fluke burden of 11·5 demonstrates immunity. This procedure appears to be highly questionable, and at higher levels of X-irradiation theoretical fluke burdens become in some instances identical with and in other instances almost identical with the means of the "challenge controls" (= normal infections). In the absence of information concerning variability of fluke burdens in normal infections single mean values of the yields of flukes from ten rats can be misleading. It will be interesting to see the results of further experiments which are promised. There may be significant differences in rats which are related to a single dose of X-irradiated cysts or to the intensity of the "vaccinating" or the "challenge" doses. It would appear to be a profitable line of investigation "to ascertain the extent to which the nutritional substrate on which young flukes feed is interfered with by the local tissue reaction of the host to the infection with X-irradiated flukes" (Dawes, 1963e). There is a suggestion that flukes of a challenge infection reach maturity at an unusually small size and that this may be a nutritional effect arising as a result of modification of the hepatic tissue on which feeding occurs by the leucocytic infiltrations of the inflammatory reaction in response to X-irradiated flukes. The rate at which the host reaction dies down after infection with X-irradiated cysts and the timing of the challenge infection may prove to be important considerations in the interpretation of results which are supposedly indicative of an immune response (Dawes, 1963e). It does appear that the prospect of immunizing the hosts of *F. hepatica* against liver fluke disease is poor in the present state of our knowledge and with techniques at present available.

Acknowledgements

The authors are grateful for permission to use many photographs on which their figures are based. Citations in the captions indicate sources. Their thanks are given to Professor Buckley, Editor of *Journal of Helminthology*, in respect of fifteen figures from the papers by Dawes (1961b, 1962a) and to the Cambridge University Press, publishers of *Parasitology*, in respect of twenty-five figures from other papers by Dawes (1963a-c, 1964). Figures borrowed from the Thesis by Hughes (1963) are published with the permission of the

University of London. Thanks are given also to Mr. J. Wells, personal technician to the senior author, whose work includes the preparation of the section on which Fig. 9 is based, which was photographed by Mr. K.W. Iles, A.R.P.S., as were many of the others, originally. To him and also to Mr. R. D. Reed of King's College, who also rendered photographical assistance, we are very grateful. Mr. T. R. Davis, A.I.S.T., kindly supplied the photographs reproduced in Fig. 12.

References

Alicata, J. E. (1938). *Bull Hawaii agric. Exp. Sta.* No. 80, 22 pp.

Arenas y Martorell, R., Espinosa, A., Padrón, E. and Martinez Andreu, R. (1948). *Rev. Kuba Med. trop.* **4**, 92–97.

Ashworth, C. T. and Sanders, E. (1960). *Amer. J. Path.* **37**, 343.

Belding, D. L. (1958). "Basic Clinical Parasitology." Appleton-Century-Crofts. New York.

Bezubik, B. and Furmaga, F. (1959). *Acta Parasit. Polonica* **7**, 591–98.

Biggart, J. H. (1937). *J. Path. Bact.* **44**, 488–89.

Brand, T. von. (1952). "Chemical Physiology of Endoparasitic Animals." Academic Press, New York.

Bueding, E. (1948). *Physiol. Rev.* **29**, 194–218.

Bugge, G. (1935). *Berl. tierärztl. Wschr.* **8**, 65.

Bürgi, K. (1936). *Mitt. Grenzgeb. Med. Chir.* **44**, 488–537.

Cameron, T. W. M. (1951). "The Internal Parasites of Domestic Animals", 2nd ed. Black, London.

Catchpole, B. N. and Snow, D. (1952). *Lancet* **2**, 711–12.

Clunies Ross, J. (1928). Pamphlet, Council of Scientific and Industrial Research, Australia No. 5, 23 pp.

Clunies Ross, J. and McKay, A. C. (1929). *Bull Coun. sci. industr. Res. Aust.* No. 43, 62 pp.

Compes, H. (1923). Vet. med. Inaug. -Diss. (Auszug), Berlin.

Culbertson, J. T. (1941). "Immunity against Animal Parasites." Columbia University Press, New York.

Dawes, B. (1946:1956). "The Trematoda-with Special Reference to British and other European Forms". Cambridge University Press.

Dawes, B. (1961a). *Nature, Lond.* **190**, 646–47.

Dawes, B. (1961b). *J. Helminthol.* R. T. Leiper Suppl., 41–52.

Dawes, B. (1961c). *Trans. roy. Soc. trop. Med. Hyg.* **55**, 310–11.

Dawes, B. (1962a). *J. Helminthol.* **36**, 11–38.

Dawes, B. (1962b). *Trans. roy. Soc. trop. Med. Hyg.* **56**, 13–14.

Dawes, B. (1962c). *J. Helminthol.* **36**, 259–68.

Dawes, B. (1962d). *Parasitology* **52**, 483–93.

Dawes, B. (1963a). *Parasitology* **53**, 109–22.

Dawes, B. (1963b). *Parasitology* **53**, 123–33.

Dawes, B. (1963c). *Parasitology* **53**, 135–43.

Dawes, B. (1963d). *Nature, Lond.* **198**, 1011–12.

Dawes, B. (1963e). *Nature, Lond.* **200**, 602-03.

Dawes, B. (1964). *Parasitology* **54**, 1–21.

Durbin, C. G. (1952). *Proc. Helminth. Soc. Wash.* **19**, 120.

Ershov, V. S. (1959). *Proc. XVIth int. vet. Congr., Madrid* **1**, 279–89.

Facey, R. V. (1959). *Brit. med. J.* **1**, 1115.

Facey, R. V. and Marsden, P. D. (1960). *Brit. med. J.* **2**, 619–25.

Flury, F. and Leeb, F. (1926). *Klin. Wschr.* **5**, 2054–55.

Gibson, T. E. (1964). *In* "Advances in Parasitology" (B. Dawes, ed.), vol. 2, p. 221–57. Academic Press, London and New York.
Gray, S. J. and Sterling, K. (1950). *J. clin. Invest.* **29**, 1604–13.
Grembergen, G. Van (1950). *Ann. Soc. Zool. Belg.* **81**, 15–20.
Gresham, G. A. and Jennings, A. R. (1962). "An Introduction to Comparative Pathology." Academic Press, London and New York.
Gresson, R. A. R. and Threadgold, L. T. (1959). *J. biophys. biochem. Cytol.* **6**, 157–62.
Healy, G. R. (1955). *J. Parasit.* **41**, No. 6 Sect. 2 (Suppl.) Abst. 42, p. 25.
Higashi, T. (1960a). *Jap. J. Parasit.* **9**, 470–79 (English summary, p. 479).
Higashi, T. (1960b). *Jap. J. Parasit.* **9**, 673–84 (English summary, p. 684).
Hsü, H. F. (1939). *Chin. med. J.* **56**, 122–30.
Hsü, H. F., Hsü, S. Y. Li. and Osborne, J. W. (1962). *Nature, Lond.* **194**, 98–99.
Hughes, D. L. (1959). M. Sc. Thesis, University of London.
Hughes, D. L. (1962a). *Nature, Lond.* **193**, 1093–94.
Hughes, D. L. (1962b). *Parasitology* **52**, 4p.
Hughes, D. L. (1963). Ph. D. Thesis, University of London.
Ibrovic, M. and Gall-Palla, V. (1959). *Veterinaria, Sarajevo* **8**, 531–37 (English summary p. 531).
Jarrett, W. F. H., Jennings, F. W., McIntyre, W. I. M. and Mulligan, W. (1957). *Vet. Rec.* **69**, 1329–40.
Jarrett, W. F. H., Jennings, F. W., McIntyre, W. I. M., Mulligan, W. and Sharp, N. C. C. (1959). *Amer. J. Vet. Res.* **20**, 527–31.
Jennings, F. W., Mulligan, W. and Urquhart, G. M. (1956). *Exp. Parasit.* **5**, 458–68.
Kendall, S. B. and Parfitt, J. W. (1962). *Brit. vet. J.* **118**, 1–10.
Kerr, K. B. and Petkovich, O. L. (1935). *J. Parasit.* **21**, 319–20.
Khaw, O. K. (1935). *Proc. Soc. exp. Biol., N. Y.* **32**, 1003–05.
Kouri, P. (1948). *Rev. Med. trop. Habana* **4**, 63–67, 77–87.
Kouri, P. and Arenas y Martorell, R. (1931a). *Vida Nueva* An. 5, 28, 389–451.
Kouri, P. and Arenas y Martorell, R. (1931b). *Vida Nueva* An. 5, 28, 553–79.
Kouri, P. and Arenas y Martorell, R. (1932a). *Vida Nueva* An. 5, 29, 231–61.
Kouri, P. and Arenas y Martorell, R. (1932b). *Bol. Mens. Asoc. Med. vet. Cuba* **1**, 52–55.
Krull, W. H. (1941). *Proc. Helminth. Soc. Wash.* **8**, 55–58.
Krull, W. H. (1956). *Cornell Vet.* **45**, 511–25.
Krull, W. H. (1958), *Cornell Vet.* **48**, 17–24.
Krull, W. H. and Jackson, R. S. (1943). *J. Wash. Acad. Sci.* **33**, 79–82.
Lagrange, E. and Gutmann, A. (1961). *Riv. Parassit.* **22**, 93–101.
Lagrange, E. and Kemele, W. (1962). *Riv. Parassit.* **23**, 167–72.
Lämmler, G. (1955). *Arzneimittel-Forschung*, **5**, 497–520.
Lämmler, G. (1956). *Z. Tropenmed. u. Parasit.* **7**, 289–311.
Lämmler, G. (1959). *Z. Tropenmed. u. Parasit.* **10**, 379–84.
Lämmler, G. and Loewe, H. (1962a). *Arzneimittel-Forschung* **12**, 15–21.
Lämmler, G. and Loewe, H. (1962b). *Arzneimittel-Forschung* **12**, 164–68.
Lapage, G. (1956). "Veterinary Parasitology", 964 pp. Oliverand Boyd, Edinburgh and London.
Lavier, G. and Deschiens, R. E. A. (1956). *Bull. Soc. Pat. exot.* **49**, 541–53.
Lavier, G. and Marchal, G. (1942). *Sang* **15**, 151.
Lederman, F. (1958). *Bull. Off. int. Epiz.* **50**, 385–421.
Leiper, J. W. G. (1938). *J. Helminth.* **16**, 173–76.
Lengy, J. (1960). *Bull. Res. Counc. Israel* **9B**, 71–130.
Leuckart, R. (1886–1901). "Die Parasiten des Menschen und die von ihnen herrührenden Krankheiten". Leipzig.
Levinam, D. A. (1950). *Klin. Med. Mosk.* **28**, 57.
Lucker, J. T. and Foster, A. O. (1957). *Fmrs' Bull. U. S. Dep. Agric.* No. 1330, 50 pp.

Lutz, A. (1893). *Zbl. W. Bakt.* **13**, 320–28.
Manson-Bahr, P. and Walton, J. (1941). *Brit. J. Surg.* **28**, 380–83.
Mansour, T. E. (1959). *Biochim. biophys. Acta* **54**, 456–64.
Marek, J. (1927). *Dtsch. tierärztl Wschr.* **34**, 513–19.
Montgomerie, R. F. (1926). *J. comp. Path.* **39**, 113–31.
Montgomerie, R. F. (1928a). *J. Helminth.* **6**, 167–74.
Montgomerie, R. F. (1928b). *J. comp. Path.* **41**, 191–98.
Montgomerie, R. F. (1931). *J. Helminth.* **9**, 209–12.
Murphy, F. D. and Pascall, K. G. (1950). *Brit. med. J.* **1**, 647.
Müller, W. (1923). *Zool. Anz.* **57**, 273–81.
Nakagawa, K. (1915a). *Tokyo Izi Shinshi.* No. 1923, 1143–44.
Nakagawa, K. (1915b). *Chiugai Izi Shimpo*, **853**, 12, 62, 1144.
Neuhaus, W. (1938). *Z. parasitenk.* **8**, 476–512.
Nöller, W. (1925). "Die Leberfäule (Leberegelkrankheit) unserer Haustiere". Jena.
Nöller, W. and Schmid, F. (1928). *S. B. Ges. naturf. Fr. Berl.* 148–54.
O'Donnell, G. P. (1949) *Bristol med.-Chir. J.* **66**, 74–76.
Oshima, T. (1956) *Jap. J. Parasit.* **5**, 404–15.
Oshima, T. and Kihata, M. (1958), *Bull. Inst. Publ. Hlth, Tokyo* **7**, 270–74.
Oshima, T., Yoshida, Y. and Kihata, M. (1958), *Bull. Inst. Publ. Hlth, Tokyo* **7**, 256–69.
Owen, D. U. (1928). *Ann. trop. Med.* **22**, 47–52.
Pantelouris, E. M. and Gresson, R. A. R. (1960), *Parasitology* **50**, 165–69.
Patterson, S. W. (1928). *Lancet* **2**, 1291–92.
Pearson, I. (1960). *Aust. vet. J.* **36**, 312.
Pearson, E. G. (1963) *Exp. Parasit.* **13**, 186–93.
Poynter, D. (1963). *In* "Advances in Parasitology" (B. Dawes, ed.), vol. 1, pp. 179–212. Academic Press, London.
Poynter, D., Jones, B. V., Nelson, A. M. R., Peacock, R., Robinson, J., Silverman, P. H. and Terry, R. J. (1960). *Vet. Rec.* **72**, 1078–90.
Railliet, A., Moussu, G. and Henry, A. (1913). *Rec. Méd. Vét.* **90**, 1–6.
Ramage, J. S. (1951). *Lancet* **1**, 146.
Reinhard, E. G. (1957). *Exp. Parasit.* **6**, 208–32.
Rial, B., Gomensoro, C. and Chappe, W. (1951). *Arch. Pediat. Urug.* **22**, 257–70, 373–84, 459–71.
Roche, M., Perez-Gimenez, M. E., Layrisse, M. and Di Prisco, E. (1957). *J. clin. Invest.* **36**, 1183–92.
Roderiguez, C. (1952). *Rev. Med. Cordoba* **40**, 9–12.
Schumacher, W. (1938). *Z. Parasitenk.* **10**, 608–43.
Schumacher, W. (1956). *Z. Parasitenk.* **17**, 276–81.
Shaw, G. W. B. and Clyne, A. J. (1941). *J. roy. Army med. Corps* **76**, 173–74.
Shaw, J. N. (1932). *J. Amer. vet. Assoc.* N. S. **34**, 76–82.
Shibanai, D., Tozawa, M., Takahashi, M. and Isoda, M. (1956). *Bull. Azabu vet. coll.* **3**, 77–86.
Shirai, M. (1927). *Sci. Rep. Gov. Inst. Inf. Dis. Tokyo* **6**, 511–23.
Sinclair, K. B. (1962). *Brit. vet. J.* **118**, 37–53.
Sinitsin, D. (1914). *Zbl. Bakt.* **74**, 280–85.
Smithers, S. R. (1962). *Nature, Lond.* **194**, 1146–47.
Soulsby, E. J. L. (1961). *Vet. Rec.* **73**, 1053–58.
Stephenson, W. (1947). *Parasitology* **38**, 123–27.
Susuki, S. (1931). *Taiwan Igakkai Zasshi* **30**, 97–102.
Taliaferro, W. H. (1929). "The Immunology of Parasitic Infections." Century Co., New York.
Taylor, A. W. (1961). *Lancet* **2**, 1334–36.
Taylor, E. L. (1949). *Rep. 14th. Vet. Congr. London* **2**, 81–87.

Taylor, E. L. (1951). Quoted by Lapage (1956). Original not traced.
Taylor, E. L. and Parfitt, J. W. (1957). *Trans. Amer. micr. Soc.* **76**, 327–28.
Thorpe, E. and Broome, A. W. J. (1962). *Vet. Rec.* **74**, 755–56.
Urquhart, G. M. (1954). *Exp. Parasit.* **3**, 38–44.
Urquhart, G. M. (1955). Ph. D. Thesis, University of Glasgow.
Urquhart, G. M. (1956). *J. Path. Bact.* **71**, 301–10.
Urquhart, G. M., Mulligan, W. and Jennings, F. W. (1954). *J. infect. Dis.* **94**, 126–33.
Vickerman, K. (1963). *In* "Techniques in Parasitology", p. 69. Blackwell Scientific Publications, Oxford.
Villella, J. B., Gomberg, H. J. and Gould, S. E. (1961). *Science* **134**, 1073–74.
Vogel, H. (1934). *Zoologica, Stuttgart* **33**, 1–103.
Ward, G. R. (1911). *Brit med. J.* **1**, 931–35.
Weinland, E. and von Brand, T. O. (1926). *Z. vergl. Physiol.* **4**, 212–85.
Wikerhauser, T. (1960). *Amer. J. vet. Res.* **21**, 895–97.
Wikerhauser, T. (1961a). *Vet. Arhiv.* **31**, 71–80.
Wikerhauser, T. (1961b). *Vet. Arhiv.* **31**, 229–36.
Williams, V. N. (1963). *Nature, Lond.* **198**, 203.
Yamaguti, S. (1943). *Jap. J. Zool.* **10**, 461–67.
Yokogawa, S. (1915a). *Tokyo Iji Shinshi* No. 1920, 197.
Yokogawa, S. (1915b). *Tokyo Iji Shinshi* No. 1922, 1083.
Yokogawa, S. (1915c). *Tokyo Iji Shinshi* No. 1934, 1742.
Yokogawa, S. (1916), *Nisshiu Ogaku* **6**, 323–70.
Yokogawa, S. (1918). *Okayama Igakkwai Zasshi* No. 336–37.
Yokogawa, S. (1919). *Rep. Government of Formosa* 289 pp.
Yokogawa, M., Yoshimura, H., Sano, M., Okura, T. and Tsuji, M. (1958). *In* "Trans. 18 Branch-Meeting Parasit, in E. Div. Parasitol. Soc. Japan," p. 12–13.
Yokogawa, S., Cort, W. W. and Yokogawa, M. (1960). *Exp. Parasit.* **10**, 81–205.
Yokogawa, M., Yoshimura, H., Sano, M., Okura, T. and Tsuji, M. (1962). *J. Parasit.* **48**, **525–31.**

The Biology of the Hydatid Organisms

J. D. SMYTH*

Department of Zoology, Australian National University, Canberra, Australia

I. INTRODUCTION

Considering the status of Hydatid disease (= hydatosis, hydatidosis, echinococcosis or echinococciasis) as a disease of world importance, surprisingly little is known regarding the general *biology* of the causative organisms. This review attempts to summarize the advances which have been made in this field, approximately covering work carried out during the last ten years. Literature earlier than this period is dealt with only where it is

* Support for these studies came in part from the National Institute of Allergy and Infectious Diseases, National Institutes of Health, U.S. Public Health Service Research Grant No. AI-4707 and from the Australian Wool Research Committee.

particularly relevant to more recent work. This survey does not consider the clinical, pathological or epidemiological aspects of hydatid disease, except where these topics bear a special relationship to a particular problem in the general biology. A survey of the geographical distribution has been given by Gemmell (1960).

II. Speciation

A. Historical

The presentation of evidence, within recent years, suggesting that at least two species of *Echinococcus* are apparently responsible for hydatid disease in man and other host animals, has revolutionized our ideas of the epidemiology, epizootiology and general biology of the causative organisms. This general conclusion—which requires further qualification in the light of recent work—has led to a general upsurge of interest in the disease and to an extension of research effort in the various centres throughout the world. Before dealing briefly with the observations which led to this conclusion, it is appropriate that the various forms which the disease can take be clearly defined, for some confusion still exists in the literature on this basic question. Euzéby (1960, 1962) has given a useful summary of the pathological picture. The following types of hydatid cysts occur:

1. *Unilocular* (= univesicular): characterized by having only one bladder, or many completely isolated bladders, each enclosed in its own envelope. This envelope, the *laminated membrane*, is solid and continuously restricts the living germinal epithelium from invading adjacent tissue.

2. *Multivesicular* (= multicystic): characterized by the presence of many adjoining bladders, but each possessing its own membranes. This form of hydatosis is sometimes referred to as "veterinary echinococcosis" because it is believed to be more common in cattle; early workers in particular, have confused it with the multilocular form (below) which it superficially resembles.

3. *Alveolar:* characterized by a malignant type of growth made up of a series of proliferating vesicles embedded in dense fibrous stroma. In the vesicles, the hydatid fluid is replaced by a jelly-like mass in older cysts, and there are few protoscoleces (see Section IV, B, 1). The laminated membrane is very thin and cells of the germinal membrane are able to send out stolon-like growths into the surrounding tissue. These growths infiltrate along the vesicular spaces of the organisms parasitized and new daughter vesicles are budded off as they go.

4. *Multilocular:* characterized by having many bladders embedded in a common adventitious membrane. According to Euzéby (1960) this form is an early stage of the alveolar type although some authors (e.g. Yamashita, 1960) use the term *multilocular echinococcosis* as being synonymous with *alveolar echinococcosis*. Since most authors (e.g. Cameron, 1960a; Schiller,

1960; Rausch, 1952; Vogel, 1955a) use the latter term, this terminology will be used in the present text.

Whether one or two species of *Echinococcus* were involved as pathogens has been a matter of conjecture since the latter part of the nineteenth century. Many well-known parasitologists such as F. Dévé and E. Brumpt favoured the single form or "Unicist" Theory which held that there was one species of parasite responsible—*E. granulosus*—which under certain unknown circumstances, gave rise to the alveolar type. This view has been held until recently. Others held the "Double or Dualist Theory" which postulated the existence of two species whose larvae had different pathological effects and whose adults may be different morphologically. The Dualist view was championed particularly by Posselt (1936) who based his arguments not only on the pathological evidence but also on the localized distribution of the alveolar form, together with some cytological and histological observations on the hook size and other minor features. The arguments for and against these rival theories now belong to history and will not be further discussed here. In general, although the so-called Dualist theory has received much experimental evidence in its support and its broad conclusion has been generally accepted, the problem of speciation, as discussed later (p. 174), may prove to be a more complex one than is generally believed when viewed against the broader background of modern genetical theory.

Over the years a number of workers have undoubtedly contributed to the solution of the problems raised by the various forms of hydatosis, but the main credit for producing evidence regarding the speciation of *Echinococcus* must go to Rausch (1953, 1954, 1956, 1958), Rausch and Schiller (1954) and Vogel (1955a, b, 1957).

The chain of events which led to the elucidation of some of the problems of speciation began with the discovery of alveolar infections in the tundra vole (*Microtus oeconomus inuitus*) on St. Lawrence Island, Alaska, by Rausch and Schiller (1951). Alveolar hydatosis was also found to be prevalent in the Eskimo population on the island. Rausch (1953) speculated that this form was probably identical with the form causing alveolar disease in man in Europe and Russia. This form of the disease was later attributed to a new species, *Echinococcus sibiricensis* (Rausch and Schiller, 1954). Natural infections in other mammals, ground squirrels (*Citellus undulatus*), voles (*Clethrionomys rutilus*) and shrews (*Sorex tundrensis*) were also reported in the same region by Thomas *et al.* (1954). Shortly after, Vogel (1955a), as a result of feeding experiments, dealt with below, concluded that the worm responsible for the European alveolar hydatid and *E. sibiricensis* were co-specific. This author also pointed out that, on the basis of priority, the valid name for this species was *Echinococcus multilocularis* Leuckart 1863, and he demonstrated that both the life cycle and the morphology differed from those of *E. granulosus*.

Vogel's conclusions were based on a careful series of observations and experiments which must be regarded as classical of their kind. He began by investigating possible natural hosts in the Serbian Alps near villages in which human alveolar echinococcosis had recently been found. He found terminal proglottides in eleven red foxes, *Vulpes vulpes*, and fed them to a wide range of possible intermediate hosts. Successful infections were obtained in Nordic burrowing voles *(Microtus oeconomus ratticeps)*, field voles *(Microtus arvalis)*, cotton rats *(Sigmodon hispidus)* and white rats. The cysts obtained appeared to correspond histologically with those of human alveolar echinococcosis. Naturally infected hosts were then discovered and one infected field vole was fed to a dog and 55 days later about 1,000 *Echinococcus* strobila were recovered. Morphological studies revealed that these specimens differed from *E. granulosus* in a number of features, the chief being the total size, the position of the genital pore, the number of testes and the number of segments. These and other differences are summarized in Table I. Thus

TABLE I

Distinguishing Characteristics of E. granulosus *and* E. multilocularis *(from Smyth and Smyth,* 1964*)*

	E. granulosus	*E. multilocularis*
Hosts—Larval stage (chief)	Ruminants (domestic animals, Cervidae) wallaby, camel, horse, pig, some primates, man	Rodents (Cricetidae), Insectivores (Soricidea)
Host—Adult worm (chief)	Dog, coyote, dingo, jackal, wolf, hyena, fisher	Dog, red fox, arctic fox
Cyst characteristics	Usually a single fluid-filled vesicle surrounded by a dense connective tissue capsule: no exogenous budding (Dissanaike and Paramananthan, 1960)	Alveolar, the small sacs and large central cavity (in old cysts) filled with gelatinous fluid made from necrosed tissue (Euzéby, 1960). Proliferates by exogenous budding (Rausch, 1956)
No. of hooks in larva	24–52 (Lubinsky, 1960a)	12–34 (Lubinsky, 1960a)
Adult: Length	4–6 mm (Euzéby, 1960) 1·5–6 mm (Rausch, 1956)	1·5–3·5 mm (Euzéby, 1960) 1·2–3·7 mm (Rausch, 1956)
No. of segments	3 (Vogel, 1957); scolex + 3 (Yamashita *et al.*, 1958b); 3–4 (Euzéby, 1960)	3–5, mostly 4 in dogs (Vogel, 1957); 3–5 mature seg. is less than ½ body length (Euzéby, 1960); scolex + 4 (Yamashita *et al.*, 1958b)

TABLE I *(continued)*

	E. granulosus	*E. multilocularis*
Size of sucker	Average 140μ (Yamashita *et al.*, 1958b)	Av. 87·3μ faster growth rate (Yamashita *et al.*, 1958b)
No. of hooks	32–40 (Yamashita *et al.*, 1958b); 28–46 (Lubinsky, 1960a)	26–36 (Yamashita *et al.*, 1958b); 14–34 (Lubinsky, 1960a)
Size of hooks: Large	33·2–39·8μ av. 36·8μ (Vogel, 1957); 30–35μ (Euzéby, 1960)	27·6–34·3μ av. 30·9μ (Vogel, 1957); 28–35μ (Euzéby, 1960)
Small	22·1–34μ av. 28·5μ (Vogel, 1957)	22·7–31μ, av. 26·9μ (Vogel, 1957)
Position of genital pore	Nearer post. end of segment. (Rausch, 1956; Vogel, 1957; Euzéby, 1960)	Near middle of segment (Rausch, 1956; Vogel, 1957); nearer ant. end of segment (Euzéby, 1960)
No. of testes	45–65, av. 56 (Rausch, 1956; Yamashita *et al.*, 1958b); 38–52 av. 44·2 (Vogel, 1957)	17–26 av. 22 (Rausch, 1956); 15–30 (Yamashita *et al.*, 1958b); 14–31 av. 22 (Vogel, 1957)
No. of testes in front of cirrus sac	9–23, av. 15·8 (Vogel, 1957). Both ant. and post. to genital pore (Rausch, 1956)	0–5 av. 2·3 (Vogel, 1957). From level of genital pore to post. end of segment (Rausch, 1956)
Form of ovary	Kidney-shaped, not acinose (Vogel, 1957; Euzéby, 1960)	Acinose, two side lobes connected by small isthmus (Vogel, 1957; Euzéby, 1960)
Gravid uterus	Lateral pouches present (Rausch, 1956; Vogel, 1957; Euzéby, 1960)	No lateral pouching (Rausch, 1956; Vogel, 1957; Euzéby, 1960)
Maturation time	40–51 days in dogs (Choquette, 1956); 57 days eggs in faeces of dog (Drežančić and Wikerhauser, 1956); 47th day eggs in faeces of dog (Clunies Ross, 1936); 48–61 days eggs in faeces of dog (Yamashita *et al.*, 1958b)	In dogs 1½–2 months. Reduces in size and fertility and no. of segments after 3rd month (Vogel, 1957); 30–35 days eggs in faeces (Yamashita *et al.*, 1958b)
Size of embryophore	29·6–44μ by 27–42·5μ (Yamashita *et al.*, 1958b)	29·5–40·5μ by 27·5–39·5μ (Yamashita, *et al.*, 1958b)

Vogel established, for the first time, what appeared to be clear morphological differences between *E. granulosus* and *E. multilocularis*.

Ironically enough, Posselt, who was a champion of the "dualist theory", had obtained similar results many years before, for Vogel (1955a) had an opportunity to examine some of Posselt's material in the Innsbruck Pathological Institute consisting of the small intestine of a dog which Posselt had fed in 1901 with human alveolar material from the Tyrol. In the bottom of this museum jar were detached strobila in a good state of preservation and examination of these showed that they corresponded morphologically with the tapeworms obtained by Vogel from naturally infected material.

Vogel further made a detailed morphological study of the St. Lawrence Island form but concluded that since there were some dissimilarities in the hooklet size, host specificity and localization of larvae, it should be regarded as a geographic race or subspecies of the European form; Cameron (1960a) used the name *E. multilocularis sibiricensis*. Vogel showed further that unlike *E. granulosus*, *E. multilocularis* matures not only in dogs but also in cats and foxes.

The general conclusion was thus emerging that there were two species of *Echinococcus* responsible for Hydatid disease: *E. granulosus*, which matured in dogs and closely allied species (see Section III, B) but not in foxes, and utilized mainly ungulates for intermediate hosts, and *E. multilocularis*, which matured additionally in foxes and cats, and utilized mainly microtine rodents as intermediate hosts. Not all animals which have been found to be successful hosts of one species of *Echinococcus* have been tested experimentally with the other species, but where experimental infection has been attempted with both species it has been found, with a few exceptions discussed below, that *E. multilocularis* and *E. granulosus* do not develop in the same intermediate hosts. The larvae of both species, will however, develop in man, but man is not a satisfactory intermediate host for *E. multilocularis* in that development is slower than in natural intermediate hosts and protoscoleces may be rare or lacking altogether (Rausch and Schiller, 1956; Yamashita *et al.*, 1958a).

It has become evident from a recent survey of infections in natural and experimental hosts (Smyth and Smyth, 1964) that the division of the causative agents of hydatid disease into two species is likely to prove to be an oversimplication of the situation, and that speciation is probably more complex than at first was evident. Thus, a number of workers have reported organisms which do not fall easily into either of these well defined categories representing *E. granulosus* and *E. multilocularis*. For example, the variety of *E. granulosus* which occurs naturally in Canada differs from *E. granulosus* found in Europe, the Middle East and Australasia, in that the larva occurs almost exclusively in the lungs of larger deer and does not produce experimental infections in cattle, sheep or pigs although these animals are the natural intermediate hosts of the classical *E. granulosus*. On the basis of these results,

Cameron (1960b) terms this a subspecies *E. granulosus canadensis*. Morphologically, the Canadian variety appears to be indistinguishable from the classical *E. granulosus* although there may be some serological differences (Cameron and Webster, 1961).

Again, Vibe (1959) has reported the development of larval *E. multilocularis* in sheep and cattle, the "natural" intermediate hosts of *E. granulosus*. Dogs fed with sheep and cattle cysts developed worms which differed morphologically from the *E. granulosus* found commonly in local dogs. If this is a "true" *multilocularis* infection it appears that in Siberia there exists a further strain or subspecies of *E. multilocularis* which is capable of using ungulates as intermediate hosts. Lukashenko and Zorikhina (1961) also reported the case of two wolves in the Barabinsk region, one infected with *E. multilocularis* and one with *E. granulosus*.

On the basis of intermediate host specificity and/or on morphological detail, Sweatman and Williams (1962, 1963), Williams and Sweatman, (1963) have further distinguished four separate subspecies of *E. granulosus* as follows: *E. granulosus granulosus*, *E. granulosus borealis*, *E. granulosus canadensis* and *E. granulosus equinus*. *E. granulosus borealis* is of particular interest, as Sweatman and Williams (1963) report its development in the red fox *(Vulpes fulva)*, in which *E. granulosus (granulosus)* does not reach maturity. The definitive and intermediate hosts are surveyed in the next section.

It is clear from this survey that speciation in *Echinococcus* is a complex matter involving perhaps at each end of a hypothetical scale well defined species with morphological, physiological and immunological characteristics clearly recognizable as belonging to *E. granulosus* and *E. multilocularis*. Between these two points are a variety of "races", "strains" or "subspecies", incorporating characteristics belonging to both species. Smyth and Smyth (1964) have speculated that these may represent "clones" the production of which could be predicted on theoretical grounds, for this parasite possesses two characteristics which particularly favour the expression of mutants and their selection and establishment in different hosts:

(i) The adult tapeworm is a self-fertilizing hermaphrodite. Thus, if a mutation occurs, it could appear in both eggs and sperm and homozygotes (e.g. double recessives) could develop.

(ii) The larva (i.e. hydatid) reproduces by polyembryony. Thus, a large population of genetically identical individuals (i.e. a clone) could be formed from a single mutant.

It is not difficult to imagine the selection of a mutation by an unusual host in which development would not normally occur. Thus, if only one egg out of several millions represented a mutation adapted to an unusual intermediate host, it could produce a clone of many thousands (perhaps millions) of protoscoleces. This multiplication of a single mutant would assure a high chance of success in the definitive host also. This reproductive pattern may account for the multiplication of "strains" or "subspecies"—which clearly represent only morphological or physiological mutations—reported throughout the world. It is to be expected, then, that as further experimental work is carried out, an increasing number of such mutants will be described. These may possess different morphological characteristics, a different range of hosts, or different physiological or immunological patterns.

B. OTHER SPECIES

In addition to *E. granulosus* Batsch, 1786 and *E. multilocularis* Leukart, 1863, and their reported varieties, ten other species of *Echinococcus* have been reported in the literature. These are *E. oligarthrus* (Diesing, 1863) Cameron, 1926; *E. cruzi* Brumpt and Joyeux, 1924; *E. felidis* Ortlepp, 1937; *E. lycaontis* Ortlepp, 1934; *E. longimanubrius* Cameron, 1926; *E. minimus* Cameron, 1926; *E. cameroni* Ortlepp, 1934; *E. intermedius* Lopez-Neyra and Solei Planas, 1943; *E. ortleppi* Lopez-Neyra and Solei Planas, 1943; *E. patagonicus* Szidat, 1960. The validity of these species have been discussed by Rausch (1953), Cameron and Webster (1961), Euzéby (1962), and Smyth and Smyth (1964).

E. oligarthus and *E. felidis* parasitize felines (jaguar and lion respectively) and, as far as is known, do not occur in canine hosts. Rausch considers *E. longimanubrius* and *E. minimus* as *species inquirendae* which may be co-specific with *E. granulosus* as also may be *E. cameroni.* Similarly Euzéby states that *E. intermedius* and *E. ortleppi* should be considered co-specific with *E. granulosus* at present. Both species have been described from the dog in Spain and North Africa respectively; the intermediate hosts are not known. Rausch regards *E. lycaontis*, from the cape hunting dog *(Lycaon pictus)* in South Africa as a separate species. Little is known of the life cycle of any of these species so that this review is entirely concerned with *E. granulosus* and *E. multilocularis.*

III. HOST SPECIFICITY

A. GENERAL

An attempt is made below (Section III, B and C) to survey recent work on the host specificity of *E. granulosus* and *E. multilocularis.* As already discussed, much of this work is based on the assumption by workers that two sharply defined species exist, for the fact that gradations of morphological, physiological and immunological characteristics can occur between these species—as indicated in the previous section—is only now becoming apparent.

Again, the importance of strain differences in definitive and intermediate hosts (especially laboratory animals) has received little attention although Yamashita *et al.* (1958a) have particularly drawn attention to this problem. These workers infected ten strains of mice with *E. multilocularis* and found that in two strains there was 100% infection and larvae developed as rapidly and as successfully as in a microtine rodent, the natural host. In all other eight strains development was slow, scolex formation was delayed and only one of these eight strains gave 100 % infection; the remainder giving 8–79 % infection.

For these reasons, therefore, many of the records quoted below may be open to reinterpretation when more evidence becomes available.

B. DEFINITIVE HOSTS

1. *Echinococcus granulosus*

a. General. A survey of the natural and experimental definitive and intermediate hosts has been made by Smyth and Smyth (1964). The dog is undoubtedly the most important host throughout the world but the worm has been recorded from a number of other species, jackal, coyote, dingo, several species of wolf and the fisher (one record; see Section III, B, g). Some notes on the main hosts are given below, and a list of natural definitive hosts is given in Table II.

TABLE II

Natural Definitive Hosts of E. granulosus *(from Smyth and Smyth,* 1964)

Host	Distribution
Dog *(Canis familiaris)*	All countries (Cameron and Webster, 1961)
Pariah dog *(C.* sp.*)*	India (Maplestone, 1933); China (Hsü-Li, 1941)
Wild hunting dog *(Lycaon pictus)*	Kenya (Nelson and Rausch, 1963)
Coyote *(C. latrans)*	Canada (Freeman *et al.*, 1961)
Dingo *(C. dingo)*	Australia (Gemmell, 1959a)
Jackal *(C. aureus)*	Near East (Hinshaw, 1937); Ceylon (Dissanaike and Paramananthan, 1960); Pakistan (Lubinsky, 1959b); Israel (Wertheim, 1957); Algeria (Jore d'Arces, 1953); Kenya (Nelson and Rausch, 1963)[1]
Wolf *(C. lupus)*	Alaska (Rausch and Williamson, 1959); Canada (Sweatman, 1952); Russia, Novosibirsk region (Lukashenko and Zorikhina, 1961)
Timber wolf *(C. lupus lycaon)*	Canada, (Freeman *et al.*, 1961); U.S.A. (Riley, 1939)
Hyena *(Crocuta crocuta)*	Kenya (Nelson and Rausch, 1963)
Fisher[2] *(Martes pennanti)*	Canada (Sweatman, 1952)
Fox[3] *(Vulpes corsac)*	Russia, Stalingrad region (Shumakovich and Nikitin, 1959)
Lion *(Leo leo)*	Africa (Ortlepp, 1937)

[1] One of 16 black-backed jackals *(Thos mesomelas)* infected; no gravid segments.
[2] Found in one of fifty-two animals examined.
[3] Only one animal found infected.

b. Dog. The dog as definitive host is too well known to need documentation. Survey lists of parasites of dogs throughout the world almost invariably show the existence of *E. granulosus*. High infections are traditionally associated with dogs which frequent abattoirs, sheep stations and farms. For example, Gemmell (1957a, 1959a) showed that in New South Wales (Australia) 15–38% of dogs in these categories were infected. This author also reported that city dogs, unassociated with the pastoral industry, can become infected. In Sydney two (3·8%) of fifty-two dogs were infected and in Melbourne, six (3·4%) of 174. Similar surveys have been made in various other countries.

c. Jackal. The occurrence of *E. granulosus* in a jackal in Ceylon was shown by Dissanaike and Paramananthan (1960). This result confirms a suggestion made earlier by Dissanaike (1958) that a sylvatic cycle was being carried out in the jungles around Colombo and could explain the infection of cattle from the Northern and North Central provinces with hydatid cysts. The intermediate hosts could be wild herbivores, including monkeys, which are known to be infected (Dissanaike, 1958).

d. Wolf. The wolf *(Canis lupus)* and the timber wolf *(C. lupus lycaon)* undoubtedly play a major part in maintaining a sylvatic cycle for *E. granulosus*. The most substantial evidence probably comes from Alaska and Canada where a number of surveys have been carried out. Rausch and Williamson (1959) have summarized the incidence in this host. In their own work these authors found 30% of 200 wolves infected but an incidence as high as 62% has been reported by Sweatman (1952). They conclude that the chief intermediate hosts in this sylvatic cycle appears to be various species of deer, moose and caribou.

e. Red fox (Vulpes vulpes). E. granulosus does not appear to reach maturity in the red fox and any early positive records of gravid worms being found may relate to *E. multilocularis*. Gemmell (1957b, 1959b) has reviewed the role of the fox as a definitive host. Mattoff and Jantscheff (1954) infected sixteen red foxes in Europe with *E. granulosus*. They found that most of the tapeworms were lost before the 35th day and in those which remained, gravid segments were never obtained. Similar experiments were carried out in Yugoslavia by Drežančić and Wikerhauser (1956) and in Hungary by Boray (personal communication to Gemmell, 1959b). In Australia, Gemmell (1957b, 1959b) found only one (non-gravid) specimen in 102 red foxes examined.

Gemmell (1959b) carried out a comprehensive experiment feeding the same material to foxes and dogs. Although some foxes became infected with light worm burdens (maximum 45 compared with dog control 14,493!) never more than one or two segments were found and ova were never produced. It can generally be concluded, therefore, that the red fox plays no part in the spread of *E. granulosus*.

Sweatman and Williams (1963) report development of what they term a subspecies *E. granulosus borealis* in a species of red fox, *Vulpes fulva*. It must be borne in mind, however, that the experimental work of Gemmell, referred to above, was carried out with *Vulpes vulpes* and specific host differences may be operating here. There is also a record from the Stalingrad region of one fox, *Vulpes corsac*, being naturally infected with mature specimens of *E. granulosus* (Shumakovich and Nikitin, 1959).

Fox bile has been shown (Smyth and Haslewood, 1963) to be only very slightly toxic to *E. granulosus* and factors other than biochemical ones—such as immunological ones—may be responsible for the resistance of this host to infection.

f. The dingo (Canis dingo). In Australia, the dingo probably plays a major part in maintaining a separate sylvatic cycle of the disease. Durie and Riek (1952) found that in Southern and Central Queensland nine out of eleven dingoes examined were infected (82%) and in Northern New South Wales ten out of twenty-one (45%) dingoes examined were infected. These workers also showed that the wallaby acts as the main intermediate host in this cycle; a survey of sixty-five of these marsupials showed a 20% infection.

g. Fisher (Martes pennanti). The listing of the marten or fisher as a definitive host is based on a single record by Sweatman (1952) who found one infected animal out of fifty-two specimens examined in Ontario. This worker has since questioned the authenticity of this record (Sweatman, personal communication).

h. Coyote (C. latrans). There are only a few records of infections in this host and this animal may prove to be a poor host (Freeman *et al*., 1961).

i. Cat. There are a few, somewhat doubtful, references to *E. granulosus* in the cat and it is likely that some of these records—if not all—may refer to *E. multilocularis*.

j. Negative experimental infections. A number of hosts have been fed experimentally with cysts of *E. granulosus* and negative results recorded. For references see Smyth and Smyth (1964).

2. *Echinococcus multilocularis*

Surveys of natural infections and experimental infections have shown that the definitive host spectrum of *E. multilocularis* is wider than that of *E. granulosus*. The most important difference being that *E. multilocularis* matures both in the fox and cat in which *E. granulosus* may sometimes develop but does not reach maturity.

A list of definitive hosts is given in Table III. Although the dog is universally recognized as definitive host for *E. multilocularis* as well as *E. granulosus*, other Canidae may be important hosts in different countries.

According to Rausch (1958) the arctic fox, *Alopex lagopus*, is the most important definitive host in North America, but the wolf *(Canis lupus)*, the

TABLE III

Natural Definitive Hosts of E. multilocularis *(from Smyth and Smyth,* 1964)

Host	Distribution
Dog *(Canis familiaris)*	Wherever *E. multilocularis* is found (Cameron and Webster, 1961)
Fox *(Vulpes vulpes)*	Siberia, Krasnoyarsk territory (Romanov, 1958); Alaska (Rausch, 1956); Rebun Island, Japan (Inukai *et al.*, 1955); Upper Bavaria (Mendheim, 1955); S. Germany (Vogel, 1957)
Arctic fox *(Alopex lagopus)*	Chukot Peninsula N. E. Siberia (Ovsyukova, 1961); Krasnoyarsk Territory Central Siberia (Mamedov, 1960); Alaska (Rausch, 1956)
Wolf[1] *(C. lupus)*	Barabinsk forest-steppe region (Lukashenko and Zorikhina, 1961)
Cat *(Felis domestica)*	Japan, Rebun Island (Ambo *et al.*, 1954)

[1] Only one animal found infected.

grey fox *(Urocyon cinereoargenteus)* and the red fox *(Vulpes vulpes)* have been infected experimentally. Rausch also claims that the red fox is "difficult" to infect and it may not be an important natural host, and that in the coyote, *Canis latrans*, strobila do not reach the egg-producing stage of maturation.

The successful infection of a cat with *E. multilocularis* was accomplished by Vogel (1955a, 1957).

C. INTERMEDIATE HOSTS

As discussed earlier, a pattern is emerging that ungulates generally act as intermediate hosts for *E. granulosus*, whereas rodents serve for *E. multilocularis* but this is a generalized statement and subject to the reservations regarding host and parasite strain differences previously stated.

Sheep, pigs and cattle act as the main intermediate hosts of *E. granulosus* in well developed agricultural countries and in the less well developed countries comparable domestic animals (e.g. goats) may be important. Wild microtine rodents, especially voles, are probably the major intermediate hosts of *E. multilocularis*. A wide range of intermediate hosts, both natural and experimental, have been reported. These have been reviewed by Smyth and Smyth (1964) and are summarized in Table IV (see Appendix); it is not intended to discuss them further here.

It is interesting to note that both dogs and cats have been reported as *intermediate* hosts but such records are extremely rare (Whitten and Shortridge, 1961). Since dogs infected with adult *E. granulosus* would be continuously in contact with eggs, a mechanism of natural immunity must be operating in most cases. Berberian (1957) has suggested that the action of intestinal juices or bile may be responsible (see Section V, A, 2). The immunological considerations of these data are discussed in Section VII, B, 1.

D. BIOCHEMICAL BASIS FOR HOST SPECIFICITY

The basis for the host specificity of the intermediate and definitive hosts of *E. granulosus* and *E. multilocularis* is not known; as with most parasites it is probably immunological. There is, however, some experimental evidence to suggest that specific compounds in bile may play a major part in determining host specificity in certain instances (Smyth, 1962b; Smyth and Haslewood, 1963). Thus, it was found that, in general, bile from herbivore hosts, (ox, sheep, rabbit) lysed the cuticle of protoscoleces of *E. granulosus in vitro* whereas that from carnivorous hosts (dog, fox, cat) had no such effect. Analysis of bile salts of herbivore hosts showed that it tended to be rich in deoxycholic acid (usually linked with glycine) and that sodium deoxycholate or sodium glycodeoxycholate above a certain concentration caused rapid lysis of the cuticle.

In carnivores, on the other hand, the level of deoxycholic acid tended to be substantially lower than that found in herbivores and when present was usually linked with taurine. It was, therefore, speculated that the cuticle of a cestode must have a molecular configuration such that it can survive the detergent-like activity of the surface active agents present in bile.

The concentration of deoxycholic acid, and possibly its type of conjugation, thus appears to be crucial from the point of view of a scolex of *Echinococcus* attempting to establish itself in the gut. Hatching and activation of eggs may also be related to the presence of specific compounds in bile (see Section V, A, 2) so that host specificity for the intermediate stages could be similarly related to the biochemistry of the bile although other factors, as well as biochemical ones, will undoubtedly operate.

IV. MORPHOLOGY, CYTOLOGY AND HISTOCHEMISTRY

A. ADULT WORM

1. *General Morphology*

The main morphological studies carried out, during the last decade, have dealt with the anatomical differences between *E. granulosus* and *E. multilocularis* but detailed studies have also been made on the scolex and the musculature. The main morphological differences between the two species are summarized in Table I.

The most important stated differences relate to the size of the strobila (*E. granulosus* being much larger), position of the genital pore in gravid segments (more posterior in *E. granulosus*) and the form of the gravid uterus (which has lateral pouches in *E. granulosus*).

There is increasing evidence that some of the stated morphological criteria may not hold for species differentiation. Thus Vibe (1959) found that the position of the genital opening in his specimens of "*E. multilocularis*" was not strictly constant and suggested that this characteristic should not be used as a criterion for species differentiation. Lubinsky (1960a) likewise found that the hook number (Fig. 2) was "extremely variable" (see Section IV, A, 3).

In view of the evidence on speciation presented earlier (Section II, A) it is likely that a range of morphological differences may become evident as further data are collected, particularly from hosts in isolated areas of infection.

2. *Scolex*

Recent studies on the scolex have lead to the discovery of a group of gland cells forming a *rostellar gland* at the extreme tip of the rostellum (Smyth, 1963b, 1964). The existence of this gland was suspected after the living scolex of a 35-day old worm from a dog was observed to release drops of secretion when examined in saline on a warm stage (Fig. 1A). This observation led

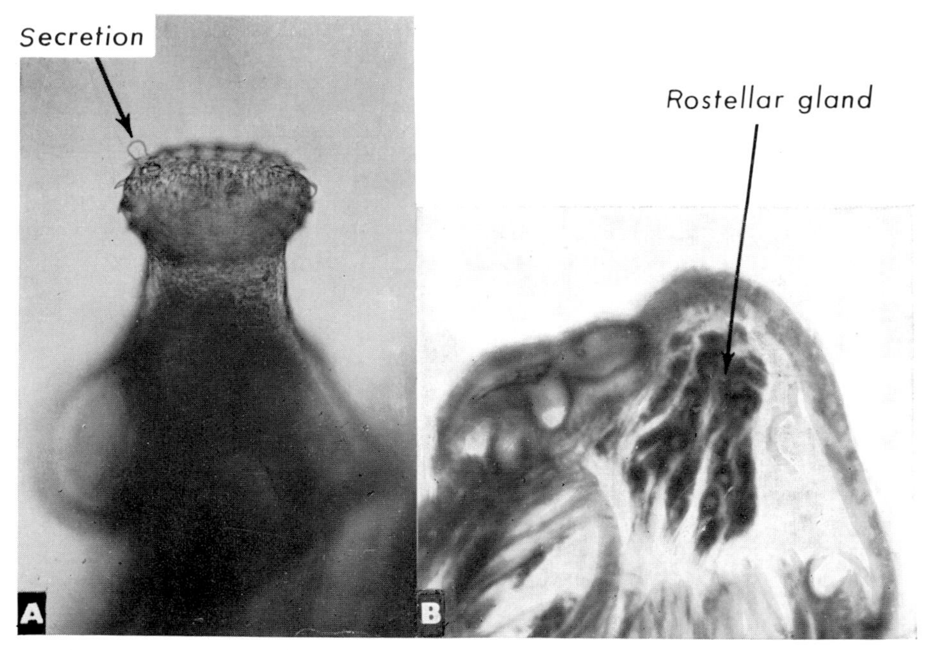

FIG. 1. A. Electronic flash photograph of everted scolex of living specimen of *E. granulosus* (35 days maturation in dog) showing secretion droplet. × 450. B. Longitudinal section of scolex of a similar worm showing cells of "rostellar gland" containing secretion droplets. × 1,000. (After Smyth, 1964.)

to a cytological investigation of the scolex in an attempt to locate the origin of the secretion. Although a group of cells was found over the rostellar pad, no evidence that these cells were secretory could be found at first. These studies were made of worms fixed while still attached to the gut; and as these were found to penetrate deeply into the base of a crypt of Lieberkühn it was realized later that fixation may have been inadequate for detailed cytological studies. When, later, worms were detached from the mucosa before fixation and fixed for long periods in formol-calcium, the gland cells were much better preserved and were revealed as long spindle-shaped cells clearly containing droplets of secretion (Fig. 1B). Cytochemical tests no this secretion have been remarkably unreactive and have not provided much evidence as to its nature. The secretion was P.A.S. negative and only doubtfully positive for protein stains. Some weak reaction was given with the P.F.A.S. test suggesting the presence of –SS– bonds. Sudan Black stained the material only after chromation, from which it was concluded that it was probably lipoprotein in nature. Whatever the nature or function of this secretion may prove to be, the secretion itself is exceptionally labile to average fixation techniques, a fact which may explain how the presence of the gland and its secretion has been overlooked for so long. Smyth (1964) further speculated that the secretion, which has not been observed in organisms of less than 32-days development, might be (a) hormonal in nature and concerned with the regulation of growth or (b) histolytic and concerned with local proteolysis in the vicinity of the scolex, a process which could have considerable nutritional significance for the worm. It was also speculated that the secretion could have antigenic properties and could be responsible for a degree of immunity well known to develop in some dogs after repeated infection. The evidence then suggests that this gland develops as the adult worm reaches maturity. In possessing this gland, the scolex of an adult worm thus differs cytologically, as well as in size (Yamashita *et al.*, 1958b), from that of the larva in a hydatid cyst. This study further emphasizes the advisibility of using another term instead of "scolex" for a larva in a brood capsule and the use of the term "protoscolex" is proposed here (Section IV, B, 1).

3. *Hooks*

Description of the sizes and numbers of hooks have been made by a number of workers. Lubinsky (1960a) has given a useful summary of the early literature. This author points out that the hook number given by Leuckart in 1886—28 to 50—for *E. granulosus* has gained wide acceptance in the international literature and has been widely cited in textbooks, whereas the earlier figures of Von Siebold (1853)—32 to 38—has been cited by relatively few authors, notably French, English and Japanese. Lubinsky (1960a) further analysed the hook sizes in six populations of *E. multilocularis sibiri-*

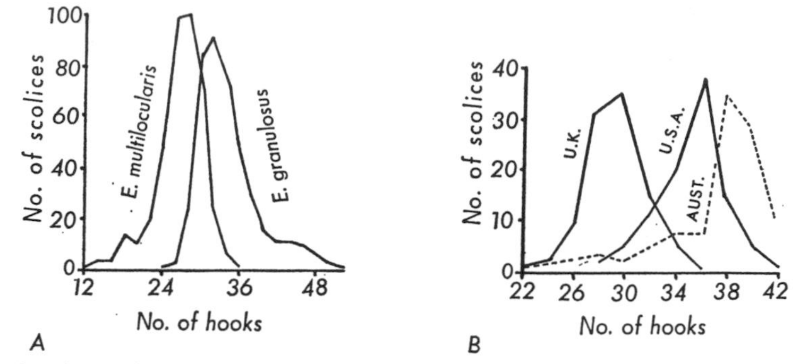

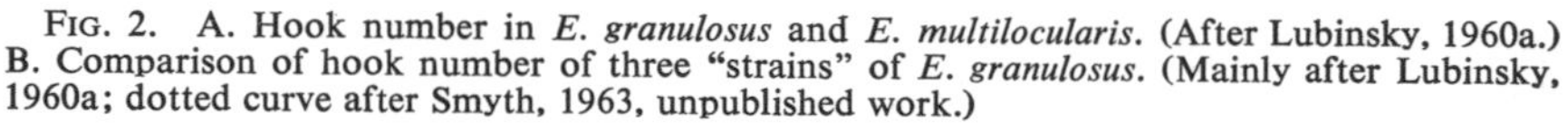
FIG. 2. A. Hook number in *E. granulosus* and *E. multilocularis*. (After Lubinsky, 1960a.) B. Comparison of hook number of three "strains" of *E. granulosus*. (Mainly after Lubinsky, 1960a; dotted curve after Smyth, 1963, unpublished work.)

censis and nine populations of the Canadian *E. granulosus*. The number of hooks was found to be extremely variable. Over 90% of the scoleces of *E. multilocularis sibiricensis* had 20–32 hooks, the mean being 28. Over 90% of the scoleces of *E. granulosus* had 28–40 hooks, the mean being 32. The difference between the populations was always statistically significant, despite a considerable overlap of the population variations curve (Fig. 2A). Lubinsky (1960a) also showed that *E. granulosus* material from widely separated areas (e.g. Mississippi and England) gave two population variation curves which showed little overlapping. These are compared in Fig. 2B, with the hook number in the Australian strain of *E. granulosus* (Smyth, 1963, unpublished work). Some variation in the number of hooks in the European *E. multilocularis* was also observed by Lubinsky (1960a) the range of the other species being 24–28. From these results, this author stresses that the use of the number of hooks of *Echinococcus* for purposes of species differentiation necessitates the examination of a considerable number of scoleces and a careful statistical and biological evaluation of the data obtained. Lubinsky (1959a) described a number of anomalies of oncotaxy which consisted of (a) those increasing the number of hooks, (b) those decreasing the number of hooks, (c) those resulting in transformation of the crown into a belt of irregular hooklets and sphaerules.

4. *Proglottis*

Apart from the general observation of gross morphological differences between the species and subspecies, few new observations on the morphology of the proglottis appear to have been made. Bacigalupo and Rivero (1949) have described the presence of a sphincter muscle around the vaginal opening in *E. granulosus*.

Kilejian *et al.* (1961) have shown that the adult of *E. granulosus* stored large amounts of glycogen in the parenchyma and lesser amounts in the vitelline glands and ova; all the reproductive organs were reported as being free of glycogen. The cuticle gave a strongly positive reaction for alkaline phosphatase but not acid phosphatase.

B. LARVAL STAGE—HYDATID CYST

1. *Protoscolex*

a. General. Most authors refer to hydatid cysts as containing "scoleces". This use of the term "scolex" for the larval stage of *Echinococcus* gives rise to some ambiguity, when describing its morphology. Thus, it is confusing to speak of a "scolex" (i.e. the head region) of a "scolex" (a larva)! For this, and other reasons (see below), the author (Smyth, 1962b, c, 1963a) has made use of the term "protoscolex" previously used by some authors (e.g. Carta and Deiana, 1960) for the larval stage in the brood capsule. This expressive term is derived from the Italian word "protoscolice". As shown in Section IV, A, 2, the true scolex of an adult worm differs markedly from the true scolex of the larva, in (a) containing well developed gland cells which secrete through the rostellum and (b) being of a substantially larger size (Yamashita *et al.*, 1958b). The term "protoscolex"—suggesting a developmental stage in the formation of the adult scolex—thus seems particularly appropriate and is used further here.

The protoscoleces have been shown to contain heavy deposits of glycogen but only minute amounts of acid and alkaline phosphatases (Kilejian *et al.*, 1961). Vanni and Radice (1950) examined the protoscoleces of *E. granulosus* by fluorescent microscopy and distinguished organisms with a blue fluorescence and other smaller ones with an orange-red fluorescence. These latter may prove to be the small contracted dead protoscoleces shown to be present in the majority of cysts from sheep (Smyth, 1962c).

b. Musculature. Coutelen *et al.* (1952) have described the musculature of the protoscolex in some detail and Fig. 3 is a composite diagram of the two deeper muscle layers based on their drawings. Briefly, the musculature may be conveniently considered in three groups:

(i) The superficial (subcutaneous) circular and longitudinal fibres, which are evaginatory in function.

(ii) Muscle fibres which control the movement of the suckers, the rostellum and the hooks.

(iii) The deep muscle fibres, which are invaginatory in function.

The subcutaneous circular and longitudinal fibres appear to cover the whole of the organism from the rostellum to the peduncle, except for a band in the region of the suckers where no circular muscle fibres can be seen. Four bundles of longitudinal fibres pass between the suckers and are inserted below the hooks on the rostellum. The subcutaneous muscle fibres of the suckers consist of a circular band on the border of each sucker and a

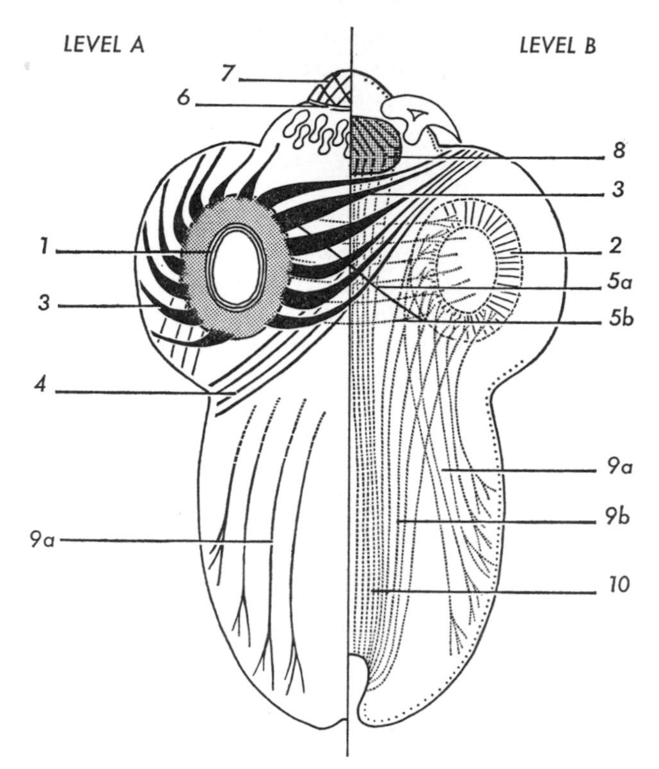

FIG. 3. Muscular system of protoscolex of *E. granulosus*; each side taken at different levels. Redrawn from Coutelen *et al.* (1952). 1. Circular muscle fibres—narrow the surface of the cup of the sucker and make the edges of the sucker jut out. 2. Radial muscle of sucker—contracts the surface of the sucker. 3. Oblique anterior sucker muscles—pull sucker in, and forwards. 4. Oblique posterior sucker muscles—pull sucker to the side and back. 5. Co-ordinating sucker muscles: (a) inserted at lateral surface—cross at midline; (b) inserted on base of sucker and parallel to each other. 6. Circular muscle of rostellum—elongates it. 7. Oblique muscle of rostellum. 8. Radial muscle fibres of the "rostellar pad"—control hooks. 9. Invaginatory muscle of sucker: (a) from base of suckers to cuticle; (b) from base of suckers to peduncle. 10. Invaginatory fibres of rostellum.

layer of circular and one of longitudinal fibres in the "cup" of each sucker.

Deeper circular and radial fibres (1, 2, Fig. 3) control the shape and depths of the sucker while anterior and posterior bands of oblique fibres (3, 4) inserted in the lateral surface of the suckers control part of the movement of the suckers. Two other groups of fibres (5a, 5b) are concerned with the movement of adjacent suckers and their position relative to one another. At the same depth, the rostellum has a band of circular fibres in front of the hooks and a double layer of criss-crossed fibres in the anterior region of the rostellum (6, 7). The rostellar pad has a system of fibres similar to those of the suckers. The radial muscles in this case control the movement of the hooks (8).

The deep muscle system consists of two posterior oblique bands of fibres

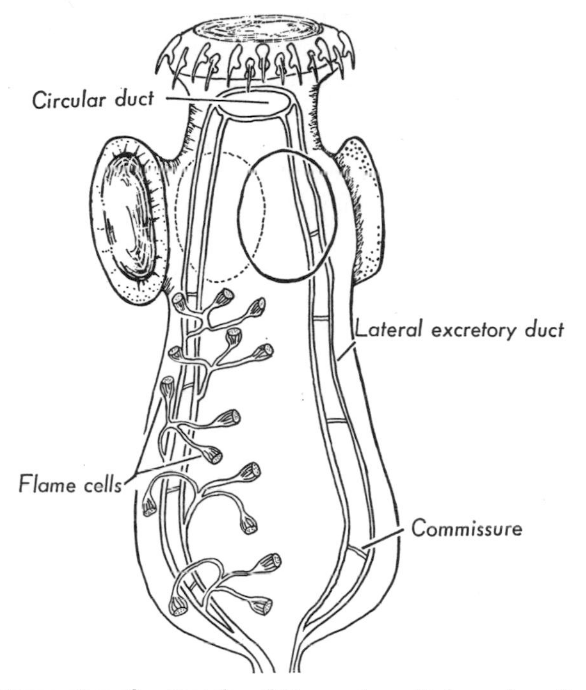

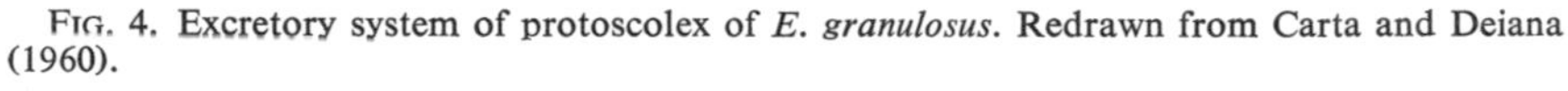

FIG. 4. Excretory system of protoscolex of *E. granulosus*. Redrawn from Carta and Deiana (1960).

from the base of each sucker (9a, 9b) which retract the suckers and help with their invagination. There is also a band of closely packed fibres which is inserted in the base of the rostellar pad and which passes directly along the anterior-posterior axis to the base of the organism (10). The contraction of these nine muscle bands invaginate the rostrum and suckers.

c. Excretory system. Carta and Deiana (1960) described the excretory system of *E. granulosus* and a figure based on their observations is given in Fig. 4. The flame cell formula, which was independent of the host origin of the material was 2(3 + 3 + 3 + 3 + 3) = 30. Some specimens possessed a flame cell in the peduncle of the protoscolex in which case the formula was 2(3+3+3+3+2+1) = 30.

2. *Germinal Membrane*

The early work of Coutelen (1938) has shown that the germinal membrane is not a syncytium but a fine granular material in which are embedded various kinds of cellular elements with clearly defined walls. Little detailed work has been carried out since on the structure of the germinal membrane and this apparently has been confined to *E. granulosus*. Since the growth pattern of a developing cyst is different in *E. granulosus* and *E. multilocularis*, it is particularly important that more information on the germinal membrane of the latter species be made available. Schwabe and Schinazi (1958) have shown that protonephridial flame cells occur both in the stalk (= peduncle of some authors) connecting the protoscolex to the brood capsule and in the brood capsule membranes themselves; these workers failed to find flame cells in the germinal membrane of the mother cyst.

Čmelik and Briski (1953) were unable to demonstrate DNA in the hydatid cyst wall of *E. granulosus*. However, Kilejian *et al.* (1961) described the presence of three types of nuclei with different amounts of DNA: (a) small vesicular, (b) large vesicular and (c) small compact. These authors give a photomicrograph of a Feulgen preparation of the germinal membrane. Examination of this and a comparison of it with aceto-orcein squashes, has led the writer to conclude that the small compact nuclei figured by Kilejian *et al.* are, in fact, chromosomes. This conclusion is based firstly on the fact that in the photomicrograph shown by Kilejian *et al.* there are eighteen small stained structures, and secondly, that one of these structures is banana-shaped. The writer has shown that in *E. granulosus* the diploid number is 18 and one pair of chromosomes is characteristically banana-shaped (Smyth, 1962a). This conclusion, of course, is open to confirmation by other workers but the evidence is suggestive. It would be particularly interesting to determine if similar structures can be readily observed in the germinal membrane of *E. multilocularis*.

Kilejian *et al.* (1961) have shown that the germinal membrane of *E. granulosus* contains heavy deposits of glycogen; these authors also recorded the presence of small amounts of acid phosphatase. Schwabe (1959) has shown that the germinal membrane is permeable to water.

3. *Laminated Membrane*

Although the chemistry of the laminated membrane has been studied in some detail (Section VI, A, 2) there has been little recent morphological or histological study on this structure. The membrane is clearly secreted by the germinal membrane, as suggested by Yamashita (1960), for protoscoleces of *E. granulosus* and *E. multilocularis* when cultured *in vitro* readily become vesiculated and form laminated membranes (Yamashita *et al.*, 1962; Smyth, 1962c; see Section VIII). Schwabe (1959) has suggested that the laminated membrane may result from the interaction of a mucopolysaccharide arising as a metabolic product of scoleces with some as yet unidentified host factor. If this result is confirmed, it would suggest that the formation *in vitro* results similarly from an interaction with some factor in the medium. Since the laminated membrane was formed *in vitro*, in a variety of media, the present author concluded (Smyth, 1962c) that it is formed from intrinsic parasite materials rather than from an interaction between intrinsic and extrinsic factors as suggested by Schwabe. It must be borne in mind, however, that since all the culture media in which a laminated membrane was formed *in vitro* in our experiments, contained materials of host origin (e.g. serum, amniotic fluid), this result is still equivocal and does not, in fact, disprove Schwabe's view.

The laminated membrane has been shown to give a strong P.A.S. reaction, a result related to its mainly mucopolysaccharide composition (see Section VI, A, 2).

Kilejian *et al.* (1961) also found that the laminated membrane did not give a positive histochemical reaction for acid and alkaline phosphatases; tests for DNA were likewise negative. Schwabe (1959) studied the permeability of the laminated membrane to inorganic and organic ions.

C. EGG

No recent work appears to have been carried out on the morphology of the egg or the embryonic formation of its membranes. The egg does not appear to differ significantly from that of other Taeniidae and has a thickened embryophore.

Lubinsky (1958) has found hexacanth embryos in the eggs of *E. granulosus* with morphologically abnormal numbers of hooks (2–12) instead of the usual six. Twelve hooked embryos were the most common anomaly occurring in the reindeer strain (i.e. from dogs infected with reindeer cysts) to the extent of 2·2 per thousand. Lubinsky suggests these may result from the fusion of two dividing eggs. An alternative explanation is that they could be tetraploids but the phenomenon clearly warrants further investigation.

Meyers (1955) used Stokes' Law to determine the weight, volume and density of *E. granulosus* eggs. Twelve determinations gave the following figures: density 1·1184; volume, $18{\cdot}30 \times 19^{-9}$ ml; weight, $20{\cdot}44 \times 10^{-9}$g.

D. CHROMOSOME NUMBER

As objects of chromosome studies, the cestodes have been somewhat neglected. This is not surprising for the chromosomes in the species studied have been found to be unusually small and often exceedingly difficult to handle by routine methods. The diploid number of chromosomes in *Echinococcus granulosus* has recently been reported as 2n = 18 (Smyth, 1962a). This figure was based on aceto-orcein squash material pretreated with hypotonic saline after mitosis had been inhibited by colchicine treatment. The set of chromosomes were found to contain two large banana-shaped chromosomes, which are particularly characteristic. It would be especially interesting to compare the chromosome patterns of *E. multilocularis* and *E. granulosus* and their varieties and determine whether chromosome modifications, i.e. polyploidy, haploidy or deficiency, are concerned in the speciation pattern.

V. Life Cycle

(A composite diagram of the life cycle is shown in Fig. 5.)

A. Egg

1. *Viability*

a. Effect of temperature. The resistance of *Echinococcus* eggs to various physical factors—especially temperature—has not been much studied. Eggs of both species can clearly withstand low temperatures for long periods; this is a position similar to that found with other cyclophyllids (Smyth, 1963a). Batham (1957) found that eggs stored in tap-water at 10°–21°C, rapidly lost their viability but some were infective after 32 days. Eggs kept dry may have been viable for 1 year but Batham regarded the evidence here as somewhat unreliable. Schiller (1955) found that the eggs of *E. multilocularis* could withstand very low temperatures for short periods. Eggs in gravid proglottides, frozen to –26°C, for 382 h were infective after 54 days but not after 65 days. Some eggs kept at –51°C, for 24 h were infective. These results are based on experiments with free proglottides but according to Schiller somewhat longer periods are required to destroy the eggs within the carcass of a fox. There is evidence (Schiller, 1954) that blowflies can transmit *E. multilocularis* eggs so that survival of carcasses may be important in the epidemiology of alveolar hydatidosis. Methods of testing viability of cyclophyllidean eggs involve using a hatching fluid. Meyers (1957) has given details of a relatively simple method using pepsin followed by trypsin. At temperatures just above zero (2°C) Thomas and Babero (1956) found that eggs of *E. multilocularis* remained infective to ground squirrels after 2½ years. Meymarian and Schwabe (1962) quoted preliminary experiments in which exposure to moist heat at 60°C, for 10 min, 70°C, for 5 min, or 100°C, for 1 min, is sufficient to kill oncospheres.

b. Effect of germicides. Meymarian and Schwabe (1962) tested the effect of a number of germicides on the viability. The effect of formalin is particularly worth commenting on: 5–20% formalin fixed the eggs but 100% hatched after 24 h. Hercus *et al.* (1962) found that eggs in specimens of *E. granulosus* may remain viable after storage in formalin for 2 weeks.

2. *Hatching*

Cyclophyllidean eggs in general fall into two groups: (a) those which require both pepsin and trypsin treatment in order to hatch, and (b) those which do not require pepsin (Smyth, 1963a). The eggs of *E. granulosus* fall into the latter group since pepsin did not seem to have any effect on the dissolution of the embryophore or the activation of the free oncospheres (Berberian, 1957; Meymarian, 1961). Meymarian found that both $NaHCO_3$

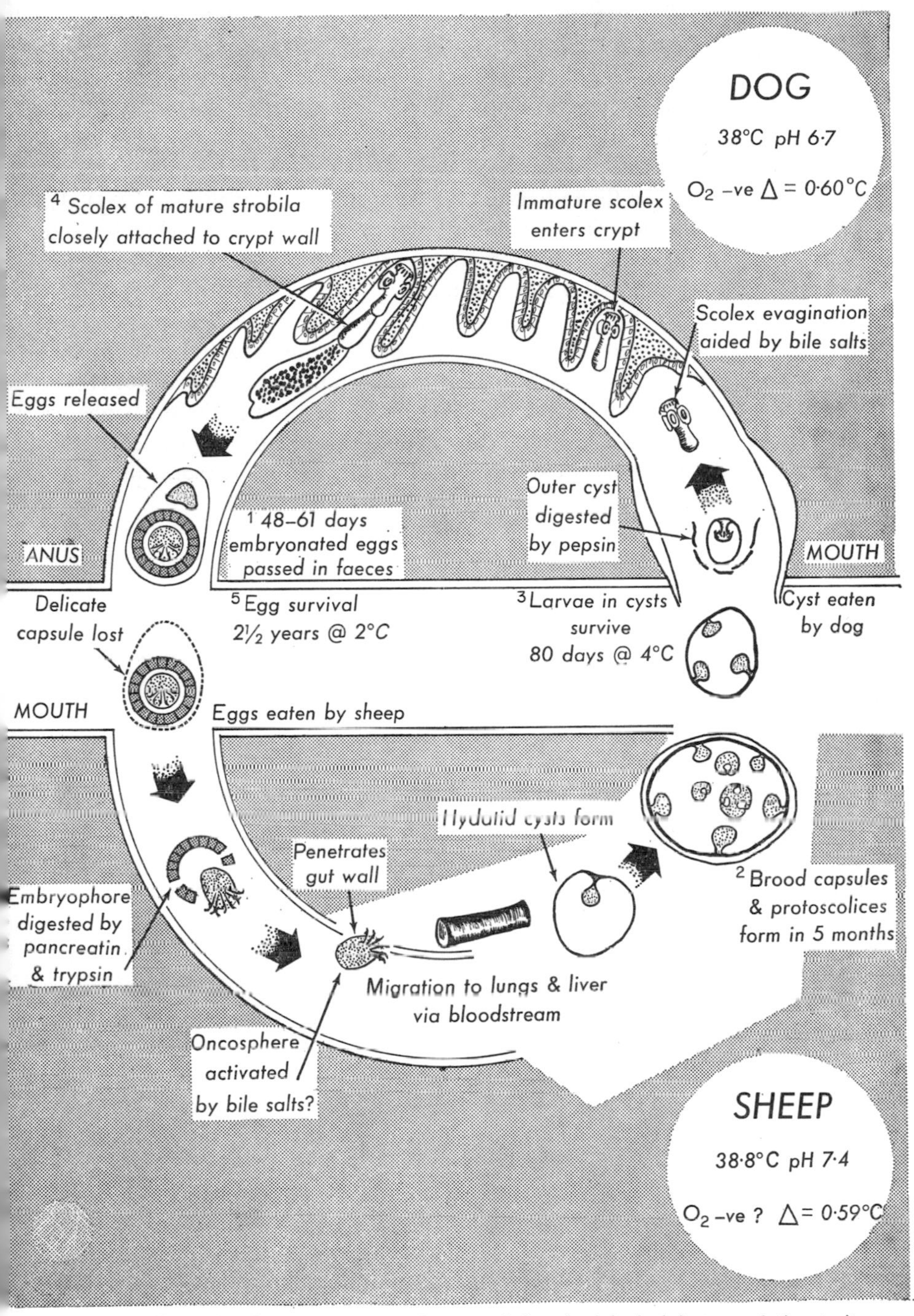

FIG. 5. Life cycle of *E. granulosus* and some of the physiological factors relating to it. Details based on [1]Yamashita *et al.* (1958b); [2]Dew (1922); [3]Gelormini (1941); [4]Smyth (1964); [5]Thomas and Babero (1956: based on *E. multilocularis*).

and NH_4OH removed shells without activation of oncospheres and furthermore that trypsin did not cause the disruption of the shells as readily as pancreatin. Both pancreatin and trypsin caused some activation of the freed oncospheres but sheep bile, bile salts (sodium glycocholate, sodium taurocholate, sodium tauroglycocholate) and cholesterol all impaired the activating effects of these enzymes. Whole bile was also shown to be a more effective synergistic agent than the individual bile salts tested. Meymarian (1961) also found that not all oncospheres could be activated and gave some evidence to support the view that oncosphere from attached proglottides (being less mature) were activated at a lower rate.

Berberian (1957) studied the effect of the digestive juices and bile of man, sheep, cattle, dogs and cats on the ova of *E. granulosus*. He found that human bile, intestinal juices of sheep and cattle, and artificial pancreatic juice caused rapid disruption of the shells and activation of the oncospheres. In dog bile or dog or cat jejunal juice hatching was poor and oncospheres were non-mobile. Berberian suggested this explained the general immunity of dogs and cats to hydatid disease (but see Section III, C).

B. LARVAL DEVELOPMENT

1. *Histogenesis of Hydatid Cyst*

The degree of development varies from an invasion of the viscera, which is followed by a rapid reaction and elimination by the host, to establishment which results in the successful development of fertile cysts. This question is basically an immunological one and is discussed in detail in Section VII, B.

The establishment of hydatid cysts in a wide range of hosts following laboratory or natural infection has been recorded (Table IV). There have been, however, relatively few studies on the histogenesis of the larval stages, this field being covered by contributions by Mankau (1956a, 1957), Rausch (1954), Yamashita *et al.* (1958a, 1960), Yamashita (1960), Lubinsky (1960b) and Webster and Cameron (1961).

Most attention appears to have been paid to the development of *E. multilocularis*; the larval development of *E. granulosus* has been described in detail by a number of earlier workers. The minimum time for protoscoleces to appear in successful infection has been recorded only in a few cases. In the tundra vole protoscolex formation has been observed as early as 30 days (Rausch, 1954). In the white mouse, protoscoleces usually take 3–4 months to develop (Mankau, 1956a) but in some cases appear earlier (Yamashita *et al.*, 1958a).

Yamashita (1960) gives a detailed account of the development of *E. multilocularis* and in particular, draws attention to the importance of strain difference in hydatid infections. In susceptible animals, echinoccocal foci become visible on the 5th day after inoculation but may be observed as early as 24 h. At the 5-day stage, the parasite takes the form of a mass of germinal cells enclosed in a thin larval membrane. Unilocular vesiculation begins in 3-4 days, a vacuolar structure appearing in the central position of the germinal mass. After 5–10 days, the larval membrane shows an indistinct appearance and the vesicle wall is made up of a thin germinal cell layer alone. Yamashita gives histopathological evidence to suggest that the parasite invades the liver by way of the portal circulation system. In the multilocular cyst, exogenous protrusion takes place very frequently and finally the cysts "manifest a complicated botryoid or coralloid aggregation of numerous minute cysts of which the lumina communicate with one another." This is a developmental picture very similar to that described *in vitro* by Rausch and Jentoft (1957). Later development involves the establishment of the laminated membrane and the germinal layer and Yamashita draws attention to the presence of fine reddish granules sometimes deposited in contact with the laminated membrane and concludes that the germinal membrane is capable of producing the substances which manufacture the laminated membrane. This is borne out by *in vitro* results (Section VIII). The initial stage in the development of a protoscolex is represented by a dense accumulation of nuclei which projects into the lumen of the capsule as a hill-like, spherical and then ellipsoidal structure. For details of the later stages in development the original papers by Rausch (1954), Mankau (1956b), Yamashita *et al.* (1958a) and Yamashita (1960) should be consulted. Webster and Cameron (1961) have also given a detailed account of primary and secondary infections of *E. multilocularis* with particular reference to the histopathology in a number of hosts. Schwabe *et al.* (1959) have shown that resistance to secondary infection by *E. multilocularis* apparently develops with age in white mice, at least with the strain of mice used in their experiments. Mice 48 days of age or younger were highly susceptible to an initial infection by intraperitoneal injection. In contrast, mice 71 days of age or older were relatively resistant to infection by the same method.

2. *Viability of Hydatid Cyst*

It has long been recognized that hydatid material will remain infective for long periods in carcasses exposed to a range of environmental conditions. Gelormini (1941) reported survival after 80 days at 4°C. Mikačić (1956) showed that from —2°C to +2°C protoscoleces remained alive for 10 days, from +10°C to +15°C for 4 days, and from 20°C to 22°C for only 2 days. At very low temperatures (—5°C to —12°C) viability was 5 days. Batham (1957) found that under normal outside winter conditions (rain, sun and wind) —2°C to +30°C in Dunedin (New Zealand) most protoscoleces in hydatid cysts in sheep viscera remained alive after 12 days, many after 21 days and a few after 24 days.

C. MATURATION IN THE DEFINITIVE HOST

Although a number of workers have fed hydatid cysts to dogs and obtained adult worms, there have been few studies of the morphogenesis from protoscoleces to sexually mature adult strobila. Yamashita *et al.* (1956, 1958b) have compared the development of *E. granulosus* and *E. multilocularis* after 15, 16, 35, 117, 135, 290 and 375 days in the dog.

In general, *E. multilocularis* develops faster than *E. granulosus*. The eggs of *E. multilocularis* appear in the faeces on the 30–35th day after infection, whereas eggs of *E. granulosus* do not appear in faeces until 48–61 days after infection (Yamashita *et al.*, 1958b). Recent studies (Smyth, 1963, unpublished work) show that strobilae of *E. granulosus*, if removed from a dog gut at 35 days, can produce shelled eggs after a further 7 days *in vitro* culture. It can, therefore, be assumed that approximately 41 days represents the minimum development time to egg-production but the subsequent appearance in the faeces requires a further period of about 7 days.

Yamashita *et al.* (1958b) have drawn attention to the manner in which the dimensions of structures such as hooks and suckers can vary during development and stresses the importance of taking the age of a specimen into account when considering the morphology of the worm. Thus on the 16th, 35th, 135th and 375th day of development the proportions by which various dimensions increase are as follows: rostellum, 1:1·4 : 1·5:2; large hook, 1:1·3 : 1·5:1·6; small hook, 1:1·1 : 1·2:1·3; sucker, 1:1·1 : 1·3:1·5. Over the the same periods, the length of the strobila increased ratios of 1:3:4:8. Attention has been already drawn to the difference in size between the larva and adult scolex when advocating the use of the term "protoscolex" (Section IV, B, 1).

VI. PHYSIOLOGY

A. CHEMICAL COMPOSITION

1. *Protoscoleces and Adults*

Although the chemical composition of the hydatid fluid (which was much studied by earlier workers) is relatively well known (see "Handbook of Biological Data") that of the larva and adult has been much less studied. With a few exceptions, much of the recent work in this field has been carried out by Agosin and his co-workers (Agosin *et al.*, 1957; Agosin, 1959).

The general chemical composition of the protoscoleces of *E. granulosus* (values as % wet weight, fresh tissues) is dry substance 14·8±1·17, in organic substances 2·0±0·33, polysaccharide 2·8±0·08, lipid 2·0±0·29 and protein 9·2±0·48. Approximately 13·6% of the dry substance consists of lipids, a figure within the range of values for this constituent given for other cestodes. In contrast, the protein content—about 62·5% of the dried substance—is substantially higher than that generally described for other helminths. The polysaccharide content is not high, amounting to some 19·8% of the dried substance—a figure substantially below that of other cestodes.

Further analysis of the gross chemical constituents of *E. granulosus* seems confined mainly to a study of the polysaccharides. Agosin (1959) studied polysaccharides isolated from the protoscoleces by electrophoretic and chromatographic techniques. By the former technique two bands were demonstrated. The first, stronger, band migrated precisely like genuine glycogen. The second band, which was weaker, moved faster than glycogen but more slowly than glucosamine or galactosamine. When subjected to more prolonged hydrolysis (i.e. 3N H_2SO_4 for 6 h at 100°C) the presence of glucose and galactose were confirmed and a new spot corresponding to glucosamine appeared. Agosin concluded that *E. granulosus* contained two polysaccharides, glycogen and a polysaccharide with at least a base of glucosamine and galactose. The nature of the slow component of the acid hydrolysis has not been identified.

The above results in general have been confirmed by Kilejian *et al.* (1962) when using Pflügers method for isolating polysaccharide. When, however, the procedure of Gary *et al.* (1958) was utilized, the infrared spectrum at 6·5μ gave an absorption band which showed a larger proportion of other constituents than in the sample isolated by Pflügers' method. A precipitate, produced by the addition of acetone in Gary's procedure, gave an infrared spectrum very closely resembling that of the mucopolysaccharide isolated from the laminated membrane (see Section VI, A, 2). When protoscoleces were treated with malt diastase in order to remove the glycogen, the infrared spectrum of the resulting precipitate with alcohol was almost identical with the precipitate from acetone.

Specimens of polysaccharide from adults gave electrophoretic and chromatographic patterns not different from those of the corresponding preparations from the protoscoleces.

Kilejian *et al.* (1962) also showed that alcohol-petroleum ether extracts of samples of protoscoleces gave spectra characteristic of lipids, which were, in general, similar to that obtained from the corresponding extract of the laminated membrane with some minor variations. Although these workers stress that such spectra are unlikely to provide detailed pictures of their composition, they point out that the spectrum obtained in the region 7·5–11μ is highly characteristic of phosphatides, rather than of cholesteryl esters, glycerides, fatty acids or cholesterol. One of the peaks in the spectrum was highly suggestive of lecithin.

2. *Laminated Membrane*

a. Carbohydrate. The laminated membrane of a hydatid cyst is relatively easy to obtain in quantity so that it is thus not surprising to find that this structure has been subjected to analysis by a number of workers.

The early work, which is reviewed by Kilejian *et al.* (1962) suggested that the cyst wall consisted of a protein carbohydrate complex which on degradation gave glucosamine and was probably related to chitin.

In general, this result confirms earlier work. Čmelik (1952a) subjected the entire cyst wall to trichloracetic acid extraction and to pepsin, and to combined pepsin and trypsin digestion. This latter procedure yielded a pure polysaccharide which was easily soluble in water and positive to the Molisch, Dische and Elson-Morgan tests but negative to Biuret, ninhydrin and sulfosalicylic acid. On analysis the polysaccharide gave 4·5%N, 0·08%P, and 3·99% ash. The principal products of hydrolysis were considered to be glucosamine and aldohexoses. Subsequently, Čmelik and Briski (1953) showed that the sugars separated from the cyst wall after sodium hydroxide digestion were glucose, galactose and glucosamine. Before hydrolysis, the presence of a protein was demonstrated; nucleoprotein was also isolated from the cyst wall. These studies, were carried out on *complete* cyst wall, i.e. without separating the laminated membrane from the germinal membrane.

Studies by Kilejian *et al.* (1962) were carried out on the laminated membrane alone. These workers showed that a mucopolysaccharide, with a characteristic infrared spectrum and containing only galactose and glucosamine as the sugar units, was the chief structural material isolated from the laminated membrane. A mucopolysaccharide, with a similar infrared spectrum was isolated from hydatid fluid and the protoscoleces. The electrophoretic mobility of the mucopolysaccharide isolated from the protoscoleces, was, however, quite different from that isolated from the laminated membrane or cyst fluid.

The infrared spectrum of the mucopolysaccharide suggested the presence of appreciable amounts of protein in the material although chromatographic analysis yielded only quite small amounts of amino acids. These authors, could not, as yet, account for this apparent discrepancy in results.

Kilejian *et al.* (1962) could not detect the presence of nucleoprotein in the laminated membrane and concluded that the nucleoprotein reported by Čmelik and Briski (1953) originated in the germinal membrane which they failed to separate from the laminated membrane.

Although chemical and histochemical methods showed the presence of large amounts of glycogen in the protoscoleces (Agosin *et al.*, 1957; Kilejian *et al.*, 1961, 1962) only a trace of glycogen is reported to be present in the laminated membrane (Kilejian *et al.*, 1961, 1962).

b. Lipid. Čmelik (1952a) found that lipids—mainly fatty acids and cholesterol,—made up only 1·25% of the dry substance of the hydatid cyst wall. Kilejian *et al.* (1962) found that the lipid material of both the protoscoleces and laminated membrane gave an infrared spectrum characteristic of lecithin in association with a variable fraction of cholesterol.

3. *Germinal Membrane*

The chemical composition of the germinal membrane has been little studied. Kilejian *et al.* (1961) recorded the presence of glycogen and DNA—the latter being clearly related to the presence of large numbers of nuclei, Krvavica *et al.* (1959) identified the following amino acids in hydrolysates of germinal membrane: aspartic acid, glutamic acid, serine, glycine, glucosamine, threonine, alanine, valine, proline, arginine (the last named in the hydrolysate of sterile layers only).

4. *Hydatid Fluid*

The chemical composition of the hydatid fluid is well known having been studied by numerous early authors ("Handbook of Biological Data", 1959). Most of the recent work has dealt with analyses of protein or related components, amino acids or nitrogen. Carbone and Lorenzetti (1957) reported a range of 17·3–227 mg protein per 100 ml in samples of hydatid fluid obtained from nearly one hundred cysts. Pozzi and Pirosky (1953) identified sixteen amino acids in the hydrolysates of protein from hydatid fluid. Magath (1959) fractionated hydatid fluids (from man) electrophoretically and reported 7·5 mg protein per 100 ml of which 44% was albumin, 39% α and β globulins and 17% was γ globulins. Cameron and Staveley (1957) reported the presence of specific anti-P inhibiting substance in the hydatid fluid. Goodchild and Kagan (1961) showed that there were striking similarities in the paper and starch electrophoretic patterns of proteins in hydatid fluid and serum from three different host species. Analyses revealed that there was 13–77 times as much nitrogen in the hydatid fluid of *E. multilocularis* as in that of *E. granulosus*.

B. RESPIRATORY METABOLISM

1. *Oxygen Consumption*

Work on the respiration of *E. granulosus* has been confined to the larva; no respiratory studies appear to have been carried out on adult worms. Agosin *et al.* (1957) have studied respiration of the protoscoleces by the conventional Warburg technique and Farhan *et al.* (1959) polarographically by means of the dropping mercury electrode.

At 38°C Agosin *et al.* (1957) obtained a figure of 258 ± 7·3 mm³/h/mg fresh tissue in Ringer's solution, which recalculated on their own figure for dried weight (14·8 ± 1·17%) gives a figure of approximately 1·9 mm³/h/g dried tissue. The corresponding figure in hydatid fluid was 2·3. Farhan *et al.* (1959) showed clearly that the oxygen consumption decreased with decreasing oxygen tension until at a tension of 0·80 ± 0·48 mm³/ml it became and remained at zero. At a tension of 3·76 mm³/ml which these authors state is the average oxygen consumption of hydatid fluid saturated with air ("optimum rate"), the Qo_2 varied within the range 0·39–1·50 mm³/h/g with a mean of 0·85 ± 0·39 optimum rate. This figure is thus substantially lower than that of Agosin *et al.* (1957) given above. Farhan *et al.* (1959) however, quote average manometric figures (unpublished) for the Qo_2 in hydatid fluid of 2·16 and 1·68-figures which closely correspond with those of Agosin *et al.* mentioned above. Smyth and Roberts (1963, unpublished work) have also obtained figures within this range (1·73 ± 0·05 mm³/h/g) based on fourteen determinations. These figures are based on pepsin-treated material—a procedure which eliminates a possible source of error due to the presence of a proportion of dead protoscoleces, usually present in sheep hydatid material.

Farhan *et al.* (1959) found that the dissolved oxygen content varied with the location of the cyst; fluid from bovine lung cysts contained 2·91 ± 0·05 mm³/ml, whereas that from the liver gave a lower figure, 1·28–2·28 mm³/ml. The respiration of protoscoleces at these tensions was found to be 96% and 54% of the mean optimum rate respectively and Farhan *et al.* (1959) concluded that the protoscoleces of *E. granulosus* are able to carry out a predominantly aerobic metabolism. These authors also show that protoscoleces possibly repay an oxygen debt following anaerobic incubation.

2. *Respiratory Quotient*

Agosin *et al.* (1957) found the R.Q. to be 0·88 but this figure does not contain a correction for bound CO_2. On account of the extremely large and somewhat variable amounts of CO_2 released from the calcareous corpuscles on acidification, this factor could not be calculated. Anaerobic production of CO_2 was given as $Qco_2 = 1{\cdot}65$.

3. *Factors Affecting Respiration*

a. Oxygen tension. Attention has already been drawn to the fact that the oxygen consumption is dependant on the oxygen tension.

b. Ionic composition of the medium. Agosin *et al.* (1957) have shown that the ionic composition of the medium has a marked influence on the respiratory rate. Respiration was highest in hydatid fluid and reasonably high in Ringer but when components of Ringer's solution were used, in concentrations higher or lower than those normally used, the oxygen consumption was lowered.

c. Temperature. Agosin *et al.* (1957) have shown that within the temperature range 28–38°C the Q_{10} has a value of 2·1. They also show that the oxygen consumption does not appear to be affected by storing protoscoleces at 5°C for 24 h.

d. pH. Agosin *et al.* (1957) found that in the pH range 4·5–8·5 in Ringer's solution, no marked difference in oxygen consumption was observed.

c. Inhibitors. Agosin *et al.* (1957) have shown that the aerobic respiration is markedly cyanide sensitive, although the degree of inhibition is less than that reported for other cestodes. This clearly indicates the dependence of the respiration on catalysis by heavy metals but does not necessarily imply that this organism possesses a cytochrome system; Bueding and Charms (1951) have shown that the nematode *Litomosoides carinii* does not contain cytochrome *c*, although its respiration is markedly inhibited by cyanide.

Agosin *et al.* (1957) have further shown that anaerobic and aerobic respiration is highly sensitive to inhibitors of glycolysis but DL-glyceraldehyde was completely ineffective. This question is discussed further below.

C. CARBOHYDRATE METABOLISM

1. *General Metabolism*

Most of the work on the carbohydrate metabolism of *E. granulosus* has been carried out by Agosin and his co-workers (Agosin *et al.*, 1957; Agosin and Aravena, 1959, 1960a, b; Agosin and Repetto, 1963). The review of Agosin (1959) gives a particularly valuable summary of the field up to that date.

As pointed out earlier, nearly 20% of the dried weight of the protoscoleces is polysaccharide and in common with many other helminth parasites the principle source of energy appears to be glycogen; it has not been possible to demonstrate a statistical difference between glycogen consumption *in vitro* under aerobic or anaerobic conditions (Agosin, 1959). Under *aerobic* conditions, the final product of metabolism was almost entirely lactic acid (1·165 units of lactic acid per unit of glycogen and 2·64 units of O_2). Acetic acid was also found to be quantitatively important (0·212 units of acetic acid per unit of glycogen). No formic acid was produced and no fatty acid higher than acetic. Small quantities of pyruvic and succinic acids were also formed, together with a little ethyl alcohol.

Under *anaerobic* conditions, the same workers found that lactic acid was again the main metabolic product but that succinic acid, which is quantitatively unimportant under aerobic conditions, was also secreted in substantial amounts. In contrast with aerobic conditions however, no pyruvic acid was excreted; acetic acid was excreted.

Agosin *et al.* (1957) further showed that in the metabolic balance about 70% of the carbon of the metabolized glycogen was recovered. In contrast, under anaerobic conditions, the carbon contents of the final products corresponded to 91% of the carbon of the used glycogen. In these calculations, carbon corresponding to the CO_2 produced has not been taken into account as it was not possible to establish what proportion was derived from inorganic sources (i.e. calcareous corpuscles).

Under anaerobic conditions, the lactic acid produced corresponded to about 50% of the utilized glycogen. This high proportion of lactic acid is somewhat similar to the position in *Hymenolepis diminuta* (Read, 1956) and *Schistosoma mansoni* (Bueding, 1949). In a number of other parasitic helminths, fatty acids form a substantial part of the final metabolic products.

Pyruvic acid is only excreted under aerobic conditions and succinic acid which, under the same conditions, is unimportant may correspond to about 30% of the consumed polysaccharide in anaerobiosis. The fact that the carbon atoms of the aerobic metabolism corresponds to a small percentage of the C-glycogen atoms suggests that in the presence of O_2 some C-atoms are completely oxidized. According to Agosin (1959), if all the oxygen used were for complete oxidation of the carbohydrates, some 40% of the polysaccharides would have been accounted for in this way. In fact, only 30% was recovered which suggests that other substrates—apart from carbohydrates—are probably utilized by the protoscoleces.

2. *Intermediary Metabolism*

a. Embden-Meyerhof cycle. The evidence suggests that not only does *E. granulosus* possess a phosphorylative glycolytic system resembling that which occurs in vertebrate tissue and yeast (Agosin and Aravena, 1960b) but also a pentose phosphate cycle and possibly a Krebs' tricarboxylic cycle. The first and last of these are not unexpected, having been reported in a number of parasites, but a pentose phosphate cycle has only been reported from a very few other helminths such as *Ascaris lumbricoides* (Entner and González, 1959).

The presence of these cycles in whole or in part should involve (a) demonstration of the enzymes involved in the reactions of the cycle, (b) demonstration that the cycle is actually operative in the intact cell.

Although not every enzyme of the Embden-Meyerhof cycle has been individually assayed, Agosin and Aravena (1959) have produced sufficient experimental evidence to show that phosphorylative glycolysis contributes substantially to glucose utilization. Several of the reactions of glycolysis were demonstrated in cell-free extract of protoscoleces making the presence of the entire Embden-Meyerhof sequence probable. Thus the presence of phosphofructokinase activity, as well as aldolase and glyceraldehyde-phosphate dehydrogenase were demonstrated in cell-free extracts.

Arising out of this work, the same authors showed the existence of four kinases, specifically catalysing the phosphorylation in position 6. The gluco-, fructo- and mannokinase activities were shown to be inhibited by glucose-6-phosphate, while mannose-6-phosphate inhibited only gluco- and mannokinases.

b. Pentose-phosphate pathway. Agosin *et al.* (1957) detected the presence of ribulose, ribose, glyceraldehyde, sedoheptulose, glucose and fructose in the extracorporeal fluid of *E. granulosus* after aerobic or anaerobic incubation. Since ribose, ribulose and sedoheptulose are pathways of the pentose phosphate pathway, these authors speculated that glucose might be metabolized also via this sequence.

De Ley and Vercruysse (1955) reported that they were unable to demonstrate the presence of glucose-6-phosphate and 6-phosphogluconic acid dehydrogenases in *E. granulosus*. Using modified techniques, Agosin and Aravena (1960b) were able to demonstrate the presence of these enzymes and in addition transketolase, transaldolase, phosphopentose isomerase, ribokinase, 3-phosphoglyceraldehyde dehydrogenase, triosephosphate isomerase, and possibly phosphoketopentose epimerase. These conclusions were based on the detection of the relevant intermediates of the pentose phosphate cycle and presents unequivocal evidence that this cycle, in addition to the Embden-Meyerhof cycle, operates *in vivo* and is probably of physiological importance. Agosin and Aravena (1960a) purified the phosphopentose isomerase over sixty times and examined its properties in detail.

c. Krebs' tricarboxylic cycle. Agosin *et al.* (1957) showed that the respiration of protoscoleces is inhibited by fluoracetate but not by malonate and from this result, suggested as a working hypothesis, the presence of a terminal sequence that could resemble an incomplete Krebs' cycle. Agosin and Repetto (1963) later showed that intact protoscoleces, as well as cell-free preparations, could oxidize several Krebs' cycle intermediates. Cell-free preparations are shown to contain a condensing enzyme, a TPN- and a DPN- dependent isocitric dehydrogenase, succinic, malic and a α-ketoglutaric dehydrogenases, as well as fumarases, aconitase pyruvic oxidase and α-carboxylase. They further demonstrated that phosphoenolpyruvic, carboxydismutase and probably "malic" enzyme were involved in CO_2-fixation reactions in *E. granulosus*, C^{14} labelled CO_2 was found to be utilized by intact protoscoleces for the synthesis of Krebs' cycle intermediates, protein, lipids and phospholipids, nucleic acids and polysaccharides. Most of the C^{14} was fixed into succinic acid and excreted into the medium. Acetate-2-C^{14} was also found to be utilized for the synthesis of various cellular materials including polysaccharides.

These results, together with the labelling patterns from $C^{14}O_2$-fixation of the amino acids obtained from protein hydrolysates, as well as of the Krebs' cycle intermediates, were considered by Agosin and Repetto (1963) to indicate the presence of the Krebs' cycle in *E. granulosus*.

D. PROTEIN AND LIPID METABOLISM

No recent work appears to have been carried out on the protein and fat metabolism of *E. granulosus* and these are fields which clearly call for further study.

VII. Immunity

A. General Considerations

Few problems in helminth immunology have been so intensely studied as those relating to the immunology of hydatid disease. A large proportion of the work carried out, however, has been directed towards the development of reliable methods for serological diagnosis of the disease. Valuable as these studies have been, the intensity of their development may have been largely responsible for the fact that the more basic problems of immunology to this parasite have been somewhat neglected. The fact that the cyst contains a fluid—hydatid fluid—which is easily obtainable in quantity—has led perhaps to an unwarranted emphasis on the chemical and biological properties of this material.

The literature on serological diagnosis is particularly voluminous, much of it dealing with the efficiency of the Casoni or other diagnostic tests (see below). As this paper is concerned with the *biology* of the hydatid organism, rather than the clinical aspects of hydatid disease, recent work in serological diagnosis will be dealt with somewhat briefly. Recent valuable reviews, which deal mainly with serological diagnosis, have been those of Pérez (1960) and Kagan (1963).

B. Immunology of the Cystic Stage

1. *Natural Resistance*

Although *E. granulosus* and *E. multilocularis* can develop in an extensive range of intermediate hosts (Smyth and Smyth, 1964) the *degree* of development achieved varies greatly. The following patterns of development and resistance can be recognized; some of these have been dealt with in some detail by Webster and Cameron (1961) and Rausch and Schiller (1956):

(i) Complete resistance; invasion of host tissue by an oncosphere is never achieved.

(ii) Invasion without establishment; the host tissue reaction is severe and the parasite is contained within fibrous tissue, forming a pseudotubercle or granuloma.

(iii) Invasion with slow development; inhibition due to host tissue reaction, usually results in infertile cysts.

(iv) Invasion with rapid development of fertile cysts.

As discussed earlier, the experimental evidence suggests that the degree of development attained is dependent on the strain of both host (see Section III, A) and parasite.

Cameron and Webster (1961) and Webster and Cameron (1961) have shown that the strain termed *E. granulosus canadensis* which develops naturally in the lungs of deer will not develop in cattle or sheep. This strain when it infects man similarly shows a predilection for the lungs. Earlier workers too, were aware of physiological differences of protoscoleces from different hosts. Thus, de Waele and de Cooman (1938), using material of horse origin, successfully infected white mice and white rats; rabbits were negative. On the other hand, Dévé (1946) found that with material of sheep origin, rabbits and white mice were receptive and rats refractory. A possible genetical explanation for the occurrence of such "strains" has been given earlier (Section II, A).

2. *Serological Diagnosis*

a. General comments. There is evidence to show that antibodies may be detected at a very early stage of an infection with larval *Echinococcus*. Thus, Rausch and Schiller (1956) found that antibodies could be detected in a young goat in which the liver showed signs of invasion but in which a cyst never became established. Bacigalupo and Franzani (1936) had earlier shown that intravenous injection of hydatid fluid into rabbits induced antibody production after 5 days.

Many workers in this field have stressed the difficulties of obtaining reliable, reproducible results with the diagnostic methods now in use. There are conflicting reports of the efficiency of various tests and antigen extraction methods (Garabedian *et al.*, 1959). Crude hydatid fluid, filtered or unfiltered extracts of protoscoleces, or cyst walls, or fractions of these have been used as antigens by various workers, but the procedures whereby these antigens have been utilized have varied greatly.

In addition to the well-known Casoni reaction (CR) and the complement-fixation test (CF), the last ten years or so have seen the introduction of two other methods—the indirect haemagglutination (HA) and the bentonite flocculation test (BF).

b. The Casoni reaction (CR). According to Meltzer *et al.* (1956) the sensitivity of the skin test developed by Casoni varies from 53·8 to 100%. A number of other workers have reported on various aspects of this test and give different opinions regarding its reliability (Magath, 1959; Bensted and Atkinson, 1953; Boccheti and Begani, 1952; Bulgakov, 1958; Lass, 1951; Pautrizel and Gosman, 1953; Sannikov, 1961). In some countries where the disease is not normally seen (e.g. the U.S.A.) difficulties arise in procuring a suitable antigen. Hydatid fluid does not retain its antigenic property for more than a few weeks although Magath (1959) points out that if passed through a Seitz filter and sealed in glass ampoules it remains suitable for the Casoni test for months or even several years. Magath has found that the addition of thimerosal (Merthiolate) in a concentration of 1:50,000 preserves the antigenic properties almost indefinitely (i.e. more than 25 years).

Several workers (Bocchetti and Begani, 1952; Pautrizel and Gosman, 1953) have shown that after injection of the antigen in the Casoni reaction the patient develops a marked eosinophilia and that this increase of eosinophiles taken together with the Casoni reaction gives a reliable diagnostic method. Bulgakov (1958), however, did not find the occurrence of eosinophilia of great diagnostic value.

c. The Complement Fixation test (CF). Meltzer *et al.* (1956) found that the CF test has a sensitivity of only 52·4% in proven cases but other workers have given lower or higher figures and have found the test giving positive results with other infections (Kagan *et al.*, 1959; Bensted and Atkinson, 1958; Jezioranska and Dobrowolska, 1956; Magath, 1959; Knierim, 1959; Knierim and Niedmann, 1961; Garabedian *et al.*, 1959). There seems to be general agreement among these workers that both this test and the Casoni are not as reliable as the alternative tests now available. Bensted and Atkinson (1958) have shown that false positive results may be given by patients previously immunized with vaccines, such as anti-rabies vaccine, containing the same or related host protein.

d. The Indirect Haemagglutination test (HA). The application of the indirect haemagglutination method of Boyden (1951) to the serology of hydatid disease has apparently proved most successful. The procedure involves the use of sheep red cells sensitized with tannic acid (1:20,000) and coated with hydatid fluid antigen. Garabedian *et al.* (1959) compared the diagnostic efficiency of this procedure with the CR and CF tests and obtained the following results: CR 88·6%; CF 77·2%; HA 87·3%. The percentages of false positives however, were: CR 18·1%; CF 5·9%; HA 0%; which revealed the specificity of the new method.

Several authors report somewhat comparable results with this method although false positives were found in some cases (Kagan *et al.*, 1959; Addis and Mandras, 1958). For a critical analysis of some of the technical problems involved, see Kagan (1963). This worker has shown that this test has such a high level of reactivity, that by adopting as criteria of a positive test a haemagglutination titre of 1:400 or higher and a positive flocculation test, results which are positive for hydatid disease can be recognized and those which are non-specific (e.g. due to various collagen diseases) can be distinguished.

e. The Bentonite Flocculation test. Norman *et al.* (1959) adapted the bentonite flocculation test described by Bozicevich *et al.* (1951) for trichinosis and like the HA test found it superior to the CF test. Kagan *et al.* (1959) found this test gave a 90% sensitivity compared with 96% for the HA test and 36·3% for CF.

3. *Larval Antigens*

Analysis and attempted isolation of the antigens, involving the immunological responses mentioned above, have received much attention in recent years. Pérez (1960) and Kent (1963) and other workers emphasize that the term "antigen" has been frequently used loosely to represent heterogenous and ill-defined mixtures and recommend the use of the plural term "antigens".

This fact is now widely recognized and increasingly precise methods for analysing the antigen mosaic are being used. These include chemical fractionation and associated methods (dialysis, gel filtration, various forms of chromatography) gel-diffusion, paper electrophoresis, agar-gel-electrophoresis (Kent, 1957, 1963; Kagan, 1961, 1963; Kagan and Norman, 1961; Bono and Pellegrini, 1956; Jimenez-Millan, 1960; Čmelik, 1952b; Kagan *et al.*, 1960).

One of the most useful of these techniques has been the double-diffusion technique of Oakley and Fullthorpe (1953) and the plate method of Ouchterlony (1958). Kagan (1961), Kagan and Norman (1961) and Kagan (1963) give an excellent account of the application of these techniques to *Echinococcus* material. The most striking result obtained by these workers, has been the demonstration of the paucity of parasite material in the hydatid antigens used. They found that of twenty-three different antigen-antibody components followed in *E. granulosus* antigens, four were of parasite origin, six of host origin, and thirteen bands were of undetermined origin. *E. multilocularis* antigens showed twenty-seven bands which could be followed. Four of these were of parasite origin, seven of host origin and sixteen could not be identified. It was also noted that although protoscoleces antigens were composed almost entirely of parasite antigenic components, these were not as reactive in serologic tests as those in the hydatid fluid although the latter was shown to contain only a small number of parasite components. Kagan (1963) speculates that metabolic products produced by the protoscoleces accumulate in the hydatid fluid and may be responsible for the serologic activity of the antigen.

Related to serologic diagnosis have been studies on the serum proteins of animals infected with hydatid disease which have been carried out by a number of workers (Goodchild and Kagan, 1961; Biondo and Beninati, 1954; Congiu and Pirlo, 1956). Detailed consideration of work in this field falls outside the scope of this review.

C. VACCINATION

Surprisingly little work appears to have been done during the last decade in this field and there appears to have been few attempts to develop vaccines against the cystic stage in man and other intermediate hosts or against the adult stages in the dog.

Early workers (Turner *et al.*, 1935, 1936) had claimed that, using antigens prepared from protoscoleces or cyst material, they were able to establish partial immunity in dogs against intestinal infection with *E. granulosus*. They emphasized, however, that some dogs seemed to have a natural resistance to infection, a fact well recognized by other workers. The basis of this natural resistance is not known, but purely physiological factors such as the variation in composition of bile during metabolic disturbance in the host could be responsible in some cases. It is more likely, however, to be immunological in origin.

Forsek and Rukavina (1959) similarly vaccinated dogs with antigens prepared from sheep, pigs and cattle using hydatid fluid, extract of protoscoleces and extract of germinal and cuticular membrane. Thirty dogs were infected with fertile hydatid cysts 60–90 days after infection. Ten dogs were used as controls. All dogs were bled 45–160 days after infection. The authors claimed that (a) the immunized dogs had much smaller worm burdens than the controls, (b) the worm burden diminished with longer intervals between infection and bleeding, and (c) the antibody titre in the serum was in constant proportion to that of the antigen tested and diminished with time.

The use of *larval* antigens against *adult* material would appear to show a lack of understanding of the biological problems involved. We must ask the question—can a circulating antibody act against a parasite established in the gut? To be effective it is likely that the host reaction must be active against either (a) the mode of attachment or (b) the metabolism of the whole worm. It is difficult to visualize an efficient mechanism for achieving the latter since strobila will be partially free in the gut, coated in mucus and bathed in semi-digested food and enzymes. It is likely then that immune reactions must be directed against the mode of attachment.

It has been shown (Smyth, 1964) that attachment to the gut is remarkably intimate, the rostellum forcing itself deep into the crypts of Lieberkühn with the hooks lightly penetrating the crypt epithelium and each sucker firmly taking a plug of mucosa within its cavity. The scolex has also been shown to contain a rostellar gland (Fig. 1) which secretes actively through the cuticle (Smyth, 1964). On theoretical grounds then, the *adult* scolex should provide better antigenic materials than the *larval* protoscolex, in which these glands are not developed.

It is interesting to note therefore, the results of Gemmell (1962) with *E. granulosus*. This worker vaccinated thirteen dogs with powdered, frozen-dried adult worms; there were nine controls. Except for one dog, in which no infection was established, some worms developed in all dogs. In control dogs, there was a rapid growth of the subterminal and terminal segments of *E. granulosus* between the 35th and 49th days. In vaccinated dogs almost all the subterminal and terminal segments failed to undergo this period of rapid growth and this worker suggested that the more important antigen-antibody reactions may be associated with inhibition of growth. It is worthwhile noting in all three dogs vaccinated with adult tapeworms this inhibition to growth was not overcome.

According to Smyth (1963b), the cells of the rostellar gland have not been seen to secrete from the scolex earlier than 32 days of development. Thus, it can be speculated that the appearance of the scolex secretion may be related to the accelerated growth phase of the worms and that this secretion may play a major part as antigenic material in the case of adult worms.

Gemmell's results are suggestive that a functional antigen is present in *adult* tapeworms but that it is not completely effective. The freeze-drying treatment used by this worker may have greatly reduced the activity of any antigenic material and fresh material may be more effective.

It is evident that, on the theoretical grounds, the scolex secretion could be expected to have antigenic properties and its nature and possible extraction and use as an antigen clearly warrant further investigation.

VIII. *In Vitro* Cultivation

There are many aspects of the biology of the hydatid organism, such as those relating to metabolism or nutrition, for example, which can only be solved satisfactorily by *in vitro* studies. Yet, strangely enough, since Coutelen (1927a) carried out some exploratory experiments, until recently, there have been no further studies in this field. Rausch and Jentoft (1957) described some preliminary experiments on *E. multilocularis*. Unfortunately, these workers provided somewhat incomplete details of the culture techniques used; the precise data on volume and composition of media and type of culture vessels not being given. A basic medium of 40% ascitic fluid in Hank's saline plus antibiotics supplemented with other nutrients, from natural sources, such as vole embryo extract was used. Some cultures were carried out in the presence of HeLa cells. Undifferentiated germinal membrane, cut from the cyst wall, was used as culture material. In the best results obtained the tissue proliferated and produced vesicles by the 29th day. In some cultures, the subgerminal membrane was present and by about 55 days up to twenty scolices were present in some vesicles. The maximum time a culture was maintained was 134 days.

In vitro cultivation of the protoscoleces of *E. multilocularis* and *E. granulosus* has recently been attempted by a number of workers: Yamashita *et al.* (1962), Smyth (1962c) and Webster and Cameron (1963). Yamashita *et al.* worked exclusively on *E. multilocularis* in Japan and Smyth exclusively on *E. granulosus* in Australia, but the results of these two groups of workers were almost identical. Yamashita *et al.* used a basic medium of 0·5% lactalbumin hydrolysate in Hank's saline reinforced with bovine serum, bovine bile and liver extract, and treated brood capsules with trypsin, before cultivation, to free the protoscoleces from the germinal membrane.

Smyth used a variety of natural media (hydatid fluid, bovine serum, bovine amniotic fluid), embryo extracts (chick and bovine) and synthetic media (Parker 199) and combinations of these as culture media; pepsin was used to free protoscoleces from the germinal membranes. Both workers found that protoscoleces became vesicular under the culture conditions provided and after a time secreted a laminated membrane. They showed that vesiculization could take place by two routes either (a) by the protoscolex itself becoming vesicular and secreting a laminated membrane, or (b) by forming a posterior "vacuole." Yamashita *et al.* described the formation of germinal cells, similar to the early stage of a brood capsule, in some cysts; Smyth, similarly reported the formation of "cells or clusters of cells within the cysts".

It is clear that in both these series of experiments the culture conditions provided approach those required for the development of the vesicular larval stage but possibly some cultural condition was abnormal or some growth factor lacking for in neither case were protoscoleces actually formed. It is particularly interesting to see the development of a laminated membrane in the experiments of Yamashita *et al.*, as no such membrane was formed in the experiments described by Rausch and Jentoft (1957) mentioned above, and Smyth (1962c)—before Yamashita's interesting results had become available—speculated that the fact that *E. granulosus* formed a laminated membrane *in vitro* might be a basic difference between *E. granulosus* and *E. multilocularis*. Although the laminated membrane of *E. multilocularis* does not develop *in vivo* to anything like the extent that it does in *E. granulosus*, it is clear that under the conditions provided *in vitro* there is little difference in the mode of formation of this membrane. It must be remembered that the protoscoleces of *Echinococcus* are essentially embryos—although somewhat exceptional ones—for they can potentially develop in either of two directions: (a) they can become vesicular and form a new cyst (as *in vitro* or in secondary hydatiasis), or (b) they can develop into adult tapeworms if taken into the dog's gut. The factors which stimulate and control development of the embryonic tissue in either of these directions it is not known, but the elucidation of this problem is likely to provide a fascinating chapter in the study of the biology of this organism. It is basically a problem in experimental embryology.

Attempts by the writer in this laboratory to stimulate development in the direction of forming a strobila have been unsuccessful to date. The elegant continuous flow method of Berntzen (1961), which has proved so strikingly successful for *Hymenolepis diminuta*, has been completely unsuccessful (at least in our hands) for *E. granulosus* and the evidence suggests that this organism may require some unusual culture conditions in order for development to occur.

In this respect, Webster and Cameron (1963) have had some encouraging results mainly with *E. multilocularis* but also with *E. granulosus*. Using an extensive range of some forty-six media, both vesicular and some degree of strobilate development were obtained. Small intact *E. multilocularis* cysts were found to produce new vesicles in culture in Morgan's medium, M150, proliferations beginning on the 4th day. Numerous vesicles usually developed but only in one instance did protoscoleces develop.

When protoscoleces were cultured *in vitro* in most media, vesiculation began and followed a course close to that described above by Yamashita *et al.* (1962) and Smyth (1962c). In certain complex media, made up mainly of synthetic materials but containing some natural media, larva became segmented forming two segments at the end of 14 days and three at the end of 3 weeks. At this stage the worms had attained a length of 0·55 mm. No further development occurred however, and worms remained in this state for a further 4 weeks before ultimately degenerating. Somewhat comparable results were obtained with *E. granulosus*. These workers concluded that more than 5% serum in any medium induced vesiculation and that a slightly acid pH (6·8) was necessary for development in the strobilate direction. Although development beyond the stage of segmentation has not been obtained, to have reached this stage is, in itself, a major achievement and these interesting results have provided valuable pointers to the direction in which further studies in this field might take.

Appendix

Table IV

Intermediate Host List

List of animals examined as possible intermediate hosts of *E. granulosus* and *E. multilocularis*. Only recent references are given. Data from Smyth and Smyth (1964).

* = species identification uncertain;
0 = no reliable information available;
1 = experimental infection;
2 = secondary infection (intraperitoneal inoculation);
3 — larva invades tissue but has limited survival;
4 = generally slow development;
5 = acephalic cysts;
+ = positive; − = negative.

(a) Host (as named by worker)	(b) *E. granulosus*	(c) *E. multilocularis*	(d) Reference
Infra class—Metatheria			
Order—Marsupialia			
Family Phalangeridae			
Trichosurus vulpecula (Aust. brush-tailed opossum)	+1	0	(b) Sweatman and Williams (1962)
Family Macropodidae			
Macropus sp.			
Wallaby	+	0	(b) Dew (1953)
Small scrub wallaby	+	0	(b) Dew (1953)

Table IV *(continued)*

(a)	(b)	(c)	(d)
Protemnodon rufogrisea (Red-necked wallaby)	+1	0	(b) Sweatman and Williams (1962)
Infra class—Eutheria			
Order—Insectivora			
Family Erinaceidae			
Erinaceus europaeus (European hedgehog)	–1	0	(b) Sweatman and Williams (1962)
Family Soricidae			
Sorex jacksonii (Shrew)	0	+	(c) Thomas *et al.* (1954)
S. tundrensis (Tundra shrew)	0	+	(c) Rausch (1958)
Order—Primates			
Sub-order—Prosimii			
Family Lemuridae			
Lemur mongoz	+	0	(b) Ratcliffe (1942)
Sub-order—Anthropoidea			
Family Circopithecidae			
Macaca sinica (Ceylon toque monkey)	+	0	(b) Dissanaike (1958)
M. sylvana (Barbary ape)	+	0	(b) Urbain (1932)
M. mulatta	+	0	(b) Allen (1957)
M. philippinensis	+	0	(b) Allen (1957)
Macaca sp.	0	+4	(c) Rausch and Schiller (1956)
Cynopithecus niger (Celebes ape)	+	0	(b) Waworoentoe and Mansjoer (1950)
Cynocephalus porcarius (S. African baboon)	+	0	(b) Heller (1923)
Family Hominidae			
Homo sapiens	+	+	(b) Yamashita (1960) (c) Yamashita (1960)
Order—Lagomorpha			
Family Leporidae			
Oryctolagus cuniculus (Rabbit)	–; –1	–1	(b) Clunies Ross (1929) (c) Sadun *et al.* (1957)

TABLE IV *(continued)*

(a)	(b)	(c)	(d)
Oryctolagus cuniculus (Rabbit)	+*	–3	(b) Arcos Porras (1951) (c) Rausch and Schiller (1956)
Oryctolagus cuniculus (Rabbit)	–1; +2	–2	(b) Dévé (1949) (c) Lubinsky (1960b)
Oryctolagus cuniculus (Rabbit)	+1	0	(b) Sweatman and Williams (1962)
Order—Rodentia			
Sub-order—Sciuromorpha			
Family Sciuridae			
Tamiasciurus hudsonicus (Red squirrel)	–1	–3	(b) Rausch and Schiller (1956) (c) Rausch and Schiller (1956)
Marmota marmota caligata (Hoary marmot)	–1	–1	(b) Rausch and Schiller (1956) (c) Rausch and Schiller (1956)
Citellus undulatus lyratus (Ground squirrel)	0	+	(c) Thomas *et al.* (1954)
C. undulatus ablusus	–1	–3	(b) Rausch and Schiller (1956) (c) Rausch and Schiller (1956)
Eutamias asiaticus (Asiatic chipmunk)	0	+1	(c) Yamashita *et al.* (1958a)
Glaucomys sabrinus (Flying squirrel)	–1	–3	(b) Rausch and Schiller (1956) (c) Rausch and Schiller (1956)
Family Geomyidae			
Thomomys bottae (Western pocket gopher)	0	–1	(c) Rausch and Schiller (1956)
Family Heteromyidae			
Dipodomys sp. (Kangaroo rat)	0	–1	(c) Rausch and Schiller (1956)
Sub-order—Myomorpha			
Family Cricetidae			
Reithrodontomys megalotis (Harvest mouse)	0	–1	(c) Rausch and Schiller (1956)
Peromyscus gossypinus (Cotton mouse)	0	+1	(c) Sadun *et al.* (1957)
P. maniculatus (Deer mouse)	–1	+1	(b) Hutchison and Bryan (1960) (c) Sadun *et al.* (1957)
P. boylii	–1	–1	(b) Rausch and Schiller (1956) (c) Rausch and Schiller (1956)
P. truei	–1	+1	(b) Rausch and Schiller (1956) (c) Rausch and Schiller (1956)

TABLE IV *(continued)*

(a)	(b)	(c)	(d)
P. leucopus	–; –1	0	(b) Gibbs (1957)
Sigmodon hispidus (Cotton rat)	+1	+1; +2	(b) Webster and Cameron (1961) (c) Lukashenko and Zorikhina (1961)
Cricetus barabensis	0	+5	(c) Lukashenko and Zorikhina (1961)
Mesocricetus auratus	–1	–3	(b) Rausch and Schiller (1956) (c) Rausch and Schiller (1956)
Mesocricetus auratus (Albino hamster)	0	–3	(c) Yamashita *et al.* (1958a)
Golden hamster	0	+2, 4	(c) Cameron (1960a)
Dicrostonyx torquatus pallus (Varying lemming)	–1	–3	(b) Rausch and Schiller (1956) (c) Rausch and Schiller (1956)
Lemmus sibiricus (Brown lemming)	–1	+1	(b) Rausch and Schiller (1956) (c) Rausch and Schiller (1956)
Clethrionomys gapperi (Red-backed vole)	–1	0	(b) Gibbs (1957)
C. rutilus mikado	0	+1	(c) Yamashita *et al.* (1958a)
C. rutilus	0	+	(c) Rausch and Schiller (1956)
C. rutilus dawsonii	–1	+1	(b) Rausch and Schiller (1956) (c) Rausch and Schiller (1956)
C. rufocannus bedfordiae	0	+1	(c) Yamashita *et al.* (1958a)
Ondatra zibethica (Musk rat)	0	+	(c) Lukashenko and Zorikhina (1961)
Odatra sp.	0	+1	(c) Rausch and Schiller (1956)
Microtus oeconomus (Tundra vole)	0	+	(c) Lukashenko and Zorikhina (1961)
M. oeconomus ratticeps	0	+1	(c) Vogel (1955b)
M. oeconomus inuitus	0	+	(c) Rausch and Schiller (1951) (See Vogel, 1955a)
M. pennsylvanicus (Field vole)	–1	+1	(b) Gibbs (1957) (c) Rausch and Schiller (1956)
M. agrestis	0	+	(c) Lukashenko and Zorikhina (1961)
M. arvalis	0	+1; +	(c) Vogel (1955a, b)
M. gredalis	0	+	(c) Lukashenko and Zorikhina (1961)
M. montebelli montebelli	0	+1	(c) Yamashita *et al.* (1958a)
M. californicus	–1	+1	(b) Rausch and Schiller (1956) (c) Rausch and Schiller (1956)
Lagurus lagurus	0	+1	(c) Lukashenko (1960)

TABLE IV *(continued)*

(a)	(b)	(c)	(d)
Meriones unguiculatus (Mongolian gerbil)	0	+1	(c) Yamashita *et al.* (1958a)
Meriones sp.	0	+2	(c) Lubinsky (1960b)
Rhombomys opimus	0	+	(c) Chun-Syun and Alekseev (1960)
Family Muridae			
Apodemus geisha	0	+1	(c) Yamashita *et al.* (1958a)
A. agrarius	0	+	(c) Merkusheva (1958)
A. speciosus ainu	0	–1	(c) Yamashita *et al.* (1958a)
Rattus sp. (White rat)	–1	+4	(b) Hutchison and Bryan (1960) (c) Vogel (1955a)
Rattus norvegicus	0	+	(c) Lukashenko and Zorikhina (1961)
R. exulani (Polynesian rat)	–1	0	(b) Sweatman and Williams (1962)
Mus musculus (Lab. mouse)	+5	+1; +4	(b) Yamashita *et al.* (1956) (c) Yamashita *et al.* (1958a)
Mus musculus (Lab. mouse)	+2	+2	(b) Dévé (1933) (c) Lukashenko (1961)
Mus musculus (Lab. mouse)	0	+1	(c) Lukashenko (1960)
Sub-order—Hystricomorpha			
Family Erethizontidae			
Erethizon dorsatum myops (Porcupine)	0	–1	(c) Rausch and Schiller (1956)
Family Caviidae			
Cavia sp. (Guinea pig)	0	–1	(c) Lukashenko (1960)
Cavia sp. (Guinea pig)	0	–2	(c) Lubinsky (1960b)
Microcavia australis	+*	0	(b) La Barrera (1948)
Family Dasyproctidae			
Dasyprocta agouti	+*	0	(b) Brumpt and Joyeux (1924) (Cameron and Webster, 1961)
Family Chincillidae			
Chincilla sp.	–3	0	(b) Cameron (1960b)
Family Capromyidae			
Myocastor coypus (Swamp beaver)	0	+	(c) Schulte (1950)
Family Octodontidae			
Octodon degus degus (Bush rat)	–; +1	0	(b) Tagle *et al.* (1956)

TABLE IV *(continued)*

(a)	(b)	(c)	(d)
Order—Perissodactyla			
Sub-order—Hippomorpha			
Family Equidae			
Equus sp. (Horse)	+	–3	(b) Corsalini (1959) (c) Rausch and Schiller (1956)
(Mule)	+	0	(b) Gallo (1955)
(Donkey)	+	0	(b) Gallo (1955)
Order—Artiodactyla			
Sub-order—Suiformes			
Family Suidae			
Sus sp. (Domestic pig)	+	–1	(b) Gemmell and Brydon (1960) (c) Rausch and Schiller (1956)
Sus sp. (Domestic pig)	+1	0	(b) Hutchison and Bryan (1960)
Sub-order—Tylopoda			
Family Camelidae			
Lama glama pacos (Alpaca)	+	0	(b) Román (1956)
	+	0	(b) Santivanex and Cuba (1949)
Camelus dromedarius (Camel)	+	0	(b) Halawani (1956)
Sub-order—Ruminantia			
Family Cervidae			
Cervus elaphus (Red deer)	+	0	(b) Jansen (1961)
C. canadensis nelsonii (Elk, Wapiti)	+	0	(b) Khaziev (1956)
Rusa unicolor unicolor (Sambhur)	+	0	(b) Paramananthan and Dissanaike (1961)
Odocoileus sp. (White tailed deer)	+	0	(b) Harper *et al.* (1955)
Odocoileus sp. (Mule deer)	+	0	(b) Cowan (1951)
Odocoileus sp. (Black tailed deer)	+	0	(b) Magath (1954)
Odocoileus sp. (Coast deer)	+	0	(b) Miller (1953)
Mazama simplicornis	+	0	(b) Nájera (1950)
Alces americana (American moose)	+	0	(b) Cameron (1960b)
Rangifer articus (Caribou)	+	0	(b) Rausch and Williamson (1959)

TABLE IV *(continued)*

(a)	(b)	(c)	(d)
Rangifer tarangulus (Reindeer)	+	–	(b) Cameron (1960b) (c) Rausch and Schiller (1956)
Family Bovidae			
Bos sp. (Domestic cattle)	+	+*	(b) Gemmell and Brydon (1960) (c) Boko and Gavez (1958)
Bison sp. (Bison)	+	0	(b) Miller (1953)
Bison sp. (Buffalo)	+	0	(b) Lubinsky (1959b)
Oryx sp. (Gemsbok)	+	0	(b) Verster (1961)
Antilope sp. (Antelope)	–	0	(b) Harper *et al.* (1955)
Antilope sp. (Antelope)	+	0	(b) Cameron and Webster (1961)
Oreamnos americanus	+	0	(b) Rausch and Williamson (1959)
Gorgon taurinus (Wildebeest)	+	0	(b) Nelson and Rausch (1963)
Capra sp. (Domestic goat)	+	–1	(b) Verster (1961) (c) Rausch and Schiller (1956)
Ovis sp. (Domestic sheep)	+	+	(b) Gemmell and Brydon (1960) (c) Vibe (1959)
Ovis sp. (Domestic sheep)	+1	0	(b) Yamashita *et al.* (1957)
Order—Carnivora			
Canis familiaris (Dog)	very rare { +	0	(b) Whitten and Shortridge (1961)
Felis domestica (Cat)	very rare { +	0	(b) Whitten and Shortridge (1961)

REFERENCES

Addis S. and Mandras, A. (1958). *Igiene mod.* **51**, 793–800.
Agosin, M. (1959). *Biologica, Santiago* **27**, 3–32.
Agosin, M., von Brand, T., Rivera, G. F. and McMahon, P. (1957). *Exp. Parasit.* **6**, 37–51.
Agosin, M. and Aravena, L. (1959). *Biochim. biophys. Acta* **34**, 90–102.
Agosin, M. and Aravena, L. (1960a). *Enzymologia* **22**, 281–94.
Agosin, M. and Aravena, L. (1960b). *Exp. Parasit.* **10**, 28–38.
Agosin, M. and Repetto, Y. (1963). *Comp. Biochem. Physiol.* **8**, 245–61.
Allen, A. M. (1957). *A. M. A. Arch. Path.* **64**, 148–51.
Ambo, H., Ichikawa, K., Iida, H. and Abe, N. (1954). *Spec. Rep. Hokkaido Inst. Publ. Hlth.* **4**, 1–19.
Arcos Porras, P. de. (1951). *Medicamenta, Madr.* **9**, 29.
Bacigalupo, J. and Franzani, O. F. (1936). *G. Batt. Immun.* **16**, 872–75.

Bacigalupo, J. and Rivero, E. (1949). *J. Parasit.* **35**, 547.
Batham, E. J. (1957). *N. Z. vet. J.* (Sept.) 74–76.
Bensted, H. J. and Atkinson, J. D. (1953). *Lancet* **1**, 265–68.
Bensted, H. J. and Atkinson, J. D. (1958). *Brit. med. J.* **2**, 203–205.
Berberian, D. A. (1957). *Tenth Annual Report of the Orient Hospital Beirut*, pp. 33–34.
Berntzen, A. K. (1961). *J. Parasit.* **47**, 351–55.
Biondo, G. and Beninati, F. (1954). *Ann. Fac. Med. vet. Torino* **4**, 209–12.
Bocchetti, G. and Begani, R. (1952). *Nuovi Ann. Igiene Microbiol.* **3**, 234–44.
Boko, F. and Gavez, E. (1958). *Vet. Glasn.* **12**, 259–65.
Bono, G. del and Pellegrini, S. (1956). *Acta med. vet.* **11**, 1–7.
Boyden, S. V. (1951). *J. exp. Med.* **93**, 107–20.
Bozicevich, J., Tobie, J. E., Thomas, E. H., Hoyem, H. M. and Ward, S. P. (1951). *Publ. Hlth Rep., Wash.* **66**, 806–14.
Bueding, E. (1949). *J. exp. Med.* **89**, 107.
Bueding, E. and Charms, B. (1951). *Nature, Lond.* **167**, 149.
Bulgakov, V. I. (1958). *Med. Parazitol. Parazit. Bol. Moscow* **27**, 152–57.
Cameron, G. L. and Staveley, J. M. (1957). *Nature, Lond.* **179**, 147–48.
Cameron, T. W. M. (1926). *J. Helminth.* **4**, 13–22.
Cameron, T. W. M. (1960a). *Parassitologia* **2**, 371–80.
Cameron, T. W. M. (1960b). *Parassitologia* **2**, 381–90.
Cameron, T. W. M. and Webster, G. A. (1961). *In* "Studies in Disease Ecology", Vol. 2, pp. 141–160. American Geographic Society, New York.
Carbone, G. and Lorenzetti, L. (1957). *Rass. med. Sarda* **59**, 519–39.
Carta A. and Deiana, S. (1960). *Parassitologia* **2**, 1–6.
Choquette, L. P. E. (1956). *Canad. J. Zool.* **34**, 190–92.
Chun-Syun, F. and Alekseev, V. K. (1960). *Med. Parasit., Moscow* **29**, 482.
Clunies-Ross, I. (1929). *Bull Counc. sci. industr. Res. Aust.* 40.
Clunies-Ross, I. (1936). *J. Counc. sci. industr. Res. Aust.* **9**, 67–68.
Čmelik, S. (1952a). *Hoppe-Seyl. Z.* **289**, 78.
Čmelik, S. (1952b). *Biochem. Z.* **322**, 456– 62.
Čmelik, S. and Briski, B. (1953). *Biochem. Z.* **324**, 104–14.
Congiu, M. and Pirlo, F. (1956). *Arch. Ital. Sci. Trip. Parassit.* **37**, 428–33.
Corsalini, T. (1959). *Vet. ital.* **10**, 644–47.
Coutelen, F. (1927a). *Ann. Parasit. hum. comp.* **5**, 1–19.
Coutelen, F. (1927b). *Ann. Parasit. hum. comp.* **5**, 239–42.
Coutelen, F. (1938). *C. R. Soc. Biol., Paris* **129**, 149–51.
Coutelen, F., Biguet, J., Doby, J. M. and Deblock, St. (1952). *Ann. Parasit. hum. comp.* **27**, 86–104.
Cowan, I. McT. (1951). *Proc. 5th Ann. B. C. Game Conv.* pp. 39–64.
Dévé, F. (1933). *C. R. Soc. Biol., Paris* **113**, 223–34.
Dévé, F. (1946). "L'echinococcose secondaire". Masson et Cie et Paul Duval, Paris.
Dévé, F. (1949). "L'echinococcose primitive (Maladie hydatique)." Masson, Paris.
Dew, H. R. (1922). *Med. J. Aust.* **2**, 381–84.
Dew, H. R. (1953). *Arch. int. Hidatid.* **13**, 319–32.
Dissanaike, A. S. (1958). *Ceylon vet. J. Sept.-Dec.* 1–3.
Dissanaike, A. S. and Paramananthan, D. C. (1960). *Ceylon vet. J.* **8**, 82–87.
Drežančić, I. and Wikerhauser, T. (1956). *Vet. Arhiv* **26**, 179–82.
Durie, P. H. and Riek, R. F. (1952). *Aust. vet. J.* **28**, 249–54.
Entner, N. and González, C. (1959). *Expt. Parasit.* **8**, 471–79.
Euzéby, J. (1960). *Rev. Hyg. Méd. soc.* **8**, 428–38.
Euzéby, J. (1962). *Rev. Méd. vet.* **113**, 111–28.
Farhan, I., Schwabe, C. W. and Zobel, C. R. (1959). *Amer. J. trop. Med. Hyg.* **8**, 473–78.

Forsek, Z. and Rukavina, J. (1959). *Vet., Sarajevo* **8**, 479–82.
Freeman, R. S., Adorjan, A. and Pimlott, D. H. (1961). *Canad. J. Zool.* **39**, 527–32.
Gallo, C. (1955). *Atti Soc. ital. Sci. vet.* **9**, 689–91.
Garabedian, G. A., Matossian, R. M. and Suidan, F. G. (1959). *Amer. J. trop. Med. Hyg.* **8**, 67–71.
Gary, N. D., Kupferberg, L. L. and Graf, L. H. (1958). *J. Bact.* **76**, 359–64.
Gelormini, N. (1941). *Boln. Fac. Agron. Vet. Univ. B. Aires* **2**, 15–24.
Gemmell, M. A. (1957a). *Aust. vet. J.* **33**, 8–14.
Gemmell, M. A. (1957b). *Yearb. Insp. Stk. N. S. W.* 81–84.
Gemmell, M. A. (1959a). *Aust. vet. J.* **35**, 450–55.
Gemmell, M. A. (1959b). *Bull. Wld Hlth Org.* **20**, 87–99.
Gemmell, M. A. (1960). *Helminth. Abstr*, **29**, 1–15.
Gemmell, M. A. (1962). *Immunology* **5**, 496–503.
Gemmell, M. A. and Brydon, P. (1960). *Aust. vet. J.* **36**, 73–78.
Gibbs, H. C. (1957). *Canad. J. comp. Med.* **21**, 287–89.
Goodchild, C. G. and Kagan, I. G. (1961). *J. Parasit.* **47**, 175–80.
Halawani, A. (1956). *Arch. int. Hidatid.* **15**, 374–75.
Harper, T. A., Ruttan, R. A. and Benson, W. A. (1955). *Trans. N. Amer. Wildl. Conf.* 20th, 198–207.
Heller, E. B. (1923). *Int. Clin.* **4**, 253–98.
Hercus, C. E., Williams, R. J. Gemmell, M. A. and Parnell, I. W. (1962). *Vet. Rec.* **74**, 1575.
Hinshaw, H. C. (1937). *Proc. Mayo Clin.* **12**, 422–23.
Hutchison, W. F. and Bryan, M. W. (1960). *Amer. J. trop. Med. Hyg.* **9**, 606–11.
Hsü-Li. (1941). *Peking Nat. Hist. Bull.* **15**, 201–203.
Inukai, T. ,Yamashita, J. and Mori, H. (1955). *J. Fac. Agric. Hokkaido Univ.* **50**, 134–39.
Jezioranska, A. and Dobrowolska, H. (1956). *Wiad. parazyt.* **2**, 119–20.
Jimenez-Millan, F. (1960). *Rev. esp. Fisiol.* **16**, 23–26.
Jansen, J. (1961). *Tijdschr. Diergeneesk.* **86**, 82–84.
Jore d'Arces, P. (1953). *Bull. Off. int. Epiz.* **40**, 45–51.
Kagan, I. G. (1961). *Proc. Helm. Soc. Wash.* **28**, 97–102.
Kagan, I. G. (1963). *Exp. Parasit.* **13**, 57–71.
Kagan, I. G. and Norman, L. (1961). *Amer. J. trop. Med. Hyg.* **10**, 727–34.
Kagan, I. G., Allain, D. S. and Norman, L. (1959). *Amer. J. trop. Med. Hyg.* **8**, 51–55.
Kagan, I. G., Norman, L. and Allain, D. S. (1960). *Amer. J. trop. Med.Hyg.* **9**, 248–61.
Kent, H. N. (1957). *In* "1st Symposium on Host Specificity among Parasites of Vertebrates." Paul Attinger, S. A., Neuchatel.
Kent, H. N. (1963). *Exp. Parasit.* **13**, 45–56.
Khaziev, G. Z. (1956). *Veterinariya* **33**, 47.
Kilejian, A., Schinazi, L. A. and Schwabe, C. W. (1961). *J. Parasit.* **47**, 181–88.
Kilejian, A., Sauer, K. and Schwabe, C. W. (1962). *Exp. Parasit.* **12**, 377–93.
Klemm, H. (1883). *München, Inaug.-Diss. von* 1883, 1–27.
Knierim, F. (1959). *Bol. Chileno Parasit.* **14**, 75–79.
Knierim, F. and Niedmann, G. (1961). *Bol. Chileno Parasit.* **16**, 6–9.
Krvavica, S., Martinčić, T. and Asaj, R. (1959). *Vet. Arhiv* **29**, 314–21.
La Barrera, J. M. de. (1948). *Arch. int. Hidatid.* **7**, 175–76.
Lass, N. (1951). *1st International Congress for Allergy, Zurich*, pp. 618–21.
Ley, J. de and Vercruysse, R. (1955). *Biochim. Biophys. Acta* **16**, 615–16.
Lubinsky, G. (1958). *Canad. J. Zool.* **36**, 883–87.
Lubinsky, G. (1959a). *Canad. J. Zool.* **37**, 793–801.
Lubinsky, G. (1959b). *Canad. J. Zool.* **37**, 83.
Lubinsky, G. (1960a). *Canad. J. Zool.* **38**, 605–12.

Lubinsky, G. (1960b). *Canad. J. Zool.* **38**, 1118–25.
Lukashenko, N. P. (1960). *Med. Parasit., Moscow* **29**, 154–57.
Lukashenko, N. P. (1961). Paper presented at Conference on Diseases in Countries with Hot Climate, Tashkent. pp. 1–6. Ministry of Health U.S.S.R.
Lukashenko, N. P. and Zorikhina, V. I. (1961). *Med. Parasit., Moscow* **30**, 159–68.
Magath, T. B. (1954). *J. Amer. vet. med. Ass.* **125**, 411–14.
Magath, T. B. (1959). *Amer. J. clin. Path.* **31**, 1–8.
Mamedov, M. M. (1960). *Med. Parasit., Moscow* **29**, 157–61.
Mankau, S. K. (1956a). *Trans. Amer. micr. Soc.* **74**, 401–406.
Mankau, S. K. (1956b). *Amer. J. trop. Med. Hyg.* **5**, 872–80.
Mankau, S. K. (1957). *J. Parasit.* **43**, 153–159.
Maplestone, P. A. (1933). *Indian med. Gaz.* **68**, 377–79.
Matoff, K. and Jantscheff, J. (1954). *Acta vet. hung.* **4**, 411–18.
Meltzer, H., Kovacs, L., Orford, T. and Matas, M. (1956) *Canad. med. Ass. J.* **75**, 121–**28**.
Mendheim, H. (1955). *Säugetierkundliche Mitteilungen* **3**, 10–12.
Merkusheva, I. V. (1958). *Dokladi Akademii Nauk. BSSR. Minsk* **2**, 134–35.
Meyers, H. F. (1955). *J. Parasit.* **41**, 1–4.
Meyers, H. F. (1957). *J. Parasit.* **43**, 322–23.
Meymarian, E. (1961). *Amer. J. trop. Med. Hyg.* **10**, 719–26.
Meymarian, E. and Schwabe, C. W. (1962). *Amer. J. trop. Med. Hyg.* **11**, 360–64.
Mikačić, D. (1956). *Vet. Arhiv* **26**, 218–24.
Miller, M. J. (1953). *Canad. med. Ass. J.* **68**, 423–34.
Nájera, L. E. (1950). *An. Med. Pública, Santa Fé* **2**, 571–93.
Nelson, G. S. and Rausch, R. L. (1963). *Ann. trop. Med. Parasit.* **57**, 136–49.
Norman, L., Sadun, E. H. and Allain, D. S. (1959). *Amer. J. trop. Med. Hyg.* **8**, 46–50.
Oakley, C. L. and Fullthorpe, A. J. (1953). *J. Path. Bact.* **65**, 49–60.
Ortlepp, R. J. (1937). *Onder. J. vet. sci.* **9**, 311–36.
Ouchterlony, O. (1958). *Ann. Allergy* **5**, 1–78.
Ovsyukova, N. I. (1961). *Med. Parasit., Moscow* **30**, 226.
Paramananthan, D. C. and Dissanaike, A. S. (1961). *Trans. R. Soc. trop. Med. Hyg.* **55**, 483.
Pautrizel, R. and Gosman, T. (1953). *Bull. Soc. Pat. exot.* **46**, 721–23.
Pérez del Castillo, C. (1960). *In* "Immuno Biologia de la Equinocococis." Montevideo: Solano Antuna 2908, Uruguay.
Posselt, A. (1936). *Proc. 3rd Int. Congr. comp. Path., Athens, Rep. Sect. Hum. Med.* 27–52.
Pozzi, G. and Pirosky, I. (1953). *Arch. int. Hidatid.* **13**, 332.
Ratcliffe, H. L. (1942). *Rep. Penrose Res. Lab.* 21.
Rausch, R. (1952). *Arctic* **5**, 157–74.
Rausch, R. (1953). *In* "Tharpar Commemoration Volume Lucknow" (J. Dayal and K. S. Singh, eds.), pp. 233–46.
Rausch, R. (1954). *J. infect. Dis.* **94**, 178–86.
Rausch, R. (1956). *Amer. J. trop. Med. Hyg.* **5**, 1086–92.
Rausch, R. (1958). *Proc. 6th int. Congr. trop. Med. Mal.* **2**, 597–610.
Rausch, R. and Jentoft, V. L. (1957). *J. Parasit.* **43**, 1–8.
Rausch, R. and Schiller, E. L. (1951). *Science* **113**, 57–58.
Rausch, R. and Schiller, E. L. (1954). *J. Parasit.* **40**, 659–62.
Rausch, R. and Schiller, E. L. (1956). *Parasitology* **46**, 395–419.
Rausch, R. and Williamson, F. S. L. (1959). *J. Parasit.* **45**, 395–403.
Read, C. P. (1956). *Exp. Parasit.* **5**, 325–44.
Riley, W. A. (1939). *J. Amer. vet. med. Ass.* **95**, 170–72.
Román, C. (1956). *Rev. Med. exp., Lima* **10**, 85–87.
Romanov, I. V. (1958). *Zool. Zh.* **37**, 1136–42.
Sadun, E. H., Norman, L., Allain, D. S. and King, N. M. (1957). *J. infect..7 Dis.* **100**, 273–7

Sannikov, Y. I. (1961). *Med. Parazitol. Parazit. Boll., Moscow* **30**, 173–76.
Santivanex, M. J. and Cuba, C. A. (1949). *Rev. Fac. Med. vet., Lima* **4**, 22–24.
Schiller, E. L. (1954). *Exp. Parasit.* **3**, 161–66.
Schiller, E. L. (1955). *J. Parasit.* **41**, 578–82.
Schiller, L. (1960). *Ann. intern. Med.* **52**, 464–76.
Schulte, F. (1950). *Berl. Münch. tierärztl. Wschr.* **2**, 29–30.
Schwabe, C. W. (1959). *Amer. J. trop. Med. Hyg.* **8**, 20–28.
Schwabe, C. W. and Schinazi, L. A. (1958). *J. Parasit.* **44**, 558.
Schwabe, C. W., Schinazi, L. A. and Kilejian, A. (1959). *Amer. J. trop. Med. Hyg.* **8**, 29–36.
Shumakovich, E. E. and Nikitin, F. F. (1959). *Byull. Nauch. Teck. Inf. Gelm. im K. I. Skryabina* **5**, 98–99.
Siebold, C. T. E. von. (1853). *Z. wiss. Zool.* **5**, 409–25.
Smyth, J. D. (1962a). *J. Parasit.* **48**, 544.
Smyth, J. D. (1962b). *Proc. roy. Soc.* B **156**, 553–72.
Smyth, J. D. (1962c). *Parasitology* **52**, 441–57.
Smyth, J. D. (1963a). *Bull. No.* 34 *Comm. Helm. Bureau* 1–38.
Smyth, J. D. (1963b). *Nature, Lond.* **199**, 402.
Smyth, J. D. (1964). *J. Parasitology* (in Press).
Smyth, J. D. and Haslewood, G. A. D. (1963). *Ann. N.Y. Acad. Sci.* **113**, 234–60.
Smyth, J. D. and Smyth, M. M. (1964). *Parasitology* (in Press).
Sweatman, G. K. (1952). *Canad. J. publ. Hlth.* **43**, 480–86.
Sweatman, G. K. and Williams, R. J. (1962). *Trans. roy. Soc. N. Z.* (Zoology) **2**, 221–50.
Sweatman, G. K. and Williams, R. J. (1963). *Parasitology* **53**, 339–90.
Tagle, I., Rivera, G. and Neghme, A. (1956). *Bol. Chileno Parasit.* **11**, 33–34.
Thomas, L. J. and Babero, B. B. (1956). *J. Parasit.* **42**, 659.
Thomas, L. J., Babero, B. B., Gallichio, V. and Lacey, R. V. (1954). *Science*, **120**, 1102–1103.
Turner, E. L., Dennis, E. W. and Berberian, D. A. (1935). *J. Egypt. med. Ass.* **18**, 536–46.
Turner, E. L., Berberian, D. A. and Dennis, E. W. (1936). *J. Parasit.* **22**, 14–28.
Urbain, P. A. (1932). *Bull. Acad. Vét. France* **5**, 138.
Vanni, V. and Radice, J. C. (1950). *Arch. int. Hidatid.* **11**, 273–83.
Verster, A. (1961). *J. S. Afr. vet. med. Ass.* **32**, 181–85.
Vibe, P. P. (1959). *Dokladie Akademii Nauk SSSR.* **129**, 471–72.
Vogel, H. (1955a). *Dtsch. med. Wschr.* **80**, 931–32.
Vogel, H. (1955b). *Rev. ibér. Parasit.* Tomo Extraordinario, 443–49.
Vogel, H. (1957). *Z. Tropenmed. u. Parasit.* **8**, 404–54.
De Waele, A. and De Cooman, E. (1938). *Ann. Parasit. hum. comp.* **16**, 121–32.
Waworoentoe, F. K. and Mansjoer, M. (1950). *Hemera Zoa* **57**, 447–56.
Webster, G. A. and Cameron, T. W. M. (1961). *Canad. J. Zool.* **39**, 877–91.
Webster, G. A. and Cameron, T. W. M. (1963). *Canad. J. Zool.* **41**, 185–95.
Wertheim, G. (1957). *Arch. int. Hidatid.* **16**, 267–70.
Whitten, L. K. and Shortridge, E. H. (1961). *N. Z. Vet. J.* **9**, 7–8.
Williams, R. J. and Sweatman, G. K. (1963). *Parasitology* **53**, 391–407.
Yamashita, J. (1960). *Parassitologia* **2**, 399–406.
Yamashita, J., Ohbayashi, M. and Konno, S. (1956). *Jap. J. vet. Res.* **4**, 113–22.
Yamashita, J., Ohbayashi, M. and Konno, S. (1957). *Jap. J. vet. Res.* **5**, 43–50.
Yamashita, J., Ohbayashi, M., Kitamura, Y., Suzuki, K. and Okugi, M. (1958a). *Jap. J. vet. Res.* **6**, 135–55.
Yamashita, J., Ohbayashi, M. and Kitamura, Y. (1958b). *Jap. J. vet. Res.* **6**, 226–29.
Yamashita, J., Ohbayashi, M. and Sakamoto, T. (1960). *Jap. J. vet. Res.* **8**, 315–22.
Yamashita, J., Ohbayashi, M., Sakamoto, T. and Orihara, M. (1962). *Jap. J. vet. Res.* **10**, 85–96.

Recent Advances in the Anthelmintic Treatment of the Domestic Animals

T. E. GIBSON

Central Veterinary Laboratory, Weybridge, England

I. Introduction

Helminths were recognized as the cause of disease in man and animals at an early stage in history. Attempts to remove the unwelcome guests followed and anthelmintics are amongst the earliest known medicaments. The earliest anthelmintics were derived from plants and substances of vegetable origin and are still widely used as anthelmintics in some parts of the world. There has, however, during the last 40 years been an increasing tendency to use synthetic organic compounds and some of our best anthelmintics are substances produced in the laboratory by the organic chemist.

The desirable qualities of an anthelmintic for veterinary use include efficacy, safety, ease of administration and cheapness. Preliminary starvation and purgation have by tradition been considered necessary to ensure the effective action of anthelmintic drugs. The alimentary canal is cleared of much of its contents and this brings the drug into closer contact with the parasite. Starvation, however, weakens the animal and also often results in increased absorption of the drug, thus enhancing toxicity. Nowadays, preliminary starvation is considered unnecessary and most anthelmintics which have become available in the last 20 years are efficacious when given on a full stomach. Whenever possible medicaments are given to poultry and pigs mixed with the food or water. Overnight starvation sharpens appetite and ensures rapid consumption of the medicated food but it is not essential for anthelmintic action.

A wide therapeutic index is a valuable attribute in an anthelmintic and this was lacking in many of the older compounds. The mass treatment of large numbers of animals using automatic drenching equipment does not allow for the calculation of the appropriate dose for individual animals. It is usual to administer to all animals in the group a dose adequate for the average animal, which means that some of the smaller or weaker animals will receive an overdose. With many of the organo-phosphorus compounds the therapeutic index is so narrow as to lead to difficulties, but compounds such as thiabendazole or bephenium hydroxynaphthoate have such a wide therapeutic index as to make overdosage in practice impossible.

Ease of administration is an important attribute of a veterinary anthelmintic. The bulky dose of phenothiazine is difficult to administer but the advent of automatic dosing equipment of various kinds has simplified the administration of large volumes of liquid. Many difficulties in the administration of bulky insoluble powders have been removed by the pharmacist's ability to produce elegant preparations which are easy to administer. Palatability is an important consideration in those anthelmintics intended for use in the food.

The cost of a product is an important economic factor in considering its use in farm animals. The drugs now used as anthelmintics are much more expensive than those in common use a quarter of a century ago but only the high economic value of the animal limits the cost of treatment. This consideration is not a factor in the treatment of small animals where sentiment, rather than economics, decides the cost of medicaments.

Most of the anthelmintics in general use 25 years ago failed to satisfy one or other of these criteria and progress in providing new veterinary anthelmintics has been slow. The first real advance was the introduction of phenothiazine about 20 years ago but further progress has been made, particularly during the past 4 or 5 years. In many pharmaceutical laboratories throughout the world a quest is being made for new compounds of potential chemotherapeutic value, and most commercial houses test their products for anthelmintic activity using small animal screens of the type described by Steward (1955) and Standen (1958). By this means a number of new chemicals are applied in the veterinary field and are further evaluated by tests on large animals. A small number of compounds are found to have sufficient value to be introduced into commerce and the number increases year by year. Because of the rapidity with which discoveries are made it is difficult to keep abreast of recent developments. This contribution is an attempt to assess the value of the more important compounds which have been introduced during the last few years and to consider to what extent they replace the older drugs with which we are familiar. For this purpose the important diseases of the domestic animals are considered in turn and the recent advances in their treatment described.

II. Parasitic Gastro-enteritis of Cattle and Sheep

A. Phenothiazine

1. *Micronised Phenothiazine*

a. Influence of particle size and purity on anthelmintic action. Phenothiazine has been the drug of choice for the treatment of parasitic gastro-enteritis in ruminants for more than two decades. Its pharmacology and anthelmintic properties have been reviewed by a number of workers but most recently by Harwood (1953) and Griffiths (1954). The drug, although a considerable advance on compounds previously in use, has a number of disadvantages, but perhaps the most important of these is indifferent action against *Trichostrongylus* spp. particularly those which inhabit the small intestine. It was shown (Anon., 1943) that action against these worms improved progressively as the dose was increased, and Gibson (1949) demonstrated that an increase of the dose rate to 800 mg/kg was necessary to achieve full efficiency against these species. It was realized as early as 1940 that particle size had some influence on the efficiency of phenothiazine, Gordon (1940) showing that coarse phenothiazine was less effective than the normal commercial product for the removal of *Oesophagostomum columbianum* from sheep. The influence of particle size on anthelmintic efficiency was more thoroughly investigated by Gordon (1956). He showed that action against *Ostertagia* spp., *Trichostrongylus* spp. and *Haemonchus contortus* was influenced by particle size to a greater degree than action against *Oesophagostomum columbianum*. He was not able to make final recommendations as to the optimum particle size for anthelmintic efficiency but concluded that preparations should not include particles greater than 30μ and that it was preferable that the majority should be smaller than 20μ. The influence of particle size on anthelmintic efficiency was confirmed by the work of Whitten (1956), Douglas *et al.* (1956), Thomas and Elliott (1957) and Kingsbury (1958), and these workers showed that for full efficiency against *Trichostrongylus* spp. the majority of the particles should be below 10μ. It was suggested by Thomas and Elliott (1957) that the ideal preparation of phenothiazine would have 70% of the particles under 5μ or 90% under 10μ with the rest up to 30μ in size. Gordon (1958b), however, reported that extreme fineness was not always an advantage for a preparation having 50% of the particles under 2μ gave the lowest efficiency recorded against oxyurids in mice. Douglas *et al.* (1959) demonstrated a linear relationship between specific surface area and anthelmintic efficiency within the range 5,000 to 25,000 cm^2/g which represents an average particle diameter of 1–10μ and a range of a 93–63% anthelmintic efficiency. These workers also demonstrated the effect of high purity on anthelmintic efficiency a property which was confirmed by Baker *et al.* (1959), who defined a suitable phenothiazine preparation as having a specific surface area of 6,000–7,000 cm^2/g if pure or 25,000 cm^2/g if of American National Formulary standard of purity. Forsyth (1959) in a preliminary communication stressed the importance of both purity and of particle size in determining anthelmintic efficiency and in a later paper Forsyth *et al.* (1961) dealt with these aspects in more detail. When purity falls below 85% anthelmintic efficiency falls sharply. It was considered that the influence of particle size could be best expressed by the term "dose area"—a product of dose in g and specific surface area in cm^2/g. High dose areas resulted in higher efficiency than lower ones, but in general it was immaterial whether this was achieved by a high dose of coarse particles or a lower dose of fine particles. Fine grinding, resulting in a bigger dose area, was beneficial against *Ostertagia* spp. and *Trichostrongylus* spp. but appeared to be of no advantage for the removal of *Chabertia ovina*. Forsyth *et al.* calculated that in order to achieve 90% efficiency against *Ostertagia* spp., *T. axei*, *Trichostrongylus* spp. in the small intestine, and *Chabertia ovina*, phenothiazine of 85% purity should have a "dose area" of 185,000 cm^2 but if of 90% purity a "dose area" of 155,000 cm^2 is sufficient.

b. Absorption and toxicity of micronised phenothiazine. Following the demonstration that purity and particle size influence efficiency most manufacturers have put on the market "micronised" preparations of phenothiazine. It was realized from the first that a likely consequence of fine grinding would be greater absorption of the drug with increased danger of intoxication. Farrington *et al.* (1962) studied the relationship between particle size and the absorption of phenothiazine by sheep and found large amounts of the finer samples to be absorbed. 73% of samples having a specific surface area of 1,000 cm^2/g was recovered from the faeces and 20·2% from the urine. With samples having a specific surface area of 21,100 cm^2/g only 33% was recovered from the faeces and 56·9% from the urine. When samples, intermediate in particle size, having a specific surface area of 11,500 cm^2/g were used, 44% was passed in the faeces and 47·3% in the urine. Farrington and Thomson (1962), however, found that if the drug was given by intra-abomasal injection little of the drug was absorbed. Using a sample having a specific surface area of 11,500 cm^2/g 78% of the dose was recovered from the faeces and only 10·3% from the urine. The route of administration, therefore, influences the amount of absorption but the figures of Farrington *et al.* (1962) are more relevant to practice where normally administration will be into the rumen. Most workers consider that intoxication from increased absorption of "micronised" phenothiazine is of relatively slight importance in sheep. Malone (1956) gave doses of micronised phenothiazine as high as 120 g to lambs without untoward effects and Arundel (1962) reports giving nine times the therapeutic dose of fine particle phenothiazine with safety. Hebden and Setchell (1962), however, record twelve cases in which losses followed the use of micronised phenothiazine. In some of these instances carbon tetrachloride had been given within 14 days of treatment, and since carbon tetrachloride interferes with kidney function it may inhibit the excretion of absorbed phenothiazine in the urine. It is noteworthy that samples of phenothiazine responsible for three cases caused no ill effects when administered to other sheep at the research station.

c. The use of micronised phenothiazine in cattle. Cattle are less favourable subjects than sheep for treatment with phenothiazine and the likelihood that ill effects would follow the use of the micronised products was considerably greater than in sheep. Most workers have avoided the use of micronised phenothiazine in cattle but Armour *et al.* (1961b) used fine particle size phenothiazine in doses of 110, 220 and 330 mg/kg in Nigerian Zebu cattle. 110 mg/kg was effective against *Haemonchus* spp., 220 mg/kg was effective also against *Trichostrongylus* spp. and *Oesophagostomum radiatum* and at 330 mg/kg *Bunostomum phlebotomum* was also removed. *Cooperia* spp. was not removed even by the highest dose rate. Two calves given 330 mg/kg became recumbent and showed muscular tremors after treatment and one died 5 days later. It is clear, therefore, that a dose rate of 220 mg/kg should not be exceeded in cattle. Anderson *et al.* (1962) treated groups of fifteen cattle with two samples of purified phenothiazine, one having a particle size of 2–3μ and the other one of 6–8μ. A dose rate of 196 mg/kg was used with complete safety. Compared with undosed controls, a 91% reduction in faecal egg count was observed in the animals receiving the sample of 2–3μ particle size and 84% reduction in those receiving the other.

2. *Synergism between Phenothiazine and other Compounds*

a. Organo-phosphorus compounds. Kingsbury (1961a) in a preliminary communication reported that small amounts of the organo-phosphorus compounds coumaphos and coroxon act synergistically when mixed with phenothiazine. 2 mg/kg of coroxon along with 200 mg/kg of phenothiazine was more active as an anthelmintic than 2 mg/kg of coroxon alone or 500 mg/kg of phenothiazine alone. More detailed work reported by Kingsbury (1961b) showed that 300 mg/kg of phenothiazine mixed with 2·0–2·5 mg/kg of coumaphos or 1·5–20 mg/kg of coroxon removed 80% or more of *Haemonchus contortus, Trichostrongylus axei*, intestinal *Trichostrongylus* spp. and *Strongyloides papillosus*. Action against *Ostertagia* spp. varied from 33% to 99% and limited data suggested efficiency was also high against *Cooperia* spp. and *Nematodirus* spp. Following this Sloan *et al.* (1961) carried out field trials with the phenothiazine-coroxon mixture and confirmed its efficiency and reported it to be safe in use. Malone (1962) made extensive studies of the toxicity of phenothiazine-coroxon mixture and found that lambs would tolerate up to four times the therapeutic dose and that 90% of the dose was excreted within 7 days. In spite of the apparent safety of this combination fatalities have followed its use in practice and care should be taken not to exceed the therapeutic dose.

b. Dichlorophen. Batte (1961) used a combination of phenothiazine and dichlorophen at approximate dose rates of 225 mg/kg and 100 mg/kg and found it to be appreciably more active in naturally infected sheep than 550 mg/kg of phenothiazine alone. Pecheur *et al.* (1962a), using egg counts as a criterion of efficiency, showed a mixture of phenothiazine at 500 mg/kg and piperazine citrate at 100 mg/kg to be extremely effective against trichostrongylid worms in sheep but not against *Strongyloides papillosus* and considered the products were acting synergistically.

c. Phenzidole. Forsyth (1962) reported a marked synergism between 2′ phenyl benzimidazole (phenzidole) and phenothiazine. 300 mg/kg of phenothiazine along with 100 mg/kg of phenzidole was markedly more active than the two constituents administered separately. It was also reported that 225 mg/kg with 75 mg/kg of phenzidole was 90% or more efficient against *Ostertagia* spp. *Trichostrongylus axei*, intestinal *Trichostrongylus* spp., *Nematodirus* spp., *Chabertia ovina* and *Oesophagostomum venulosum.*

B. ORGANO-PHOSPHORUS COMPOUNDS

A number of these cholinesterase inhibitors have been tested as anthelmintics in cattle and sheep. Those most widely used have been fenchlorphos, trichlorphon, coumaphos and ruelene.

1. *Fenchlorphos O, O*-dimethyl *O*-(2, 4, 5-trichlorophenyl) phosphorothioate

Fenchlorphos was tested by Gordon (1958a) using 110 mg/kg in sheep and found to be ineffective against *Trichostrongylus axei*, *T. colubriformis* and *Oesophagostomum radiatum.* At 220 mg/kg it was effective against *Haemonchus contortus.* Schad *et al.* (1958) carried out critical tests on three sheep, using a dose rate of 110 mg/kg, but were unable to find any useful activity. Dorney and Todd (1959), however, recorded useful action against species of *Haemonchus*, *Ostertagia*, *Strongyloides* and *Nematodirus* using dose rates of 200, 400 and 600 mg/kg. Symptoms of intoxication were observed at the higher dose levels and they considered that 200 mg/kg should not be exceeded. Gibson (1960) found the drug at 110 mg/kg to be without effect on *Trichostrongylus axei*, and later Gibson (1961) found it to be 31% effective against *Haemonchus contortus*. In cattle, Herlich and Johnson (1957) reported some activity against several species of trichostrongylid worm at a dose rate of 110 mg/kg. Likewise Drudge *et al.* (1961) found egg counts to be markedly reduced after treatment. Worley (1957), however, could demonstrate little effect on the egg count of thirty-eight yearling cattle.

2. *Trichlorphon* *O*, *O*-dimethyl 1-hydroxy-2-trichloroethyl phosphonate

Gordon (1958a) found trichlorphon given at a dose rate of 60 mg/kg to be highly effective against *Haemonchus contortus* and 110 mg/kg to be 100% efficient. Action against *Trichostrongylus* spp. and *Oesophagostomum columbianum* was variable. Doses of 165 mg/kg caused death in a proportion of treated animals. Stampa (1959) in a field trial found trichlorphon to be effective against species of *Haemonchus* and *Nematodirus* but doses of 130 mg/kg produced toxic symptoms in some sheep. In some districts doses of 90 mg/kg produced intoxication. Gibson (1960) found 110 mg/kg to be ineffective against *Trichostrongylus axei* but 100 mg/kg was 100% efficient against *Haemonchus contortus* (1961). Southcott (1961) found 55 mg/kg to be highly efficient against *Haemonchus contortus*. Some effect was seen against *Ostertagia* spp. but even at 110 mg/kg no useful action was seen against *Trichostrongylus* spp. or *Oesophagostomum* spp. At this higher dose level signs of toxicity were frequent and sometimes death was caused. Galvin *et al.* (1962), however, were able to give doses as high as 200 mg/kg without signs of intoxication. Useful anthelmintic activity was, however, observed against only *Haemonchus* spp. and *Trichostrongylus axei*. Riek and Keith (1958) tested the drug under field conditions in cattle and at 44 mg/kg removed the majority of *Haemonchus placei* and *Oesophagostomum radiatum*. At 100 mg/kg the drug was also effective against *Trichostrongylus axei*, *Cooperia* spp. and *Bunostomum phlebotomum*. If given directly into the abomasum it was active against immature stages of these species at a dose rate of 110 mg/kg. In toxicity studies Riek and Keith gave dose rates of up to 2,175 mg/kg but observed little correlation between dose rate and severity of symptoms. At 110 mg/kg salivation, restlessness and looseness of the bowel occurred in some animals but these symptoms passed off in 1 or 2 h. Banks and Michel (1960) carried out a small controlled trial on calves infected with *Ostertagia ostertagi*. At a dose rate of 66 mg/kg 64% efficiency was recorded against 21-day-old worms and 96% against month-old worms. When the dose rate was raised to 110 mg/kg the corresponding figures were 65% and 98·5%. At the higher dose rate severe reactions were seen and in some animals death occurred within 30 min. Atropine, if given as soon as symptoms were observed, was effective as an antidote. Banks and Mitton (1960) used the drug at a dose rate of 80 mg/kg to treat acute ostertagiasis in the field with dramatic results. Alarming symptoms of poisoning were observed in many animals but spontaneous recovery was the rule.

3. *Coumaphos O*, *O*-diethyl *O*-(3 chloro-4-methyl-7-coumarinyl) phosphorthioate

Gordon (1958a) found 22 mg/kg to be 100% effective against *Oesophagostomum columbianum*. 9–11 mg/kg was very effective against *Trichostrongylus colubriformis* and 9 mg/kg was highly efficacious against *Trichostrongylus axei*. 7 mg/kg killed two sheep and 11 mg/kg killed one of six sheep but other sheep survived doses of 22 mg/kg. Stampa (1959) found the drug effective in a field trial with sheep affected with parasitic gastro-enteritis and lost only two sheep of 400 given 17·6 mg/kg. Knight *et al.* (1960) however, used 15 mg/kg in field tests and killed eight of forty-five lambs. Their trials indicated that the drug has little action against immature forms.

Herlich and Porter (1958) found 25 mg/kg extremely effective in calves against species of *Haemonchus*, *Ostertagia*, *Trichostrongylus*, *Cooperia*, *Trichuris* and *Capillaria* as well as *Strongyloides papillosus*. Toxic symptoms were mild and transitory. Galvin *et al.* (1959) also found a dose of 25 mg/kg effective in calves and without untoward effects but the same authors (1960b) reported intoxication in calves receiving from 12·5 to 50 mg/kg with death in three calves which received 25, 37·5 and 50 mg/kg.

4. *Ruelene* 4-tert-butyl-2-chlorophenyl methyl methylphosphoramidate

Douglas and Baker (1959) carried out controlled tests in sheep and found 200 mg/kg to be 90% effective against *Ostertagia* spp., *Trichostrongylus axei*, *Trichostrongylus vitrinus* and *Nematodirus* spp. and 55% effective against *Trichostrongylus colubriformis*. Immature worms were not removed and no intoxication was observed. Gibbs and Pullin (1961) carried out controlled and critical tests on naturally infected sheep and also reported high activity and no evidence of intoxication at a dose rate of 200 mg/kg. Landram and Shaver (1961), however, reported good results in field trials on naturally infected lambs using a dose rate of 75 mg/kg. Skerman (1962), in extensive controlled trials in sheep, found 50 mg/kg to be effective against *Haemonchus contortus* and 100–150 mg/kg to be effective against *Trichostrongylus colubriformis*, *Ostertagia* spp., *Trichostrongylus axei* and *Cooperia* spp. The lowest dose which caused death was 350 mg/kg and in field trials many sheep received double the therapeutic dose of 100–150 mg/kg without ill effects. Weidenbach *et al.* (1962) killed one of two sheep given 300 mg/kg but 150 mg/kg was quite safe. Brown (1962) carried out field trials using a dose rate of 125 mg/kg and reported good anthelmintic effects and no untoward symptoms in any of the treated lambs.

Alicata (1960) found 39 mg/kg almost 100% effective in a critical test on calves artificially infected with *Cooperia punctata*. No untoward effects were seen, either in these animals, or in seven others treated under field conditions. Herlich *et al.* (1961) used doses of 40–60 mg/kg in cattle with good effect against *Haemonchus placei*, *Cooperia punctata* and *Oesophagostomum columbianum*. At 60 mg/kg it was also active against *Ostertagia ostertagi*, *Trichostrongylus axei* and *Trichostrongylus colubriformis* but at this dose rate one of the three steers died. Shelton (1962) in controlled critical tests on artificially infected animals found 40 mg/kg to be highly effective against species of *Haemonchus*, *Ostertagia* and *Cooperia*. Landram and Shaver (1961) used doses of 20–100 mg/kg in calves and found doses of 40 mg/kg to be the least which gave adequate anthelmintic effect. 150 mg/kg produced severe symptoms of intoxication but no deaths. Weidenbach *et al.* (1962) found 50 mg/kg to be a safe dose but 100 mg/kg produced signs of intoxication.

5. *Haloxon O, O* di-(2-chloroethyl) *O*–(3-chloro-4-methyl coumarin-7yl) phosphate

Armour *et al.* (1962) reported on laboratory and field trials with this compound using a dose rate of 30–55 mg/kg in sheep. High efficiency was reported against *Haemonchus contortus*, *Ostertagia* spp., *Trichostrongylus axei*, intestinal *Trichostrongylus* spp., *Cooperia curticei*, *Nematodirus* spp., and *Strongyloides papillosus*. The fourth larval stages and immature adults of *H. contortus*, intestinal *Trichostrongylus* spp. and *C. curticei* were also adequately removed. Immature adults of *Ostertagia* spp. and *Nematodirus* spp. were removed and partial control was effected of fourth larval stages of these latter nematodes. In these studies no intoxication was observed in 600 sheep and a further 29,000 sheep have been treated in the field without ill effect. The minimum oral toxic dose is stated to be 250 mg/kg some five times the therapeutic dose.

6. *Conclusion*

Although most of these organic phosphorus compounds have anthelmintic activity those with the highest activity are in general also the most toxic to the host. The therapeutic and toxic doses are so near that they have little advantage over established anthelmintics. An exception to this is the new compound haloxon in which serviceable anthelmintic activity appears to be combined with an unusually wide therapeutic index amongst the organophosphorus compounds.

C. METHYRIDINE 2 (β methoxyethyl) pyridine

1. *Anthelmintic Efficiency*

Walley (1961) described the use of methyridine as an anthelmintic in cattle and sheep. The drug is active as an anthelmintic whether given orally or by subcutaneous injection. When given by the latter route it is excreted into the alimentary tract where it exerts its anthelmintic effect; the concentration in the stomach being somewhat higher by the oral route. The optimum dose is 200 mg/kg. After experiments involving over 400 cattle and 400 sheep Walley concluded that high efficiency was obtained against *Ostertagia* spp., *Trichostrongylus* spp. *Cooperia* spp. and *Nematodirus* spp. in the small intestine and against *Trichuris* spp. in the caecum. In the abomasum *Trichostrongylus* spp. were completely eliminated, *Ostertagia* spp. almost completely eliminated and *Haemonchus* eliminated from many animals. Activity against *Haemonchus* is however erratic and in some animals worm burden is unaffected. Good action is claimed against immature worms. Gibson (1962) carried out controlled tests on sheep carrying monospecific infections of *Haemonchus contortus*, *Trichostrongylus axei*, and *Nematodirus filicollis*. At a dose rate of 200 mg/kg efficiency was 91, 100 and 100% respectively. Gibbs and Pullin (1962) carried out critical tests on naturally parasitized sheep and confirmed the high efficiency of the drug, when used at a dose rate of 200 mg/kg given subcutaneously, against worms in the small intestine. Against *Haemonchus* and *Ostertagia* in the abomasum action was, however, generally poor. Groves (1961) summarized the results of field trials involving the successful treatment of 2,500 cattle and sheep and further evidence of the efficiency of methyridine in cattle and sheep under conditions of general practice was given by Gracey and Kerr (1961), Hamilton (1961), Young (1961), Macrae (1961), Watt *et al.* (1961) and Pouplard *et al.* (1961). Walley (1962) showed the drug to be equally effective when given intraperitoneally.

2. *Toxicity*

Walley (1961) reported the drug to be well tolerated in therapeutic doses although local reactions sometimes followed large doses especially in cattle. Groves (1961) carried out toxicity tests on 1,630 cattle and sheep. A dose of 200 mg/kg was well tolerated but one-third of 50 ewes and 100 lambs given twice the therapeutic dose showed incoordination 30–60 min later. Seven of the poorer animals became recumbent and 4 h later four of these were dead. The following day the three sick ewes were better and the rest of the flock was walking normally. In cattle local reactions at the site of injection were more common than in sheep. Doses of 60 ml or more at a single site resulted in oedema but no abscess formation or ulceration occurred. Thorpe (1962) describes the histology of the local reaction to the subcutaneous injection of methyridine and points out that healing is complete at 32 days, dense fibrous tissue having replaced the damaged areolar tissue and necrotic muscle. Johnston (1962) points out the danger of inadvertent injection of methyridine intramuscularly. Severe inflammation, fibrosis and abscess formation takes place in the muscle at the site of injection. Dale *et al.* (1962) reported four deaths in a flock of lambs after treatment with methyridine. These animals were weak from inadequate nutrition, recent castration and pneumonic disease for which the animals had recently been treated. The drug may not be entirely to blame in these cases but care must be taken in treating animals debilitated by other disease conditions. Walley (1962) found methyridine to be slightly more toxic when given intraperitoneally than subcutaneously but considered the hazard small providing the live weight of the treated animal was estimated with reasonable accuracy. Reeves (1962), however, reported incoordination and fast respirations in two sheep, one of which died, following intraperitoneal injection and Hiscock (1962) records similar untoward effects in a heifer which was able to stand again within 20 min. Harrow (1962) reports that methyridine is incompatible with diethylcarbamazine and the two drugs should never be given together. Within 10 min of administration of the two drugs the affected animal shows signs of distress. The pulse rate rises to 200 or more per min and heart sounds are audible at several feet. There is severe purgation, respiration is laboured and the limbs extended in spasm. Recovery or death may occur within 2 h. This phenomenon is not of constant occurrence; the two drugs having been used together in many cases without ill effect.

D. THIABENDAZOLE 2-(4′thiazolyl)-benzimidazole

1. *Anthelmintic Action in Sheep*

Brown *et al.* (1961) reported briefly on this compound which was stated to have anthelmintic activity on a wide range of parasites in a wide range of hosts. It is active against immature worms as well as adults and is well tolerated. Subsequently Hebden (1961) carried out laboratory and field trials in sheep using a dose rate of 50 mg/kg and recorded 100% efficiency against *Trichostrongylus axei*, *Trichostrongylus vitrinus*, *Trichostrongylus colubriformis* and *Oesophagostomum* spp., 90% against *Haemonchus contortus* and *Ostertagia* spp., 82% against *Nematodirus* spp., and 89% against immature *Nematodirus* spp. Gordon (1961), also using a dose rate of 50 mg/kg, obtained similar results and also reported high activity against *Cooperia curticei* and *Chabertia ovina* but there was no effect on *Trichuris* spp. Cairns (1961) carried out controlled trials on some 200 sheep with results comparable to those quoted but with action against *Strongyloides papillosus* and *Bunostomum trigonocephalum* at a dose rate of about 75 mg/kg. No action was recorded against *Capillaria* spp. Bell *et al.* (1962) carried out controlled tests on sheep confirming the efficiency of the drug against gastrointestinal parasites of sheep, when used at a dose rate of 70–100 mg/kg. High activity was also recorded against unidentified larvae both in the abomasum and small intestine. Dunsmore (1962) found 62·5 mg/kg to be 99% active against immature *Ostertagia* spp. A number of other workers have recorded similar results in laboratory and field trials, and all agree that therapeutic doses of 50–100 mg/kg are well tolerated by sheep. Bell *et al.* (1962) gave doses of 400, 800, 1,200 and 2,000 mg/kg in order to assess the toxic level of the drug. Signs of intoxication were seen at 800 mg/kg, depression and anorexia being present for 4 days. At 1,200 mg/kg the animal was weak, depressed, incoordinated and anorectic 24 h after treatment and it died 14 days later. A sheep given 2,000 mg/kg was more severely affected and died within 36 h of treatment.

2. *Anthelmintic Action in Cattle*

Thiabendazole has not been so widely tested in cattle as in sheep. Bailey *et al.* (1961) treated five cows at the rate of 55 mg/kg and noted a marked fall in faecal egg count following treatment. Two animals were re-treated 28 days later at a higher dose rate of 88 mg/kg. Following autopsy of five treated and three control animals it was concluded that the drug showed promise for the treatment of parasitic gastritis in cattle. Using faecal egg counts as a criterion of efficiency Reinecke and Rossiter (1962) found 25 mg/kg highly effective in removing *Haemonchus placei*, *Cooperia* spp., *Bunostomum phlebotomum* and *Trichostrongylus* spp. from naturally infected calves. When the dose rate was increased to 50 mg/kg *Oesophagostomum radiatum* was also effectively removed. Herlich (1962), however, in controlled tests on artificially infected calves found 55 mg/kg effective against 7-day-old *Trichostrongylus axei*, *Trichostrongylus colubriformis*, *Cooperia punctata*, and *Cooperia pectinata* but less efficient against *Oesophagostomum radiatum*. Bell *et al.* (1962) carried out a controlled test on naturally infected calves using dose rates of 100, 200 and 400 mg/kg. Action was poor in one calf which received 100 mg/kg but in the rest *Haemonchus* spp. and *Trichostrongylus* spp. were completely removed. *Ostertagia* spp. and *Cooperia* spp. were less effectively removed, the latter species being completely removed only by 400 mg/kg. The number of unidentified larvae was greatly reduced at all dose levels. In a group of calves used to compare the action of thiabendazole given as a bolus or as a drench, superior action was observed when the drug was administered as a bolus.

Reinecke and Rossiter (1962) found doses up to 800 mg/kg well tolerated by calves. A dose of 1,000 mg/kg was without permanent ill effect but the dosed calf was recumbent for 2 days. Bell *et al.* (1962) found calves given 100 mg/kg to have slightly elevated temperatures on the 4th day post treatment. One which received 200 mg/kg had laboured breathing, excessive salivation and mucoid diarrhoea and one given 400 mg/kg also showed excessive salivation and rapid breathing. These symptoms passed off within 24 h.

E. BEPHENIUM COMPOUNDS

Copp *et al.* (1958) reported briefly on the anthelmintic activity of the bephenium series of compounds which possess a wide range of anthelmintic activity being especially effective against mucosa dwelling forms of intestinal parasites. Two compounds have been used in veterinary practice: bephenium embonate and bephenium hydroxynaphthoate.

1. *Bephenium Embonate*

Rawes and Scarnell (1958) carried out laboratory and field trials with bephenium embonate on sheep infected with *Nematodirus filicollis*. At 250 mg/kg the drug was highly efficient against the adult worms and also against the parasitic larval stages. At 125 mg/kg action against larval states was much reduced. Gibson (1959a) carried out a controlled test using lambs, artificially infected with *Nematodirus battus*, and at 125 mg/kg found 94% of adult worms and 91% of larval forms to have been removed. When the dose rate was increased to 250 mg/kg the respective figures were 99·6% and 90%. Marquardt *et al.* (1960) found 250 mg/kg to be 100% effective against *Nematodirus spathiger* in sheep. Dunsmore (1960) and Galvin *et al.* (1960a) found the drug to be of little value against other trichostrongylid worms in sheep. Rawes and Scarnell (1958) did not observe any signs of intoxication even when doses of 2,000 mg/kg were used.

2. *Bephenium Hydroxynaphthoate*

Bephenium hydroxynaphthoate is also effective against *Nematodirus* spp. but also is effective against some other species of trichostrongylid nematode in sheep. Rawes and Scarnell (1959) carried out controlled tests and field trials on naturally infected sheep. Using a dose rate of 250 mg/kg the following efficiencies were recorded: *Haemonchus contortus* 96%, *Ostertagia* spp. (adult and immature) 99%, *Trichostrongylus axei* 93%, *Cooperia curticei* 100% and *Nematodirus* spp. 100%. Gibson (1959b) carried out a small controlled test on sheep, artificially infected with *Nematodirus battus*, and found 250 mg/kg to be 91% efficient against immature forms and 100% efficient against adult worms. The same dose rate was reported by Gibson (1960) to be 74% effective against *Trichostrongylus axei*. Dunsmore (1960) also found it effective against *Ostertagia* spp. and *Trichostrongylus axei* in sheep. Banks and Korthals (1960), however, found its action to be inferior to fine particle phenothiazine against *Ostertagia* spp. and *Trichostrongylus* spp. but reported 100% removal of *Nematodirus* spp. 500 mg/kg was given by Rawes and Scarnell (1959) without ill effect.

3. *The Use of Bephenium Compounds in Cattle*

Bephenium embonate has not been used in cattle. Bephenium hydroxynaphthoate was tested by Rubin (1960) and 250 mg/kg was found to be highly effective against *Nematodirus helvetianus.* Armour and Hart (1960) carried out critical tests on six cattle and high activity was observed against *Haemonchus* spp., *Cooperia* spp., *Oesophagostomum radiatum* and *Bunostomum phlebotomum* when the drug was given at a dose rate of 225 mg/kg and these results were later confirmed in field trials. Later Armour *et al.* (1961a) showed the same dose rate to be very effective against the fifth larval stages of *Haemonchus* spp., *Oesophagostomum radiatum* and the fourth and fifth larval stages of *Cooperia* spp. Some action was observed against the fifth larval stage of *Trichostrongylus axei* and the fourth larval stage of *Oesophagostomum radiatum* but against the fourth larval stages of *Haemonchus* spp. and *Trichostrongylus axei* action was poor. Eisa and Rubin (1961) carried out controlled tests on naturally infected cattle and found 85% of *Ostertagia ostertagi* to be removed by a dose of 250 mg/kg. Efficiency against other species were as follows: 87% for *Trichostrongylus axei*, 100% *Nematodirus helvetianus*, 99% for *Cooperia oncophora*, 100% for *Oesophagostomum radiatum* and 99% for *Chabertia ovina.* Rubin and Eisa (1961) carried out controlled tests on six further cattle using a dose rate of 225 mg/kg with similar results.

F. CONCLUSION

In sheep there is little doubt that thiabendazole will replace phenothiazine as the anthelmintic of choice for parasitic gastro-enteritis. Its wide range of activity, including activity against immature stages, and its wide therapeutic index make it extremely suitable for routine use. It has been suggested that its activity against immature stages may interfere with the development of immunity if used for routine dosing. Such evidence as is available suggests that this is unlikely. It is not absolutely certain that thiabendazole will prove as efficient as bephenium hydroxynaphthoate for the treatment of acute nematodiriasis. The latter drug should therefore be used until conclusive evidence is forthcoming.

For cattle methyridine is the first satisfactory anthelmintic which has become available. Experience with thiabendazole in cattle is as yet not very extensive and maybe further work will show it to be as useful in cattle as in sheep. It is clear, however, that a higher dose rate than that used for sheep will have to be employed in cattle.

III. Parasitic Bronchitis

A. INTRODUCTION

The standard treatment for parasitic bronchitis has, for many years, been the intratracheal injection of various anthelmintic mixtures or the inhalation of volatilized iodine or sulphur. There is no critical evidence that these methods are effective and the very few controlled tests which have been carried out indicate that they are ineffective. Recently Enigk (1953) described a new method of inhalation therapy in which an air compressor was used to pass an air stream over atomisers containing anthelmintic mixtures, the medicated air then being passed on to face masks attached to cattle. Fourteen anthelmintics were tested in the apparatus and Enigk concluded that ascaridol-santonin mixture was the most efficacious. The method can fairly readily be used with dairy cattle but is impracticable for store cattle and calves which would have to be restrained whilst treatment was being given. Two drugs have been described recently which have a specific action on lungworms. These are cyanacethydrazide and diethylcarbamazine.

B. CYANACETHYDRAZIDE

1. *The Use of Cyanacethydrazide in Cattle*

The drug was tested by Walley (1957) and by means of a form of critical test it was shown that it would remove the majority of mature lungworms from artificially infected cattle. It was shown that oral administration was almost as effective as subcutaneous injection. A dose rate of 15 mg/kg is used when the drug is given by injection and this is increased to 17·5 mg/kg when given orally. Walley confirmed his results using fifty-six naturally infected animals. Rubin (1959) described in abstract experiments recorded in detail by Rubin and Tillotson (1960). A controlled test was carried out on artificially infected calves and showed that treatment at 15 mg/kg on the 14th, 15th or 16th days after infection was without effect. Treatment given 20, 21 or 22 days after infection was 73% efficient and treatment after patency was 86% efficient. These results demonstrate the relative inefficiency of cyanacethydrazide against immature lungworms. Swanson *et al.* (1959) carried out laboratory studies on eight naturally infected and twenty-two artificially infected calves and field studies on 200 calves but were unable to demonstrate any benefit from the use of doses of 15 mg/kg given on 3 successive days. O'Donoghue (1958) reported clinical improvement in ten calves given 15 mg/kg orally or subcutaneously repeated after a week and again after a further 2 weeks. Rosenburger and Heeschen (1959) carried out a field trial on twenty-six farms involving about 500 animals and concluded that severely affected cattle did not respond to treatment but mildly affected animals ceased to excrete larvae and improved clinically. Zettl (1959) obtained satisfactory results in treating field cases although badly affected animals responded less well than those more lightly infected. Enigk and Düwel (1959), however, found response unsatisfactory in 200 field cases although response was better in more lightly infected animals. Enigk and Düwel (1961) in a critical trial on artificially infected calves, comparing cyanacethydrazide with other drugs, found cyanacethydrazide to remove 69% of the burden of mature lungworms present but not to have appreciable effect on immature worms. Walley (1957) reported that twice the therapeutic dose produces depression, inappetence and mild convulsions and three times the therapeutic dose caused fatal convulsions if given subcutaneously but produced only mild convulsions and depression if given orally. Groves (1958) treated 100 dairy cows and found 18·5 mg/kg and 15 mg/kg doses caused a slight depression in milk yield lasting 12–24 h. Both Walley and Groves agree that no reaction occurs at the site of injection. Rosenburger and Heeschen (1960) claim that the levulinic acid hydrazone of cyanacethydrazide is better tolerated than cyanacethydrazide itself. A 30% solution was given orally at a dose rate of 60 mg/kg on two successive days to forty-eight artificially infected animals and to 1,037 field cases. Clinical improvement followed in about 70% of cases.

2. *The Use of Cyanacethydrazide in Sheep*

Walley (1957) carried out critical tests on lungworm infected lambs using doses of 15–25 mg/kg given orally, subcutaneously or intratracheally and found the drug to be active against *Dictyocaulus filaria* and *Protostrongylus rufesceus* but not against *Muellerius capillaris* and *Neostrongylus linearis.* Further experiments on naturally infected sheep enabled him to fix the oral dose at 17·5 mg/kg and the subcutaneous dose at 15 mg/kg. Pierotti (1958) agreed that the drug is effective against *D. filaria* and *P. rufescens* but not against *M. capillaris, N. linearis* and *Cystocaulus ocreatus.* Enigk and Federmann (1958) also reported that three successive daily doses of the drug were effective in removing *D. filaria* but not against the small lungworms. Jungmann (1960) gave three daily doses of 15 mg/kg to fifty-three sheep and found action good against *D. filaria* and *P. rufescens* but poor against *M. capillaris* but Favati and Della Croce (1961) found action good against *M. capillaris* but negative against *N. linearis* and *Cystocaulus ocreatus.* These results are in agreement that the drug is effective against *D. filaria* and this conclusion is supported by Wikerhauser *et al.* (1959) and Vodrážka *et al.* (1959, 1960), although Wikerhauser *et al.* (1959) produced evidence that action on mature worms is superior to that on immature ones. Cyanacethydrazide has been used successfully in the field by O'Donoghue (1958), Hiepe *et al.* (1959) who treated 1,592 sheep and Jordan (1960). These workers all reported clinical improvement after treatment. Walley (1957) and Vodrážka *et al.* (1960) reported slight irritation following injection and the symptoms produced by two and three times the therapeutic dose were similar to those in cattle. Simunek (1961) also reported three times the therapeutic dose to be toxic in about half the animals treated.

3. *The Use of Cyanacethydrazide in Pigs*

In pigs, cyanacethydrazide was tested by Walley (1957) using dose rates of 15 mg, 20 mg or 25 mg/kg given on 3 successive days. Compared with control pigs the worm burdens of the treated pigs were low, efficiency being 97%. Dick (1958) treated 2,000 clinical cases of lungworm disease in pigs using a dose of 15 mg/kg and reported clinical improvement in 95% of cases. Wikerhauser *et al.* (1959), however, observed clinical improvement in fifty-eight pigs following treatment but at autopsy it was found that their worm burden had not been reduced. Sen *et al.* (1960) also found activity against *Metastrongylus* spp. to be indifferent. In artificially infected pigs 16 mg/kg given on three successive days reduced faecal egg count to a low level but many immature worms were present at autopsy. In a further experiment in which treatment was given 7, 21 and 42 days after administration of the larvae all three groups developed similar faecal egg counts but these diminished more rapidly than the counts of a control group. Worms were found in each of the treated groups at autopsy. Colglazier and Enzie (1961) saw no effect on worm burdens of artificially infected pigs given 15–20 mg/kg.

C. DIETHYLCARBAMAZINE

1. *The Use of Diethylcarbamazine in Cattle*

Parker (1957) tested the effect of this drug on lungworms in calves. Using artificially infected animals treatment was given with 55 mg/kg for 5 days beginning 14–18 days after infection. Faecal larval counts were depressed, deaths prevented and weight gain continued. Treatment at a later stage in the disease was less effective and death was delayed rather than prevented. In further laboratory trials Parker and Roberts (1958) showed that five daily doses of 22 mg/kg beginning at the 14th day after infection was effective in artificially infected animals but challenge on the 44th day after infection showed that treatment had interfered with the development of immunity. In three controlled field trials, in which three daily doses of the drug were given at a rate of 22 mg/kg, Parker *et al.* (1959) obtained satisfactory clinical response, including increased weight gain, in two outbreaks. In the third, which was complicated by "virus pneumonia", clinical response was unsatisfactory. Parker and Vallely (1960), using artificially infected animals, compared the effect of a single dose of 44 mg/kg injected intramuscularly with three daily doses of 22 mg/kg. The single dose was not as effective in removing worms as the series of doses but it was considered that its effect would be adequate in practice. Hollo (1961) used doses of 20, 50 and 100 mg/kg in a field trial involving three herds of cattle. In two herds, treatment was repeated once and in a third it was repeated twice. On the basis of faecal examinations it was found that 55–80% of moderately infected animals in different herds were freed from infection. In the rest of the animals the reduction in worm burden varied from 62% to 87%. In heavily infected animals 13–57% were freed from infection, and in the rest infection was reduced 59% to 88%. The best results were obtained using 20 mg/kg repeated once or twice on successive days or using 50 mg/kg repeated once. Rubin and Tillotson (1962), working with artificially infected calves, found 22 mg/kg given for 5 days beginning on the 14th day after infection successful in reducing worm burden but treatment given for 5 days beginning the day before infection or for 5 days beginning on the 5th day of patency was less effective in reducing worm burden. Jarrett *et al.* (1962) carried out similar experiments using artificially infected calves and found a 3-day course of treatment given on the 15th day after infection to almost completely remove infection. Treatment given to calves after the worms had reached patency was, however, without effect; the clinical, parasitological and gross pathological findings being similar in these calves to those in untreated controls. Enigk and Düwel (1961) in critical tests on artificially infected calves found immature stages to be completely destroyed but only 42% of mature worms were killed. In field trials good results were obtained in early cases but in established infections the drug was of little value. All workers have reported that therapeutic doses are well tolerated. Parker and Vallely (1960) carried out toxicity tests and found 440 mg/kg given by intramuscular injection to be without ill effect even though repeated on a 2nd and 3rd day. The concentrated (40%) solution used to administer these large doses caused muscle necrosis at the site of injection. Nervous symptoms followed in about 4 h when the same dose rate was given orally but recovery was rapid without treatment. Since 440 mg/kg represents twenty times the therapeutic dose there is a wide margin of safety.

2. *The Use of Diethylcarbamazine in Sheep*

Ozerskaya (1955, 1959) has used diethylcarbamazine citrate and phosphate for the treatment of parasitic bronchitis in sheep. The citrate was given by subcutaneous injection as a 1:3 aqueous solution at the rate of 100–200 mg/kg to eighteen sheep and excretion of larvae ceased 10–12 days after treatment. No worms were found at autopsy. The phosphate, given at 50 mg/kg, freed one sheep from worms and reduced the burden by 80% in another three. 100 mg/kg cleared the majority of the animals treated. Kassai (1958) also used the phosphate giving a dose rate of 200 mg/kg subcutaneously repeated twice. Excellent results were obtained against *Dictyocaulus filaria* and *Protostrongylus* spp. but *Cystocaulus* spp. was only partially affected by treatment. Kurtpinnar and Kalkan (1960) gave 20 mg/kg intramuscularly for 3 successive days to naturally infected sheep. Treatment was found to be effective against *Dictyocaulus filaria* but not against other types of lungworm. Egorov and Morozov (1960) used 100–200 mg/kg and freed 64% of animals from *Muellerius* infection. Against *Dictyocaulus filaria* infection 76% of cures was observed. Ivanov (1961) found 100 mg/kg to be tolerated without ill effect but doses of 500 and 1,000 mg/kg were toxic.

D. CONCLUSION

None of the drugs yet tested is entirely satisfactory for the treatment of parasitic bronchitis in cattle. Two specific drugs have recently been described and of these cyanacethydrazide is active against mature lungworms whilst diethylcarbamazine displays most activity against immature worms 14 or 15 days old. The majority of authors agree that diethylcarbamazine has little or no action against mature worms and it has also been shown to be inactive against very early parasitic stages. Diethylcarbamazine is undoubtedly the drug of choice for the treatment of early cases of parasitic bronchitis. In established cases, on theoretical grounds, there is a good case for using both drugs together but there are no reports of this having been tried. In sheep both drugs are active mainly against *Dictyocaulus filaria*. Most species of the smaller lungworms of sheep are not appreciably affected by either drug.

IV. Fascioliasis

A. PARENTERAL ADMINISTRATION OF CARBON TETRACHLORIDE

Although parenteral administration of carbon tetrachloride was investigated in cattle by Zunker *et al.* (1927) and in sheep by Nöller and Schmidt (1927) these routes of administration have not been commonly used until the last decade during which many workers have investigated their use.

1. *Parenteral Administration of Carbon Tetrachloride to Cattle*

The drug has been injected alone or mixed with various vehicles. Slanina *et al.* (1955) found injections of 10–20 ml of carbon tetrachloride given subcutaneously to produce marked excitement lasting about 8 min. A large painful swelling developed at the site of injection which could be reduced by dividing the dose and injecting at several sites. Virtual elimination of the flukes was reported. Komjathy (1957) gave subcutaneous doses of 5–15 ml/100 kg. Local irritations persisted for 10 min following injection and the maximum dose caused giddiness for 24 h. Sterile abscesses developed at the injection site. Pearson and Boray (1961) gave the drug intramuscularly to cattle without causing local reactions and with good anthelmintic effect. Mixed with liquid paraffin or sunflower oil the drug was given by subcutaneous injection by Winterhalter and Delak (1956) who reported the presence of flukes in animals slaughtered 4–15 days after injection. Extensive necrosis occurred at the injection site and the rapid formation of fibrous tissues prevented absorption and reduced efficiency. Horváth (1958), however, successfully used a mixture of equal parts of liquid paraffin and carbon tetrachloride given intramuscularly at the rate of 10 ml/100 kg to cattle. Efficiency based on faecal examinations was 94·2% and body weight and milk production increased after treatment. Kovacs (1958) used 4 ml/100 kg of equal parts of liquid paraffin and carbon tetrachloride intramuscularly on 193 experimental cattle and 60,000 animals in the field. Restlessness followed injection and efficiency was stated to be 90%. Kovacs (1959) used equal volumes of liquid paraffin and carbon tetrachloride with 0·5% of lignocaine base in order to minimize pain on injection. Doses of 40 ml were used and 90% of the flukes were removed. 150,000 cattle were treated with losses of 0·02–0·03%. Delak *et al.* (1961) also found that the addition of local anaesthetics to carbon tetrachloride or to carbon tetrachloride–liquid paraffin mixture diminished excitement after injection in cattle. Winterhalter (1961) added to a 150 i.u. of hyaluronidase a 2:1 mixture of carbon tetrachloride and vegetable oil. This formulation was better tolerated and produced less severe reactions than a similar mixture without hyaluronidase. Hyaluronidase also increased efficiency. Veselova and Velikovskaya (1959) used 20 ml doses of carbon tetrachloride in an equal volume of vaseline injected intramuscularly in 350 cows and claimed 85% of cures and that the treatment was well tolerated. Matevosyan and Kryukova (1961) also used the drug intramuscularly mixed with soft paraffin and reported high efficiency. Cieleszky and Kovacs (1958) investigated the excretion of carbon tetrachloride in the milk of dairy cows following the injection of 40 ml of a mixture of equal parts of carbon tetrachloride and liquid paraffin containing 1% of procaine base. A significant drop in milk yield was seen the day after treatment followed by recovery. 1·56–0·09 mg of carbon tetrachloride per litre was found in the milk and the taint persisted for 3 days. On the 8th day traces of carbon tetrachloride could be detected in one or two cows. In the butter fat the amount of carbon tetrachloride per kg was 3·83 mg on the 1st day, 2·33 mg on the 2nd and 1·37 mg on the 3rd. These amounts are considered so small as not to constitute a hazard if the milk is diluted by incorporation in bulk supplies.

2. *Parenteral Administration of Carbon Tetrachloride to Sheep*

In sheep the subcutaneous administration of carbon tetrachloride has been widely used in some countries for Obitz and Wadowski (1939) refer to the use of Nöller and Schmidt's method in 3,000 sheep. In more recent years many reports mainly from Russia and Eastern European countries of the parenteral use of carbon tetrachloride have been received. Donigiewicz (1951) gave 1 ml doses to lambs and 2 ml to adult sheep by intrarumenal injection and obtained good results in 4,170 sheep with only 0·5% of losses. Demidov (1954) administered doses of 2 ml and 3 ml subcutaneously and found the drug as effective as when given orally or by intrarumenal injection. No complications arose in the treatment of 1,018 sheep. Kagramanov (1955) claimed 100% efficiency following the administration of 2 or 3 ml to 1,728 sheep but reported after effects including rapid movements accelerated respiration and heart beat lasting ½–1½ h in some sheep. Yakovlev (1955) reported lameness lasting 2–5 min in sheep receiving carbon tetrachloride subcutaneously. Winterhalter and Delak (1955) described degenerative changes of the liver following subcutaneous injection of a 3:1 mixture of liquid paraffin and carbon tetrachloride. Regeneration began on the 4th day. Oedema appeared at the injection site during the first 3 days and the odour of carbon tetrachloride could be detected. Pregnant sheep were treated without ill effects and only dead flukes were found in the liver at autopsy 3 or 4 days after treatment. Vitenko (1956) used carbon tetrachloride by four different methods of parenteral administration: subcutaneously, subcutaneously as a 1:1 mixture with fish fat, intramuscularly mixed with Vaseline and intramuscularly mixed with fish fat. The second method gave the best results in 800 sheep. Gavrilyuk (1956) and Boray (1956a, b) both used the drug intramuscularly with good results and without toxic effects. Egyed and Nemeseri (1957) used various parenteral routes of administration of carbon tetrachloride in sheep. After the subcutaneous injection of 2 or 3 ml of the pure drug, excitement resulted from irritation at the injection site. Stupor and dizziness was seen in some animals but these ill effects passed off in 1 or 2 h. A marked limp was seen immediately after the intramuscular injection of 2–3 ml of the pure drug but this passed off in 1 or 2 h. Similar effects were seen after subcutaneous or intramuscular injection of 6–8 ml of a 1:1 mixture of carbon tetrachloride and liquid paraffin. The intramuscular route was favoured for routine use and was reported to be 90% efficient. Delak and Marzan (1959) reported high anthelmintic efficiency following the use of carbon tetrachloride as a 3:1 mixture with paraffin giving doses of 3–4 ml of the mixture but local muscle necrosis occurred at the site of injection. No permanent lameness resulted. Kendall and Parfitt (1962) describe the serious muscle necrosis which follows intramuscular injection of carbon tetrachloride and carbon tetrachloride–liquid paraffin mixture in sheep and also the ulceration which sometimes follows the subcutaneous injection of such mixtures. They state that the subcutaneous injection of carbon tetrachloride in arachis oil is, however, without ill effect.

3. *Conclusion*

The parenteral routes of administration of carbon tetrachloride appear to have no advantages over the oral route except, perhaps, ease of administration. This cannot be an important factor in sheep but may be a recommendation in intractable cattle. In cattle, however, other anthelmintics are preferred to carbon tetrachloride in most parts of the world.

B. HEXACHLOROPHENE bis (2-hydroxy-3, 5, 6-trichlorophenyl) methane*

1. *Anthelmintic Effect in Cattle*

Dorsman (1959) gave an alcoholic solution of hexachlorophene by subcutaneous injection to cattle at a dose rate of 40 mg/kg and freed them from

* See footnote on p. 149.

liver fluke; but one animal died and signs of intoxication were produced in two others. The subcutaneous administration of 5 mg/kg in olive oil was ineffective but 10 mg/kg given subcutaneously or orally gave better results as judged by faecal egg counts. 15 mg/kg given to eight cows effectively reduced the faecal egg count to zero. In these latter animals the fluke burden at autopsy was 0–3 compared with 71–225 in five untreated controls. The drug is absorbed slowly and swellings occur at the site of injection which persist for a few weeks after treatment. No effect on appetite was observed in any of the treated animals. Federmann (1959) gave oral doses of 10–20 mg/kg to cattle and reported good results assessed on egg count data and autopsy findings on ten animals. Osinga (1960), however, did not consider the drug, when given by injection or orally at dose rates of 10–15 mg/kg, to be any more effective than hexachloroethane. Painful swellings were seen after injection and digestive upset, depressed milk yield and even death have followed oral treatment. Bosman *et al.* (1961), however, used doses of 7·5–18·0 mg/kg in nineteen cattle and at slaughter all were found to be free from flukes. In a further twenty-five cattle, treated at the rate of 15–20 mg/kg, twenty had negative faecal egg counts after treatment. In more extensive field trials 535 cattle were successfully treated using dose rates of 7·5–20 mg/kg.

2. *Anthelmintic Effect on Sheep*

Federmann (1959) gave doses ranging from 10 to 20 mg/kg orally to 911 sheep and subcutaneously to 100 sheep. Six weeks after treatment faecal examinations demonstrated 89% of cures in the orally treated group and 76·6% in those treated by injection at a dose rate of 15 mg/kg. Local reactions were seen at the site of injection but these disappeared within 14 days. Osinga (1960) administered 10–20 mg/kg orally as a 10% solution in olive oil with better clinical effect than with 1 ml of carbon tetrachloride. Bosman *et al.* (1961) treated twenty-eight sheep at dose rates of 7·5–20 mg/kg and found all to be free from *Fasciola hepatica* at autopsy. Some sheep were dosed at rates from 7·5 to 30 mg/kg and in only eighteen were *F. hepatica* eggs seen after treatment. In field trails 8,865 sheep were treated at 10–15 mg/kg with good effect and with only seventeen casualties, which were attributable to causes other than treatment. Kendall and Parfitt (1962) treated ewes carrying immature flukes at a dose rate of 40 mg/kg and found a partial effect on flukes 2 weeks old and complete removal of flukes 3 and 4 weeks old. At this dose rate the drug has, therefore, great potentialities in the control of acute fascioliasis. A dose of 15–20 mg/kg is recommended for routine use and is generally well tolerated, although Osinga (1960) reported digestive upset following treatment. Federmann (1959) found 40 mg/kg to be well tolerated but 60 mg/kg caused the death of four of twenty sheep. Guilhon and Graber (1961) found doses of 10–30 mg/kg well tolerated but three of four sheep given 40 mg/kg died. 60 mg/kg was without effect on three sheep but 100 mg/

kg caused death or severe symptoms of intoxication in all sheep treated. Some risk therefore attends the use of 40 mg/kg in acute fascioliasis and treatment should be applied with caution until wider experience has been gained.

C. HETOL 1, 4-bis-trichloromethyl-benzol

Lämmler (1960) after preliminary experiments in white rats used doses of 60–150 mg of hetol in fifteen cattle and found 125 mg/kg successful in reducing the count of *Fasciola hepatica* eggs in the faeces to zero for 28 days after dosing. Enigk and Düwel (1960) used the drug on 864 naturally infected cattle at a dose rate of 160 mg/kg. Faecal egg counts and post-mortem examinations showed the drug to be efficacious against mature flukes but not against immature ones. Lämmler (1960) investigated the acute and chronic toxicity and found 390 mg/kg to be well tolerated. In Enigk and Düwel's (1960) tests sick animals tolerated treatment well and no effect was seen on milk yield in lactating cows.

Lämmler (1960) also treated fifteen sheep using dose rates from 90 to 170 mg/kg and recommended 150 mg/kg as the therapeutic dose. 880 sheep treated under field conditions responded satisfactorily. Behrens (1960) treated 3,383 sheep in the field with satisfactory results. Lämmler (1960) also studied the acute and chronic toxicity in sheep and concluded that ten times the therapeutic dose was without ill effect. Eikmeier and Kamel (1961) found therapeutic doses to be well tolerated.

D. FREON 112 difluorotetrachloroethane

Demidov *et al.* (1959) treated naturally infected cattle at a dose rate of 0·2 ml/kg injected directly into the abomasum with excellent results. Demidov (1958) used the drug dissolved in difluorodichloromethane (Freon 12) at a dose rate of 0·3–0·4 ml of the mixture per kg. Administration was by capsule or directly into the rumen and controlled tests on sixty cattle indicated high efficiency.

Demidov (1955) treated 122 sheep with doses of 0·3–0·6 ml/kg and concluded that the drug was effective and suitable for further field trial. 1,120 sheep were treated by Demidov (1959) with doses of 200–300 mg/kg by intrarumenal injection. Faecal examination 10 days after treatment failed to reveal fluke eggs in treated animals. Boray and Pearson (1960a, b) carried out laboratory trials using doses from 220 up to 1,650 mg/kg given orally except for two sheep in which the dose was given intra-abomasally and intrarumenally. The drug was mixed with 50% of liquid paraffin before use. The results were assessed by means of faecal egg counts and some 90% efficiency was found against mature flukes but immature flukes were not removed. For field trials successful treatment was carried out using doses of 220, 330 or 660 mg/kg but again immature flukes were unaffected. Demidov (1959) reported transient nervous symptoms in a few sheep following doses of 200 or 300 mg/

kg but Boray and Pearson (1960b) did not notice any ill effects in treated sheep.

V. Dicrocoelium dendriticum

Carbon tetrachloride, hexachloroethane, antimony compounds and many other substances have been used in attempts to remove *D. dentriticum* from sheep without success. Guilhon (1962) reported an experiment involving twelve infected sheep which were given doses of 50–500 mg/kg of thiabendazole. This reduced faecal egg count to zero and at post-mortem examination many treated animals to be free from flukes. This preliminary experiment indicates that thiabendazole merits further investigation as an agent for the removal of *D. dendriticum* from sheep.

VI. Tapeworms

A. Tin Arsenate

Tin arsenate was widely tested as a taeniacide in Russia during the years 1948–57 and the experiences during this period have been summarized by Tchoubarrie (1958). He states that the drug is highly efficient against *Moniezia* spp. and *Thysanosoma* spp. acting against immature as well as adult stages when given in doses of 0·3–0·4 g per animal. 18 h fast before treatment and 3 h after is recommended for maximum efficiency. Reduction of the length of fast to 12 or 14 h necessitates an increase of dose if maximum anthelmintic activity is to be achieved. Therapeutic doses are without ill effect and the toxic dose is five or seven times the therapeutic dose; Castel *et al.* (1960) used the drug in tropical Africa. They found that if the drug was given suspended in water to an unstarved sheep it was highly toxic. Given in capsules without starvation at the dose rate 250–500 mg per sheep no toxic symptoms were observed and complete elimination of *Moniezia expansa* was achieved. Given in capsules at the same dose rate after 20 h starvation the drug was safe and was completely effective against *Moniezia* spp., *Stilesia globipunctata* and *Avitellina centripunctata*. For the removal of *Moniezia* spp. a dose of 200 mg per sheep given in capsules after 20 h starvation is recommended. If starvation is omitted water should be withheld 12 h before and 4 h after treatment. For *Avitellina* spp. and *Stilesia* spp. a dose of 350 mg per sheep given after 20 h starvation is advised. Alternatively a dose of 500 mg per sheep can be given without starvation if water is withheld before treatment. Castel and Graber found the toxic dose to be about six times the therapeutic dose; a figure similar to that reported by the Russian workers. Toxic doses produce typical signs of arsenical poisoning. Animals treated with 250 or 350 mg may be safely used for human consumption 6 or 7 days after treatment although it is best to discard the intestine and spleen.

Manganese and calcium arsenates have also been used in Russia as taeniacides in ruminants.

B. YOMESAN *N*-(2′-chloro-4-nitrophenyl)-5-chlorsalicylamid

Yomesan has been used for the removal of tapeworms from man. In South Africa Stampa and Terblanche (1961) tested it in lambs, calves and kids. 50 mg/kg was found to be effective for the removal of species of *Moniezia*, *Thysaniezia* and *Avitellina* except in young lambs in which a dose of 1 g is required irrespective of weight. Doses of ten times greater than the therapeutic level were given without evidence of toxicity. The drug was ineffective against *Stilesia hepatica*. Zettl (1962) carried out a field trial in Germany involving 1,400 lambs. 50 mg/kg was given to sheep in moderately infected flocks and 75 mg/kg in heavily infected flocks. Good results were reported, scoleces as well as segments being removed. Mild diarrhoea was noted in most animals 3–4 h after treatment but this passed off in 5–6 h.

C. BITHIONOL 2, 2′-thio-bis-(4, 6-dichlorophenol)

Fukui *et al.* (1960) used bithionol at a dose rate of 7 mg/kg and bithionol acetate at a dose rate of 10 mg/kg in the horse and reported satisfactory removal of *Anoplocephala* spp. Allen *et al.* (1962) carried out tests with bithionol in sheep infected with *Thysanosoma actinoides*. A dose rate of 220 mg/kg freed the majority of sheep from tapeworms and 175 mg/kg substantially reduced infestation. The drug was generally well tolerated but the higher doses caused diarrhoea for a few days in some animals. Old animals tended to lose weight immediately after treatment; but young lambs quickly began to regain weight.

VII. ASCARIASIS

A. PIPERAZINE COMPOUNDS

1. *The Use of Piperazine Compounds in Horses*

The piperazine compounds were first used as ascaricides in horses by Sloan *et al.* (1954) who treated six horses with doses ranging from 250 to 400 mg/kg. Using egg counts as a criterion of efficiency 250 mg/kg was found to be highly effective and 400 mg/kg to be completely effective. Downing *et al.* (1955) carried out a series of critical and sub-critical tests and confirmed these results. Poynter (1955a) gave doses of 220 mg/kg by stomach tube to fourteen horses and almost 100% efficiency was indicated by egg count data. He also showed (Poynter, 1955b) that the drug could be effectively given in a bran mash. Poynter (1956a) compared the efficiency of four piperazine compounds (adipate, citrate, phosphate and carbodithioic acid) as ascaricides in the horse and found all equally efficient if the dose rate was adjusted to supply 200 mg/kg of piperazine base. Later, Poynter (1956b) treated twelve horses with piperazine-1-carbodithioic acid at a dose rate of 100 mg/kg and found, on the basis of egg count data, that all Ascarids had been eliminated. Drudge *et al.* (1957a) carried out critical tests with this latter compound and found 75, 100, 150 and 200 mg/kg to be 100% efficient against mature ascarids. All the immature ascarids were removed from two animals, one of which received treatment at 75 mg/kg and the other at 150 mg/kg. In field trials Drudge *et al.* (1957b, 1960) and Clark and Connor (1959) confirmed the ascaricidal value of piperazine-1-carbodithioic acid against mature and immature parasites in horses. The piperazine compounds have a wide margin of safety for Sloan *et al.* (1954) gave doses of 1,250–1,500 mg/kg in 5 litres of water by stomach tube without ill effect except an increase in the fluid content of the faeces for 24 h.

2. *The Use of Piperazine Compounds in Cattle*

In cattle, Lee (1955) carried out critical tests on two calves giving a dose rate of 220 mg/kg of piperazine adipate and found the drug to be 100% efficient. Using faecal egg counts as a criterion of efficiency excellent results were obtained in four calves given 440 mg/kg and in seven others given 220 mg/kg. Lee (1956) demonstrated that the strategic use of a dose of 220 mg/kg piperazine adipate at an age of 21 days controlled ascariasis in calves. Dass *et al.* (1961) treated thirty-eight calves with piperazine citrate at 220 mg/kg and found large numbers of ascarids to be voided in the faeces after dosing. On the criterion of faecal egg counts the treatment was shown to be 100% efficient.

3. *The Use of Piperazine Compounds in Pigs*

Sloan *et al.* (1954) carried out tests with piperazine adipate in pigs and recommended a dose of 300–400 mg/kg for field use. Leiper (1954) tested the compound piperazine-1-carbodithioic acid and in critical tests found a dose rate of 75 mg/kg to be 83% efficient and 100 mg/kg to be 100% efficient. A critical test on a group of six pigs to which the drug was administered in the food at 100 mg/kg revealed 89% efficiency and a further test on three pigs using a dose rate of 125 mg/kg demonstrated 90% efficiency. In subsequent tests on 116 pigs on six different farms, faecal egg count data demonstrated an efficiency of 83–100% when dose rates of 100–150 mg/kg were used. Riedel and Larson (1956) reported similar results in critical tests using dose rates of 110 and 125 mg/kg. Guthrie (1956) tested the efficiency of piperazine hexahydrate and piperazine sulphate given in the drinking water for a 24 h period. With the first drug intakes of 60–190 mg/kg resulted in the removal of 45–98% of the Ascarids. Piperazine sulphate at 103 mg/kg was found to be 93% efficient. Shumard and Eveleth (1956) found piperazine citrate to be effective in clearing pigs of Ascarids. Enzie *et al.* (1958) carried out critical tests using piperazine adipate and found greater efficiency in individually treated pigs than in those treated as a group, a fact already recorded by Leiper (1954).

Comparative tests with piperazine citrate, dihydrochloride, sulphate, adipate and a mixture of mono- and dihydrochloride given in the drinking water at the rate of 100 mg of piperazine base per kg revealed efficiencies of 100%, 86%, 69% and 36% respectively. It is generally agreed that therapeutic doses are well tolerated and Sloan *et al.* (1954), Leiper (1954) and Guthrie (1956) have given five times the therapeutic dose with no more significant effect than transient loss of appetite.

4. *The Use of Piperazine Compounds in Dogs and Cats*

Sloan *et al.* (1954) tested piperazine adipate in a number of dogs infected with Ascarids and found all the worms had been removed from one dog which was submitted to autopsy. Faecal egg counts were negative after treatment in other dogs which received dose rates from 25 to 200 mg/kg. In three cats given 100 mg/kg faecal examinations were negative after treatment. Mann *et al.* (1955) gave a series of daily doses of piperazine citrate to eleven cats and two dogs. Treatment was given at the rate of 100 mg/kg for 10 days, but as all the worms had been passed by the 4th day there was no advantage in continuing therapy after that time. Bradley *et al.* (1956) carried out critical tests with piperazine citrate. Various dose rates were used and 176 mg/kg as a single dose was 100% effective. Mahmoud *et al.* (1958) carried out critical tests on a number of dogs given piperazine adipate at several different dose rates after a 12 h fast. In three dogs given 100 mg/kg efficiency was 87%. Six dogs given 200 mg/kg and two others given 300 mg/kg were completely cleared of worms. Erhardt (1956) tested a dose range from 50 to 200 mg/kg of piperazine adipate against *Toxocara cati* in cats and calculated that the effective therapeutic dose would be 430 mg/kg. It is noteworthy, however, that most cats receiving 200 mg/kg vomited. Sprent and English (1958) found 200 mg/kg of piperazine adipate to be effective in removing immature worms from puppies and advocated its use 1 or 2 weeks after birth for the prevention of visceral larva migrans in man. Hayes and McDaniel (1959) investigated the possibility of preventing the prenatal infection of puppies by administering piperazine adipate during gestation. No effect on the establishment of infection in the pups was observed when 100 or 200 mg/kg were given daily or 200 mg/kg every 10 days during gestation. The drug is generally well tolerated by dogs and cats; Sloan *et al.* (1954) gave doses representing an intake of 2,170 mg/kg over 18 weeks without ill effect. Greenberg *et al.* (1958), however, reported vomiting following high doses in puppies, and that kittens were less sensitive than puppies.

VIII. Equine Strongylosis

A. Piperazine Compounds

Phenothiazine had been the standard remedy for equine strongylosis for some years when it was realized that the ascaricidal piperazine compounds would also remove strongylid worms.

Critical tests were carried out by Downing *et al.* (1955) and they found 220 mg/kg of piperazine adipate 100% efficient against *Trichonema* spp., 80% against *Triodontophorus* spp. and 33% against *Strongylus vulgaris. Strongylus edentatus* and *Strongylus equinus* were unaffected. Critical tests by Gibson (1957) confirmed the high efficiency against *Trichonema* spp. and *Triodontophorus* spp. Drudge *et al.* (1957a) carried out critical tests using piperazine carbodithioic acid on seven horses and found the drug to be highly efficient against small strongyles, about 50% efficient against *S. vulgaris*, 30% against *S. edentatus* and without effect on *S. equinus*. Poynter (1956a) using faecal egg counts and differential larval counts as a criterion of efficiency showed that the four piperazine compounds, adipate, citrate, phosphate and carbodithioic acid given at a dose rate of 200 mg/kg of piperazine base were highly efficient against strongylid worms except *S. edentatus*. Field trials by Drudge *et al.* (1957b) using 197 horses and by Clark and Connor (1959) with forty-seven foals confirmed the efficiency of piperazine-1-carbodithioic acid against strongylid worms.

The efficiency of the piperazine compounds against strongylid worms is of the same order as that of phenothiazine. As they also remove Ascarids and are very safe to use they are the anthelmintics of choice for routine use. Their expense has, however, meant that they have not replaced phenothiazine as widely as they might otherwise have done. Poynter and Hughes (1958) have, however, used both compounds together and found the combination to be more efficient than either drug given alone particularly against *S. vulgaris* and *S. edentatus*. The mixture used was phenothiazine at 66 mg/kg and piperazine adipate at 220 mg/kg or piperazine phosphate at 182·6 mg/kg.

B. Thiabendazole

Drudge *et al.* (1962) carried out critical tests with thiabendazole on two horses giving a dose rate of 25 mg/kg, on one horse using a dose rate of 40 mg/kg and on a fourth at 100 mg/kg. The small strongyles were completely removed at all three dose rates and high, but not complete efficiency, was also recorded against immature stages. All three dose rates were also highly efficient against *S. vulgaris*, *S. equinus* and *S. edentatus*. Egerton *et al.* (1962) administered thiabendazole to 105 horses in various ways—in the food, in capsules, in boluses and in aqueous suspensions. Egg count data indicated high efficiency, not only against adult worms, but also against immature forms. Turk *et al.* (1962) gave doses of 100 mg/kg to seventeen horses and the faecal egg counts of all of them had been reduced to zero within a week of treatment. A controlled test was carried out on five horses and confirmed that the majority of the small strongyles, *S. vulgaris*, *S. equinus* and *S. edentatus*, had been removed. Toxicity tests were carried out on four horses which were given 200 or 400 mg/kg without ill effects. A fifth horse was treated with 200 mg/kg weekly for 4 weeks without the development of any signs of intoxication. As Drudge *et al.* (1962) also reported the drug to be effective against ascarids and *Oxyuris equi* thiabendazole has great promise as an anthelmintic for horses.

IX. Hookworm in Dogs

A. Bephenium Compounds

Burrows (1958) tested the efficacy of bephenium chloride, bromide, iodide, hydroxynaphthoate and a mixture of embonate and hydroxynaphthoate in dogs infected with *Ancylostoma caninum*. Various dose rates were used and doses over 20 mg/kg, were 99·1 % efficient and doses of 20 mg/kg were 94·1 % efficient. Doses of 20–25 mg/kg rarely caused emesis but emesis was common in dogs which received 50 mg/kg or over. Senviratna (1960) treated twenty dogs with bephenium chloride at a dose rate of 20 mg/kg. Egg count data and clinical improvement indicated high efficiency. The dogs used were suffering from intercurrent disease but tolerated treatment without ill effect. One or two dogs vomited after treatment. Rawes (1961) used various dose rates of bephenium hydroxynaphthoate against hookworms and concluded that 50 mg/kg given morning and evening removed almost completely *Ancylostoma caninum* and *Uncinaria stenocephala*. A single dose of 100 mg/kg was less effective. Emesis was common in dogs given 50 mg/kg morning and evening but was less frequent when the dose was reduced to 25 mg/kg. At this level, however, anthelmintic efficiency was reduced, but Brown (1962a) found two doses of 25 mg/kg to be 83 % efficient for the removal of *A. caninum* from dogs.

B. Thenium Compounds

Thenium *p*-chlorobenzene sulphonate was tested by Rawes and Clapham (1961) and it was found that doses of 200–250 mg morning and evening removed 90 % or more of the *A. caninum* and 80 % of *U. stenocephala*. Some activity was also observed against *Toxocara canis*. Vomition occurs in some animals, unweaned puppies being especially prone. Burrows and Lillis (1962) found that doses of 25–35 mg/kg twice in one day or once daily for 2 days removed 90–100 % of the hookworms present. 8 % of dogs and 12 % of cats given less than 50 mg/kg vomited. The presence of tapeworms seemed to interfere with the complete removal of hookworms. Rawes and Clapham

(1962) and Brown (1962b) used the thenium compound combined with piperazine for the simultaneous removal of hookworms and *Toxocara canis*. 250 mg of thenium with 500 mg of piperazine given morning and evening resulted in 96% removal of *A. caninum* and 98% of *T. canis*. Vomiting was not seen except in young puppies. In unweaned puppies the dose should be reduced and milk withheld from the evening before treatment to 3 h after the second dose. This reduces the incidence of vomition and any vomition which does occur is not severe and does not significantly affect efficiency.

C. DISOPHENOL 2, 6-di-iodo-4-nitrophenol

Wood and Pankavich (1961) found 28-day-old hookworm infections to be efficiently removed by disophenol in doses of 7·5 mg/kg. Seven-day-old infections were, however, unaffected. Further experiments revealed infections up to 14 days old not to be affected by the drug and results were variable up to 20 days. Infections 21 days old or more were completely removed. Wood *et al.* (1961) reported further tests on mature hookworms and found 7·5 mg/kg given by subcutaneous injection to be highly efficient. The drug could also be given orally or by intramuscular injection. *Uncinaria stenocephala* was less easily removed than *Ancylostoma caninum* and the dose rate had to be increased to 10 mg/kg to achieve full efficiency. By the parenteral routes 36 mg/kg was lethal but orally 100–200 mg/kg has to be given to cause death. 50 mg/kg orally, however, frequently caused emesis. Koutz and Groves (1961) agreed that the drug is less efficient against *U. stenocephala* than against *A. caninum*. Using 7·7 mg/kg nine of twelve dogs were cleared of hookworms without side effects.

X. CAPILLARIASIS IN POULTRY

A. METHYRIDINE

No satisfactory treatment for this condition has been reported until recently when several workers have investigated the value of methyridine. Hendriks (1962) found 200 mg/kg given by subcutaneous injection to artificially infected birds to be effective in removing *Capillaria obsignata* 12 and 24 days old. Three times the dose produced narcosis and prostration but recovery occurred within 24 h. Five times the dose caused similar symptoms but caused death in one of eight birds. Thienpont and Mortelmans (1962) also found 200 mg/kg effective for the control of capillariasis. They found 500 mg/kg to be toxic but found that dilution of the injection reduced the incidence of side effects. In pigeons even subtherapeutic doses caused transient nausea and vomiting. Pecheur *et al.* (1962b) reported similarly but Geevaerts (1962) found 200 mg/kg well tolerated by pigeons and to be effective against *Capillaria* spp. Geevaerts was able to give 400 mg/kg without side effects and the lethal dose was 825 mg/kg. Cotteleer (1962) incorporated doses of 300–1,000 mg/kg in the drinking water. No untoward symptoms were seen but no action against *Capillaria* spp. was observed.

Hendriks (1963), however, found methyridine to be effective when administered in the drinking water. He used 0·5% and 1% solutions of a preparation containing 47·9% methyridine. The medicated water was made available to the animals for a period of 8 h and was then replaced by ordinary drinking water. During this time, on the average, each bird consumed the therapeutic dose of methyridine. Experiments carried out on artificially infected birds showed the 0·5% solution to be 65% effective and the 1% solution to be 100% effective against mature parasites. Trials were not made on birds carrying immature worms. Litjens (1963) likewise reports favourably on the use in practice of the 1% solution on flocks of naturally infected birds.

XI. Summary and Conclusions

The number of anthelmintics available for use in the domestic animals has increased enormously in recent years. An especially large number have been tested for use in the treatment of parasitic gastro-enteritis in sheep and cattle. The recognition of the influence of particle size and of purity on the anthelmintic action of phenothiazine has resulted in the marketing of superior formulations of this compound for use in sheep. These "pure micronised" preparations of phenothiazine are, however, somewhat dangerous to use in cattle. Some of the organo-phosphorus compounds, originally studied for war-like purposes, have been tested as anthelmintics and also as systemic insecticides. Those which show serviceable anthelmintic activity are usually toxic to the host in doses only slightly higher than the therapeutic dose and they seem unlikely to be of sufficient value to replace established compounds. Some of these compounds are, however, used to control warble fly in cattle and it would clearly be preferable to use an insecticide which also had an anthelmintic effect to one without. The recently described organo-phosphorus compound haloxon seems to be exceptionally safe in having a toxic dose of about five times the therapeutic dose. Thiabendazole, which has a wide range of activity against mature and immature parasites in ruminants and also a wide therapeutic index, is without doubt the anthelmintic of choice for parasitic gastro-enteritis in sheep and further experience may show it to be equally effective in cattle at a higher dose range. It has been suggested that its high activity against immature stages will interfere with the development of immunity in treated animals but there is little likelihood that it will be used in practice in such a way that immunity mechanisms will be affected. Bephenium hydroxynaphthoate has an established place in the treatment of nematodiriasis in sheep but thiabendazole seems likely to prove equally effective.

Although cyanacethydrazide and diethylcarbamazine have a specific action on lungworms, neither of these compounds is the ideal for the treatment of parasitic bronchitis. For use in cattle a drug is required which will kill all stages of the life cycle of *Dictyocaulus viviparus* from parasitic third stage to adult worm, and for use in sheep and goats a drug is required which will remove the small lungworms as well as *Dictyocaulus filaria.* If a drug active against all stages of the life cycle of the parasite were available it could probably be used in a system of strategic treatment to prevent occurrence of the disease.

Several new drugs have recently been introduced for the treatment of fascioliasis and some of these have advantages over the well established carbon tetrachloride. A non-toxic drug active against the immature stages of the fluke is still lacking and hampers the treatment of acute fascioliasis. Hexachlorophene is the best drug yet discovered for this purpose but the dose (40 mg/kg) which will remove immature flukes approaches the toxic level and will cause some losses in the field. The catastrophic losses occasioned by acute fascioliasis in sheep may well justify the risk of using hexachlorophene for its treatment.

There has been welcome progress in the provision of new taeniacides in the last few years. Some of the new products show great promise but it is too early yet to be quite certain of their value. There are still a number of tapeworm infections, e.g. *Stilesia hepatica,* for which no cure is available and progress in this field would be welcome.

The piperazine compounds have an established place as ascaricides in the domestic animals. Their action against the strongylid worms increases their value for use in horses. The thenium compounds and disophenol have provided new, useful treatments for hookworms in dogs.

It seems very likely that the spectacular progress, which has been seen in recent years, will continue in the future. The domestic animals frequently harbour multiple worm infestations, and anthelmintics for veterinary use which have a wide range of activity and low toxicity have obvious possibilities. There are still many worm infections for which such drugs remain to be provided.

References

Alicata, J. E. (1960). *Amer. J. vet. Res.* **21**, 410–15.

Allen, R. W., Enzie, F. D. and Samson, K. S. (1962). *Amer J. vet. Res.* **23**, 236–40.

Anderson, D. J., Heaton, J. W. and Higgs, B. F. T. (1962), *Sth west Vet.* **15**, 140–42.

Anon. (1943). *Imperial Agricultural Bureau Joint Publication* No. 4, 21.

Armour, J. and Hart, J. A. (1960). *Vet. Rec.* **72**, 306–309.

Armour, J., Hart, J. A., Lee, R. P. and Ross, J. G. (1961a). *Vet. Rec.* **73**, 234–37.

Armour, J., Hart, J. A. and Ross, J. G. (1961b). *Vet. Rec.* **73**, 485–89.

Armour, J., Brown, P. R. M. and Sloan, J. E. N. (1962). *Vet. Rec.* **74**, 1454–57.

Arundel, J. H. (1962). *Aust. vet. J.* **38**, 307.

Bailey, W. S., Diamond, D. L. and Walker, D. F. (1961). *J. Parasit.* **47**, (Suppl.) 40–41.

Baker, N. F., Allen, P. H., Longhurst, W. M. and Douglas, J. R. (1959). *Amer. J. vet. Res.* **20**, 409–13.

Banks, A. W. and Korthals, A. (1960). *Aust. vet. J.* **36**, 383–86.
Banks, A. W. and Michel, J. F. (1960). *Vet. Rec.* **72**, 135–36.
Banks, A. W. and Mitton, R. L. (1960). *Vet. Rec.* **72**, 241–45.
Batte, E. G. (1961). *Vet. Med.* **56**, 281–84.
Behrens, H. (1960). *Dtsch. tierärztl. Wschr.* **67**, 467–70.
Bell, R. R., Glavin, T. J. and Turk, R. D. (1962). *Amer. J. vet. Res.* **23**, 195–200.
Boray, J. (1956a). *Acta vet. hung.* **36**, 469–73.
Boray, J. (1956b). *Mag. állator. Lapja* **11**, 40–42.
Boray, J. C. and Pearson, I. G. (1960a). *Nature, Lond.* **186**, 252–53.
Boray, J. C. and Pearson, I. G. (1960b). *Aust. vet. J.* **36**, 331–37.
Bosman, C. J., Thorold, P. W. and Purchase, H. S. (1961). *J. S. Afr. vet. med. Ass.* **32**, 227–33.
Bradley, R. E., Crimmins L. T. and Zweigart, T. F. (1956). *J. Amer. vet. med. Ass.* **128**, 393–94.
Brown, C. G. D. (1962a). *Vet. Rec.* **74**, 259–62.
Brown, C. G. D. (1962b). *Vet. Rec.* **74**, 787–91.
Brown, F. G. (1962), *Vet. Rec.* **74**, 524–27.
Brown, H. D., Matzuk, A. R., Ives, I. R., Peterson, L. H., Harris, S. A., Sarett, L. M., Egerton, J. R., Yakstis, J. J., Campbell, W. C. and Cuckler, A. C. (1961). *J. Amer. chem. Soc.* **83**, 1764–65.
Burrows, R. B. (1958). *J. Parasit.* **44**, 607–10.
Burrows, R. B. and Lillis, W. G. (1962). *Amer. J. vet. Res.* **23**, 77–80.
Cairns, G. C. (1961). *N. Z. vet. J.* **2**, 147–52.
Castel, P., Graber, M., Gras, G. and Chhay-Hancheng. (1960). *Rev. Élev.* **13**, 57–74.
Cieleszky, V. and Kovács, F. (1958). *Mag. állator. Lapja* **13**, 279–81.
Clark, D. T. and Connor, N. D. (1959). *Amer. J. vet. Res.* **20**, 452–58.
Cole, D. G., Gibbs, H. C. and Jeffers, H. (1962). *Canad. J. comp. Med.* **26**, 264–67.
Colglazier, M. L. and Enzie, F. D. (1961). *Proc. helm. Soc. Wash.* **28**, 86–91.
Copp, F. C., Standen, O. D., Scarnell, J., Rawes, D. A. and Burrows, R. B. (1958). *Nature, Lond.* **181**, 183.
Cotteleer, C. (1962). *Vlaams. diergeneesk. Tijdschr.* **31**, 279–82.
Dass, N., Panda, S. N. and Biswal, G. (1961). *Indian vet J.* **38**, 194 97.
Delak, M. and Marzan, B. (1959). *Vet. Arhiv* **29**, 357–62.
Delak, M. Drežančić, I. and Marzan, B. (1961). *Vet. Arhiv* **31**, 286–96.
Demidov, N. V. (1954). *Veterinariya, Moscow* **31**, (4), 16–18.
Demidov, N. V. (1955). *Veterinariya, Moscow* **32**, (4), 29–32.
Demidov, N. V. (1958). *Bull. Informatsii vsesoyuz. Inst. Gelmint.* No. 4, 36 38.
Demidov, N. V. (1959). *Trud. gelmint. Lab.* **9**, 89–90.
Demidov, N. V., Deripasko, P. G. and Kovalev, G. V. (1959). *Trud. vesesoy. Inst. Gelmint.* **6**, 216–20.
Dick, J. R. (1958). *Vet. Med.* **53**, 413–15, 455.
Donigiewicz, K. (1951). *Méd. vét., Varsovie* **7**, 237–39.
Dorney, R. S. and Todd, A. C. (1959). *J. Amer. vet. med. Ass.* **135**, 336–338.
Dorsman, W. (1959). *Tijdschr. Diergeneesk.* **84**, 100–103.
Douglas, J. R. and Baker, N. F. (1959). *J. Amer. vet. med. Ass.* **135**, 567–69.
Douglas, J. R., Baker, N. F. and Longhurst, W. M. (1956). *Amer. J. vet. Res.* **17**, 318–23.
Douglas, J. R., Baker, N. F. and Longhurst, W. M. (1959). *Amer. J. vet. Res.* **20**, 201–205.
Downing, W., Kingsbury, P. A. and Sloan, J. E. N. (1955). *Vet. Rec.* **67**, 641–44.
Drudge, J. H., Leland, Jr., S. E., Wyant, Z. N., Elam, G. W., Smith, Jr., C. E. and Dale, Jr., E. (1957a). *Amer. J. vet. Res.* **18**, 792–97.
Drudge, J. H., Leland, Jr., S. E., Wyant, Z. N. and Hutzler, L. B. (1957b). *J. Amer. vet. med. Ass.* **131**, 231–33.

Drudge, J. H., Leland, Jr., S. E., Wyant, Z. N., Elam, G. W. and Hutzler, L. B. (1960). *Amer. J. vet. Res.* **21**, 397–402.
Drudge, J. H., Haws, D. A., Leland, Jr., S. E., Lyons, E. T. and Rust, J. W. (1961). *Vet. Med.* **56**, 135–38.
Drudge, J. H., Szanto, J., Wyant, Z. N. and Elam, G. W. (1962). *J. Parasit.* **48**, (Suppl.) 28.
Dunsmore, J. D. (1960). *Vet. Rec.* **72**, 573.
Dunsmore, J. D. (1962). *Aust. vet. J.* **38**, 42.
Egerton, J. R., Cuckler, A. C., Ames, E. G., Bramel, R. G., Brightenback, G. E. and Washko, F. V. (1962). *J. Parasit.* **48**, (Suppl.) 29.
Egorov, Y. G. and Morozov, I. G. (1960). *Trud. nauchno-issled. vet. Inst. Minsk.* **1**, 174–78.
Egyed, N. and Nemeseri, L. (1957). *Acta vet. hung.* **7**, 345–50.
Eikmeier, H. and Kamel, S. H. (1961). *Tierärztl. Umsch.* **16**, 79–81.
Eisa, A. M. and Rubin, R. (1961). *Amer. J. vet. Res.* **22**, 708–12.
Enigk, K. (1953). *Mh. prakt. Tierheilk.* **5**, 14–22.
Enigk, K. and Düwel, D. (1959). *Dtsch. tierärtzl, Wschr.* **66**, 379–82.
Enigk, K. and Düwel, D. (1960). *Dtsch. tierärtzl. Wschr.* **67**, 535–39.
Enigk, K. and Düwel, D. (1961). *Dtsch. tierärtzl. Wschr.* **68**, 601–607.
Enigk, K. and Federmann, M. (1958). *Mh. Vet. Med.* **13**, 705–709.
Enzie, F. D., Wilkens, E. H. and Colglazier, M. L. (1958). *Amer. J. vet. Res.* **19**, 19–24.
Erhardt, A. (1956). *Arzneimittel - Forschung.* **6**, 496–97.
Farrington, K. J. and Thomson, B. M. (1962). *Aust. vet. J.* **38**, 103–104.
Farrington, K. J., Thomson, B. M. and Whitlock, H. V. (1962). *Aust. vet. J.* **38**, 53–58.
Favati, V. and Della Croce, G. (1961). *Ann. Fac. Med. vet. Pisa* **14**, 294–304.
Federmann, M. (1959). *Dtsch. tierärtzl. Wschr.* **66**, 526–29.
Forsyth, B. A. (1959). *Aust. vet. J.* **35**, 99.
Forsyth, B. A. (1962). *Aust. vet. J.* **38**, 398.
Forsyth, B. A., Scott, M. T. and Bainbridge, J. R. (1961). *Vet. Rec.* **73**, 67–75.
Fukui, M., Kaneko, C. and Ogawa, A. (1960). *Jap. J. Parasit.* **2**, 217–23.
Galvin, T. J., Bell, R. R. and Turk, R. D. (1959). *Amer. J. vet. Res.* **20**, 784–86.
Galvin, T. J., Bell, R. R. and Turk, R. D. (1960a). *Sthwest. Vet.* **13**, 197–202.
Galvin, T. J., Turk, R. D. and Bell, R. R. (1960b). *Amer. J. vet. Res.* **21**, 1054–57.
Galvin, T. J., Bell, R. R. and Turk, R. D. (1962). *Amer. J. vet. Res.* **23**, 191–94.
Gavrilyuk, P. Y. (1956). *Veterinariya, Moscow* **33**, (5), 45.
Geevaerts, J. (1962). *Vlaams diergeneesk. Tijdschr.* **31**, 105–13.
Gibbs, H. C. and Pullin. J. W. (1961). *Canad. J. comp. Med.* **25**, 171–77.
Gibbs, H. C. and Pullin, J. W. (1962). *Canad. J. comp. Med.* **26**, 163–66.
Gibson, T. E. (1949). *Brit. vet. J.* **105**, 309–12.
Gibson, T. E. (1957). *Brit. vet. J.* **113**, 90–92.
Gibson, T. E. (1959a). *Vet. Rec.* **71**, 431–34.
Gibson, T. E. (1959b). *Vet. Rec.* **71**, 949–50.
Gibson, T. E. (1960). *Vet. Rec.* **72**, 343–44.
Gibson, T. E. (1961). *Vet. Rec.* **73**, 230–31.
Gibson, T. E. (1962). *Vet. Rec.* **74**, 224–25.
Gordon, H. Mcl. (1940). *J. Coun. sci. industr. Res. Aust.* **13**, 85–86.
Gordon, H. Mcl. (1956). *Aust. vet. J.* **32**, 258–68.
Gordon, H. Mcl. (1958a). *Aust. vet. J.* **34**, 104–10.
Gordon, H. Mcl. (1958b) *Aust. vet. J.* **34**, 376–81.
Gordon, H. Mcl. (1961). *Nature, Lond.* **191**, 1409–10.
Gracey, J. F. and Kerr, J. A. M. (1961). *Vet. Rec.* **73**, 171–72.
Greenberg, J., Seymour, W. E. and McEwen, A. F. (1958). *Vet Med.* **53**, 609–10.
Griffiths, R. B. (1954), *J. Pharm., Lond.* **6**, 921–43.

Groves, T. W. (1958). *Vet. Rec.* **70**, 219–21.
Groves, T. W. (1961). *Vet. Rec.* **73**, 196–201.
Guilhon, J. (1962). *Bull. Acad. vét. Fr.* **35**, 271–74.
Guilhon, J. and Graber, M. (1961), *Bull. Acad. vét. Fr.* **34**, 119–24.
Guthrie, J. E. (1956). *Vet. Med.* **51**, 235–38.
Hamilton, J. (1961). *Vet. Rec.* **73**, 169–171.
Harrow, W. T. (1962). *Vet. Rec.* **74**, 1433–34.
Harwood, P. D. (1953). *Exp. Parasit.* **2**, 428–55.
Hayes, F. A. and McDaniel, H. T. (1959). *J. Amer. vet. med. Ass.* **134**, 565–67.
Hebden, S. P. (1961). *Aust. vet. J.* **37**, 264–69.
Hebden, S. P. and Setchell, B. P. (1962). *Aust. vet. J.* **38**, 399.
Hendriks, J. (1962). *Tijdschr. Diergeneesk.* **7**, 314–22.
Hendriks, J. (1963). *Tijdschr. Diergeneesk.* **88**, 418.
Herlich, H. (1962). *J. Parasit.* **48**, (Suppl.) 29.
Herlich, H. and Johnson, J. M. (1957). *J. Parasit.* **43**, (Suppl.) 19.
Herlich, H. and Porter, D. A. (1958). *Vet. Med.* **53**, 343–48, 350.
Herlich, H., Porter, D. A. and Isenstein. R. S. (1961). *Vet. Med.* **56**, 219–21.
Hiepe, T., Heide, D. and Lippmann, R. (1959). *Berl. Münch. tierärztl. Wschr.* **72**, 315–19.
Hiscock, R. H. (1962). *Vet. Rec.* **74**, 1140.
Hollo, F. (1961). *Mag. állator. Lapja* **16**, 309–11.
Horváth, J. (1958). *Mag. állator. Lapja* **13**, 191–94.
Ivanov, I. V. (1961). *Sborn. Inf. vesesoyuz. Inst. Gel'mint.* No. 7/8, 26–28.
Jarrett, W. F. H., McIntyre, W. I. M. and Sharp, N. C. C. (1962). *Amer. J. vet. Res.* **23**, 1183–91.
Johnston, R. W. (1962). *Vet. Rec.* **74**, 285.
Jordan, H. E. (1960). *J. Amer. vet. med. Ass.* **136**. 508–10.
Jungmann, R. (1960). *Arch. exp. Vet Med.* **14**, 589–604.
Kagramanov, A. K. (1955). *Veterinariya, Moscow* **32**, (4), 39–40.
Kassai, T. (1958) *Mag. állator. Lapja* **13**, 9–13.
Kendall, S. B. and Parfitt, J. W. (1962). *Brit. vet. J.* **118**, 1–10.
Kingsbury, P. A. (1958). *Vet. Rec.* **70**, 523–28.
Kingsbury, P. A. (1961a). *Vet. Rec.* **73**, 183–84.
Kingsbury, P. A. (1961b). *Res. vet. Sci.* 265–71.
Knight, R. A., McGuire, J. A. and Walton, L. (1960). *Vet. Med.* **55**, 71–74.
Komjathy, K. (1957). *Mag. állator. Lapja* **12**, 235–36.
Koutz, F. R. and Groves, H. F. (1961). *Speculum* **14**, No. 2, 35–37.
Kovacs, F. (1958). *Mag. állator Lapja* **13**, 118–24.
Kovacs, F. (1959). *Proc. XVI th. Int. Vet. Congr. Madrid* **2**, 605–607.
Kurtpinnar, M. and Kalkan, A. (1960). *Türk vet. Hekim. dern. Derg.* **30**, 770–76.
Lämmler, G. (1960). *Dtsch. tierärtzl. Wschr.* **67**, 408–13.
Landram, J. F. and Shaver, R. J. (1961). *Amer. J. vet. Res.* **22**, 893–98.
Lee, R. P. (1955). *Vet. Rec.* **67**, 146–49.
Lee, R. P. (1956). *Bull. Epizoot. Dis. Afr.* **4**, 61–63.
Leiper, J. W. G. (1954). *Vet. Rec.* **66**, 596–99.
Litjens, J. B. (1963). *Tijdschr. Diergeneesk.* **88**, 435.
Macrae, R. R. (1961). *Vet. Rec.* **73**, 193–95.
Mahmoud, A. H., Fahmy, M. A. and Selim, M. K. (1958). *Zbl. Vet. Med.* **5**, 366–72.
Malone, J. C. (1962). *Res. vet. Sci.* **3**, 18–33.
Malone, P. H. (1956). *N. Z. vet. J.* **4**, 125–26.
Mann, P. H., Harfenist, M. and De Beer, E. J. (1955). *J. Parasit.* **41**, 575–78.
Marquardt, W. C., Fritts, D. H., McAlpin, N. R. and Hawkins, Jr., W. W. (1960). *J. Parasit.* **46**, 42.

Matevosyan, E. M. and Kryukova, K. A. (1961). *Sborn. Inf. vsesoyuz. Inst. Gel'mint.* No. 7/8, 34–42.
Nöller, W. and Schmidt, F. (1927). *Tierärztl. Rdsch.* **38**, 851–57.
Obitz, K. and Wadowski, S. (1939). *Wiad. weteryn.* **18**, No. 233, Suppl., 96–102.
O'Donoghue, J. A. (1958). *Canad. J. comp. Med.* **22**, 237–39.
Osinga, A. (1960). *Tijdschr. Diergeneesk.* **85**, 529–33.
Ozerskaya, V. N. (1955). *Trud. XL. Plenum. vet. Sekt. Akad. Selskokhoz. Nauk im. Lenina.* 237–46.
Ozerskaya, V. N. (1959). *Trud. gel' mint. Lab.* **9**, 208–10.
Parker, W. H. (1957) *J. comp. Path.* **67**, 251–62.
Parker, W. H. and Roberts, H. E. (1958). *J. comp. Path.* **68**, 402–10.
Parker, W. H. and Vallely, T. F. (1960). *Vet. Rec.* **72**, 1073–77.
Parker, W. H., Roberts, H. E., Vallely, T. F. and Brown, F. T. (1959). *Vet. Rec.* **71**, 509–11.
Pearson, I. G. and Boray, J. C. (1961). *Aust. vet. J.* **37**, 73.
Pecheur, M., Pouplard, L. and Gregoire, C. (1962a). *Ann. Méd. vét.* **106**, 101–108.
Pecheur, M., Pouplard, L. and Gregoire, C. (1962b). *Vlaam. diergeneesk. Tijdschr.* **31**, 114–16.
Pierotti, P. (1958). *Atti Soc. ital. Sci. vet.* **12**, 442–45.
Pouplard, L., Pecheur, M. and Gregoire, C. (1961). *Ann. Méd. vét.* **105**, 285–65.
Poynter, D. (1955a). *Vet. Rec.* **67**, 159–63.
Poynter, D. (1955b). *Vet. Rec.* **67**, 652–26.
Poynter, D. (1956a). *Vet. Rec.* **68**, 291–97.
Poynter, D. (1956b). *Vet. Rec.* **68**, 429–31.
Poynter, D. and Hughes, D. L. (1958). *Vet. Rec.* **70**, 1183–88 .
Rawes, D. A. (1961). *Vet. Rec.* **73**, 390–92.
Rawes, D. A. and Clapham, P. A. (1961). *Vet. Rec.* **73**, 755–58.
Rawes, D. A. and Clapham, P. A. (1962). *Vet. Rec.* **74**, 383–85.
Rawes, D. A. and Scarnell, J. (1958). *Vet. Rec.* **70**, 251–55.
Rawes, D. A. and Scarnell, J. (1959). *Vet. Rec.* **71**, 645–50.
Reeves, R. J. C. (1962). *Vet. Rec.* **74**, 1111–12.
Reinecke, R. K. and Rossiter, L. W. (1962). *J. S. Afr. vet. med. Ass.* **33**, 193–99.
Riedel, B. B. and Larson, E. J. (1956). *J. Amer. vet. med. Ass.* **129**, 156–59.
Riek, R. F. and Keith, R. K. (1958). *Aust. vet. J.* **34**, 93–103.
Rosenburger, G. and Heeschen, W. (1959). *Dtsch. tierärztl. Wschr.* **66**, 169–73.
Rosenburger, G. and Heeschen, W. (1960). *Dtsch. tierärztl. Wschr.* **67**, 403–405.
Rubin, R. (1959). *J. Parasit.* **45**, (Suppl.) 57.
Rubin, R. (1960). *Amer. J. vet. Res.* **21**, 178–80.
Rubin, R. and Eisa, A. M. (1961). *Vet. Med.* **56**, 487–89.
Rubin, R. and Tillotson, A. J. (1960). *Amer. J. vet. Res.* **21**, 1040–45.
Rubin, R. and Tillotson, A. J. (1962). *Amer. J. vet. Res.* **23**, 42–48.
Schad, G. A., Allen, R. W. and Samson, K. S. (1958). *Vet. Med.* **53**, 533–34, 554.
Sen, H. G., Kelley, G. W. and Olsen, L. S. (1960). *J. Amer. vet. med. Ass.* **136**, 366–368.
Senviratna, P. (1960). *Vet. Rec.* **72**, 200–203.
Shelton, G. C. (1962). *Amer. J. vet. Res.* **23**, 506–509.
Shumard, R. F. and Eveleth, D. F. (1956). *Vet. Med.* **51**, 515–17.
Simunek, J. (1961). *Sborn. vys. Šk. zemědělsk. Brno* Ser. B. **9**, 35–40, 247–260.
Skerman, K. D. (1962). *Aust. vet. J.* **38**, 324–34.
Slanina, L., Popluhár, L. and Vrzgula, L. (1955). *Sborn. čes. Akad. Zemědělsk. Ved., vet. Med.* **28**, 923–40.
Sloan, J. E. N., Kingsbury, P. A. and Jolly, D. W. (1954) *J. Pharm., Lond.* **6**, 718–24.
Sloan, J. E. N., Wood, J. C. and Rawes, D. A. (1961). *Vet. Rec.* **73**, 432–35.
Southcott, W. H. (1961). *Aust. vet. J.* **37**, 55–60.

Sprent, J. F. A. and English P. B. (1958). *Aust. vet. J.* **34**, 161–71.
Stampa, S. (1959). *J. S. Afr. vet. med. Ass.* **30**, 19–26.
Stampa, S. and Terblanche, H. J. J. (1961). *J. S. Afr. vet. med. Ass.* **32**, 367–71.
Standen, O. D. (1958). *E. Afr. med. J.* **35**, 439–45.
Steward, J. S. (1955). *Parasitology*, **45**, 242–54.
Swanson, L. E., Wade, A. E., Senseman, V. F. and Djafar, M. I. (1959). *Amer. J. vet. Res.* **20**, 777–83.
Tchoubarie, I. T. (1958). *Bull Off. int. Epiz.* **49**, (bis), 633–39.
Thienpont, D. and Mortelmans, J. (1962). *Vet. Rec.* **74**, 850–52.
Thomas, P. L. and Elliott, D. C. (1957). *N. Z. vet. J.* **5**, 66–69.
Thorpe, E. (1962). *J. comp. Path.* **72**, 29–32.
Turk, R. D., Ueckert, B. W. and Bell, R. R. (1962). *J. Amer. vet. med. Ass.* **141**, 240–42.
Veselova, T. P. and Velikovskaya, Y. A. (1959), *Veterinariya, Moscow* **36**, (7), 39–41.
Vitenko, S. N. (1956), *Veterinariya, Moscow* **33**, (5), 45.
Vodrážka, J., Berecký, I. and Sokol, J. (1959). *Vet. Čas.* **8**, 255–67.
Vodrážka, J., Sokol, J. and Berecký, I. (1960). *Vet. Rec.* **72**, 404–408.
Walley, J. K. (1957). *Vet. Rec.* **69**, 815–24, 850–53.
Walley, J. K. (1961). *Vet. Rec.* **73**, 159–68.
Walley, J. K. (1962). *Vet. Rec.* **74**, 927–32.
Watt, J. H., Nicholson, T. B. and Macleod, N. S. M. (1961). *Vet. Rec.* **73**, 567–72.
Weidenbach, C. P., Radeleff, R. D. and Buck, W. B. (1962). *J. Amer. vet. med. Ass.* **140**, 460–63.
Whitten, L. K. (1956). *N. Z. vet. J.* **4**, 63–68.
Wikerhauser, T., Žuković, M., Vražić, D., Richter, S. and Mažgon Z. (1959). *Vet. Arhiv* **29**, 363–68.
Winterhalter, M. (1961). *Vet. Arhiv* **31**, 55–70.
Winterhalter, M. and Delak, M. (1955). *Vet. Arhiv.* **25**, 68–74.
Winterhalter, M. and Delak, M. (1956). *Vet. Arhiv* **26**, 307–12.
Wood, I. B. and Pankavich, J. A. (1961). *J. Parasit.* **47**, (Suppl.), 20.
Wood, I. B., Pankavich, J. A., Wallace, W. S., Thorson, R. E., Burkhart, R. L. and Waletsky, E. (1961). *J. Amer. vet. med. Ass.* **139**, 1101–05.
Worley, D. E. (1957). *J. Parasit.* **43**, 632.
Yakovlev, Y. (1955). *Veterinariya, Moscow* **32**, (4), 40.
Young, J. (1961). *Vet. Rec.* **73**, 192–93.
Zettl, K. (1959) *Dtsch. tierärztl. Wschr.* **66**, 319–23.
Zettl, K. (1962) *Vet.-med. Nachr.* No. 1. 19–33.
Zunker, M., Bach, V. and Jordan, W. (1927). *Berl. tierärztl. Wschr.* 121–23.

Snail Control in Trematode Diseases: the Possible Value of Sciomyzid Larvae Snail-Killing Diptera

CLIFFORD O. BERG

Department of Entomology and Limnology, Cornell University, Ithaca, New York, U.S.A.

I. Introduction

This report concerns a family of true flies or Diptera, the Sciomyzidae or marsh flies, whose larvae kill and feed upon pulmonate snails. Larvae of more than 100 species in this family have destroyed freshwater and terrestrial gastropods in laboratory trials. More than eighty species have been reared from hatching to pupation solely on snails, introduced alive into their rearing dishes. Starvation tests to determine whether the larvae could be induced to attack each other or other invertebrates found in their habitats have proved negative. The snails killed include many trematode hosts, and predation on them has been observed also in nature. Laboratory trials against important hosts of human blood flukes from various parts of the world indicate that the hosts of *Schistosoma mansoni* Sambon are very vulnerable to attacks by these larvae, that the *Bulinus* species which transmit *S. haematobium* (Bilharz) are less so, and that *Oncomelania*, hosts of *S. japonicum* Katsurada, are invulnerable. Laboratory and field tests of Sciomyzidae experimentally introduced into Hawaii, Guam, and Australia for control of fascioliasis also are discussed.

The discovery of this food habit and of the fact that it is widespread, perhaps even universal, in a family that had been virtually unknown biologically has elicited considerable interest among entomologists, malacologists, and ecologists. The larvae range from overt predators to highly specialized parasitoids, with many species of intermediate habits forming a continuum between the two extremes. As this continuum may illustrate the evolution of predator-prey or parasite-host relations, sciomyzid biology may be of some academic interest also to parasitologists.

However, these academic and theoretical considerations are quite incidental to the major thesis presented here. This is (1) that the sciomyzid flies are one of several groups that may be of practical value in the biological control of fluke-bearing snails, and (2) that much more effort should be expended in the quest for other natural enemies of snails and in research into their potential value as agents of biological control.

Resistance to these ideas, experienced before, is anticipated again; it seems to come primarily from much emphasis on the advantages of molluscicides with little consideration of their disadvantages. Secondarily, there is reticence to get into the ecologically involved and complex matter of biological control, based on recognition (1) of this complexity, (2) of the uncertainty that biological control will provide a solution for any given problem, and (3) of the limited financial provision made for biological control studies in a period in which chemical control has predominated.

It is deemed necessary to challenge both the complacency with which some parasitologists seem to regard molluscicides as the full and complete answer to all snail control problems, and the defeatism, expressed by a few in writing and by more in conversation, concerning biological control schemes on which they have no first-hand information. If these attitudes and the policies that result from them were not questioned at the outset, snail control workers would see no practical reason for reading any further.

This inquiry and challenge is focused on attitudes and policies of snail control. Criticism of individuals and groups is not intended. Indeed, there is a tendency not even to question policies because of the possibility of reflecting indirectly on individuals identified with them. However, the resulting temptation to delete these questions has been resisted. The future of snail control is far more important than anyone's personal sensitivities, and this cause is served better by clear, forthright objectivity than by the obscurities of diplomatic language.

II. NEED FOR NEW SNAIL CONTROL METHODS

Parasitologists seem agreed that control of the snail intermediate host is one effective means of reducing the transmission rate of trematode diseases. However, there is considerably less agreement concerning the most practical methods of snail control. It is hoped that the following comments and suggestions on this matter will prove useful.

A. COST OF CHEMICAL CONTROL

Chemical control—the application of toxicants known as molluscicides—has received more attention in the literature than any other snail control method. Indeed, it has been the subject of more recent papers and reports than all other methods put together. Yet the comprehensive review of this technique by Paulini (1958) mentions some serious drawbacks and characterizes it as "a temporary measure and relatively expensive." Others (Pesigan, 1953; Meleney, 1954; Schwetz, 1954; Van der Schalie, 1958) have stated unequivocally that this method of control, with the techniques used by them, is too expensive to be practical. Van der Schalie summarized: "With the high cost of the chemical and the additional expenditure on labour, the costs involved in keeping even a 5,000-acre area under control with molluscicides would be prohibitive. Clearly, therefore, further research on methods of control is needed... Malaria has been controlled in many regions through attack on the mosquito vectors; there is hope that bilharziasis* can also be reduced to a low level through control measures directed against the intermediate hosts of the parasite. The difference between the two cases is that much is now known about the biology of the malaria vectors and about the means of reducing or eliminating them, whereas medical malacology stands today roughly where medical entomology stood a quarter of a century ago. Unless a greater effort is made and more funds are available for research, there will be no hope of achieving complete control of the vector snails, and the staggering costs of attempts to effect even a partial reduction in snail populations with molluscicides will effectively prevent the widespread application of such a control measure."

Improved molluscicides, formulations, and application methods have since reduced the cost, but the application of molluscicides onto extensive breeding areas, repeated at regular intervals as required, will inevitably continue to be an expensive control method.

* The name "bilharziasis" is used by the World Health Organization for the diseases caused by the blood flukes of man and more commonly known as "schistosomiasis" in some countries. Because WHO literature is quoted freely here, the use of the term "bilharziasis" throughout avoids a confusing duplication of terms.

B. PESTICIDES AND POPULATION DYNAMICS

A disadvantage of molluscicides that is even more serious than any economic consideration is the probability of indiscriminate poisoning of most of the animals exposed to them. Such great disturbances of the ecosystem, with simultaneous elimination of many natural enemies, usually result in more frequent and more serious outbreaks of the pest species. Studies of these so-called "side effects" have not progressed nearly as far on molluscicides as on insecticides, and comparatively little information is available on this subject. The Expert Committee on Bilharziasis of the World Health Organization (1961) reported, "So far little is known and little work has been done to study the immediate and long-term effects of molluscicide on the fauna, including fish, frogs, shrimps, etc., and on the microfauna and microflora of the habitat." Inasmuch as molluscicides have been developed more recently, and all phases of study related to them are less advanced than the corresponding investigations on insecticides, this situation is quite understandable. But it follows that medical malacologists probably should be studying our experiences with insecticides and other pesticides, learning from these studies certain general principles that would probably apply also to molluscicides, and thus avoiding costly mistakes.

The critical analyses by Brown (1961) of four extensive insect control campaigns are well worth reading in this connection. Brown reported that both operations based on the general, indiscriminate application of stable, potent, broad-spectrum insecticides resulted in ineffective control of the target species and in much destruction of other species associated with them. He commended the other two campaigns, in which thorough research on the biology of the pest species resulted in techniques which avoided destruction of associated species and in far more effective control. He emphasized the value of "new approaches" and predicted accelerated development of control measures alternative to chemical methods.

In his exhaustive review of the giant African land snail problem, Mead (1961) wrote, "...in the vast majority of cases, after repeated chemical control measures are finally discontinued, the giant snail population builds up to the point where it is just as bad if not worse than it was before... just as soon as comprehensive behavioral study can be made of these snails, there will be not only an increase in the efficacy of the existing control measures but new and more effective means of control can be conceived... If there is a formula for producing an eradication of this snail pest, it probably will be found in the realm of multiple unfavorable environmental factors." This problem is vastly different from that of controlling the aquatic and amphibious snails that harbor dangerous trematode parasites. No claim is made that specific details of the two situations are comparable nor that extrapolation from one to the other would be valid. The case of the giant African snail is simply another instance in which years of attempted control by chemical methods have led to the conclusion that intensive studies of the pest species would almost certainly show the way to more effective control methods.

The trend of contemporary thought concerning pest control methods clearly is to place increasing emphasis on principles of ecology and population dynamics. Well qualified to express this point of view, Geier and Clark (1961) wrote: "Since modern control began, we have become confronted by an increasing number of pest problems, the fundamental nature of which continues to be ignored in our attempts at solving them. An endless sequence of *ad hoc* programmes of destruction has kept the available trained personnel constantly mobilized in practical pursuits, with little time and energy for thinking about fundamentals.

"This largely explains why the concepts which generally inspire pest control practices are so much out of step with modern ecological ideas. It must be conceded that the current approach to pest control has not progressed far beyond the original idea of striking down noxious populations by a series of blows delivered almost blindly but with a maximum of violence. This [a program of applying a modern insecticide indiscriminately] may be described as the 'fly-swat approach'...

"Since eradication of the noxious species is not usually achieved by such means, their success is necessarily of short duration and, sooner or later, repeated application becomes indispensable. The fly-swat approach bears within itself the cause of its limited success—it can only reduce symptoms. It leaves unaffected the circumstances which induce pest-level densities."

The summary of a symposium that should be studied carefully by everyone concerned with pesticide work of any sort (International Union for Conservation of Nature and Natural Resources, 1961) also makes the point that many problems can be solved by non-chemical means, and that such solutions often prove to be the best ones, although they seem more complicated at first because they require understanding of the ecology of the pest species. This recommendation for more searching into the possibilities of biological control is an endorsement of Franz (1961), who urged "fullest use of any weapon provided by nature," and warned against reliance on "the continuation of present conditions with regard to the susceptibility of pests to chemical insecticides since this has already been shown to be fallacious." He predicted that one of the future trends of sound pest control practices will be to use chemical methods only as emergency measures where other possibilities cannot be discovered, and in such a way that the ecosystem is disturbed as little as possible.

Although the problems and control methods discussed by the writers cited above differ greatly from those in endemic areas of the most serious trematode diseases, the fundamental ecological principles are identical. Their timely suggestions and criticisms therefore seem just as applicable to molluscicide programs as to other pesticide operations.

C. RESISTANCE PROBLEMS

Franz's warning against reliance on the continuation of present susceptibility levels refers to the constantly increasing "resistance" of pests to pesticides. Natural populations are characteristically heterogeneous, and they often contain small numbers of individuals resistant to a particular chemical toxicant. Relieved from the competitive pressure of nonresistant members of the population by the pesticide, the resistant individuals have quickly repopulated treated areas with resistant progeny. Each successive application that has fallen short of complete eradication has exerted another strong selective pressure toward greater resistance, and the resulting genetic changes have continued until some pest species now are quite invulnerable to dosages more than one hundred times the strength of those originally effective (Metcalf, 1952; Brown, 1958). Whether dosages are greatly increased or other toxicants are substituted, new hazards are created for all associated organisms, including the natural enemies.

This process is especially characteristic of species with high biotic potentials and short life cycles, and is best known in insects, protozoa, and bacteria. Inasmuch as the genetic mechanisms of snails are essentially similar to those of insects, some snails probably are developing resistance to molluscicides. There seems to be no evidence of it, but is this because it does not exist or because no one has thought to look for it? Clearly demonstrable cases can remain undetected in all snail populations not subjected to periodic tests of susceptibility, and such tests probably are not being performed on any population. The comprehensive review of molluscicides cited above (Exp. Comm. Bilharz., 1961) does not even mention resistance. Although the meeting reported in that paper was devoted exclusively to an exhaustive discussion of molluscicides and their use, the impression is left that this important question was not even raised.

By contrast, another Expert Committee charged with the evaluation of other pesticides concluded "that resistance is at present the most important single problem facing vector-control programmes" (Exp. Comm. Insecticides, 1958). It urged more attention to ecological aspects of control problems, just as it had previously urged study of the possibilities of biological control (Exp. Comm. Insecticides, 1957).

D. STABLE RESIDUES

Another matter of great and rapidly growing concern that has received little or no attention in snail-control operations is the accumulation of pesticide residues in nature (U.S. Agric. Res. Serv., 1960; Gunther, 1962–63). This dangerous situation builds up insidiously. The killing power of toxic residues is not apparent until they have accumulated in the water and soil to the extent that there is widespread mortality of other animals as well as the target species. Although all biologists should then see the danger of long-term effects on man and domestic animals, they have not always done so.

The residue problem was recognized by many entomologists (e.g. Wigglesworth, 1945; Pickett, 1949; Brown, 1951; Solomon, 1953; Linsley, 1956), and serious efforts were made to solve it, long before Carson (1962) brought it forcibly to public attention. Entomologists had admitted the real dangers from residues of stable, broad-spectrum toxicants, and had already developed spectacularly successful alternative methods (e.g. the screwworm and medfly campaigns discussed by Brown (1961)). In the United States, the Miller Amendment to the Food, Drug, and Cosmetic Act set up special procedures to regulate pesticide residues in 1954. By the time Carson's popular book was published, minimum allowable tolerances had been set for hundreds of agricultural insecticides, and many pesticide residue laboratories were at work to assure that these minima were not exceeded. These regulations apply specifically to foodstuffs, but the transfer of stable chemical substances through food chains prevents any complacency about the presence of potent toxicants anywhere in the ecosystem.

It is not only a problem of such transfer, but often of storage and tremendous concentration in the relatively resistant species. This magnifies greatly the danger to organisms higher in the food chain. At Clear Lake, California, waters containing only 0·02 p.p.m. of DDD produced plankton containing 5 p.p.m., which in turn produced fish with fat containing up to 2,500 parts per million. Grebes that fed on the fish died although their fat contained somewhat smaller residues than that of the fish (Hunt and Bischoff, 1960; U.S. Pres. Sci. Advis. Comm., 1963).

However, medical malacologists have shown little concern about long-term effects of toxic residues, even in such intensively cultivated areas as rice paddies. The comprehensive review of molluscicides (Exp. Comm. Bilharz., 1961) contains no warning about toxic residues. Indeed, molluscicides with residual activity are welcomed for use as "chemical barriers." It is ironical that a group already at work to solve the problems she discussed bore the brunt of Carson's sharp criticism while another group, apparently oblivious to such problems and still demanding stronger and more stable toxicants, was completely overlooked. Except for their relative concealment from public view in remote, tropical regions, it seems that the medical malacologists would inevitably have attracted the most vehement criticism.

If snails are becoming resistant to molluscicides, toxic residues probably are doing far more harm than good even in uncultivated regions. Resistance to insecticides is known to develop fastest when (1) a high proportion of the total population is exposed, as to a toxicant generally dispersed throughout the area, and (2) the exposure is continuous over several successive generations. Insecticide residues in nature therefore are widely regarded as major forces inducing resistance in insect pests. Molluscicide residues must also exert continuous evolutionary pressure and result in selection for resistant properties.

E. MOLLUSCICIDE TOXICITY TO OTHER ANIMALS

The fact that so little is known about molluscicide toxicity to other animals should, in itself, sound a note of warning. An observer who must remain anonymous told of mass mortality of fish resulting from experimental application of molluscicide to irrigation ditches in Egypt. It took very little time and trouble to pick up enough poisoned fish for a large meal. The mollusciciding team had no authority to prevent this, and local residents ignored their warnings and ate well. No acute illnesses were reported, but practically nothing is known about the storage of molluscicides in mammalian tissues and their long-term effects on the consumer and his progeny.

Most members of mollusciciding teams probably have had similar experiences. The fact that local residents in endemic areas of human trematode diseases have no conception of the dangers of chronic toxicity should not surprise anyone. Information on this subtle but potentially great danger is woefully inadequate throughout the world, even among those best informed in the biological and medical sciences. The U.S. President's Science Advisory Committee Panel on the Use of Pesticides (1963) wrote, "The Panel is convinced that we must understand more completely the properties of these chemicals and determine their long-term impact on biological systems, including man. The Panel's recommendations are directed toward these needs, and toward more judicious use of pesticides or alternate methods of pest control..."

The African people who are fortunate enough to escape disastrous long-term effects of chronic toxicity due to molluscicides may still face increased problems of malnutrition because of them. Since these people are critically dependent on fish to supply minimum protein requirements in their diet, widespread application of fish-killing molluscicides would seem inevitably to intensify the serious malnutrition problem. The report of the Expert Committee on Bilharziasis (1961) does not hold forth much hope in this matter. It indicates that most molluscicides are quite toxic to fish. Then, without suggesting any alternative source of essential proteins, it concludes, "Where there is a vested interest in the fish in any locality the people concerned must be prepared for what will happen." Diseases due to malnutrition are outside the sphere of responsibility of a committee on bilharziasis, but there may be no net gain in public health if the control method selected for bilharziasis aggravates problems of malnutrition.

There is a striking contrast in points of view between those specifically charged with control of intermediate-host snails and the biologists interested in more general aspects of pest control. The former is well expressed by the Expert Committee on Bilharziasis (1961). The statement that little is known about the effects of molluscicides on the biota is followed by, "It seems important that more should be known of the practical advantages or even disadvantages of what happens to the fauna and flora. It may well be that the immediate lethal effects of a molluscicide may be prolonged and enchanced by the destruction or poisoning of those items which form the food supply of the snail intermediate hosts. The converse may also be so, and snail predators may be destroyed." The propriety of emphasizing the "practical advantages" of poisoning the associated fauna and flora unselectively is open to grave doubts. Indeed, this seems a more shocking example of pesticidal "scorched-earth" policies than the campaign which prompted Brown's (1961) use of this term.

The subordinate position given to the possibility of "even disadvantages" is consistent with the fact that the section on disadvantages of molluscicides contains no mention of their toxicity to other animals. (That section dwells principally on the relatively high cost, the unsolved problem of reinfestation and necessity for repeated treatments, and the difficult problem of training personnel in effective application methods.) Although toxicity to higher animals is recognized, it apparently is not considered a disadvantage—at least not a sufficiently important one to warrant listing it in that section. Under "future developments," the need is stressed for a "highly effective molluscicide" (evidently one with prolonged and enhanced lethal effects).

In sharp contrast to this, Janzen (1961) expressed alarm at the ever-increasing dependence of agriculture and public health on chemical control of pests and the growing number of "ever more potent pesticides," and Geier and Clark (1961) urged that the most highly selective pesticides, and the least stable ones, be given first consideration if pesticides must be used at all.

None of the foregoing discussion is intended as a denial of the value of molluscicides, which unquestionably are effective for emergency control in areas of limited size and high endemicity. If it seems to stress the limitations rather than the advantages of molluscicides, it simply reflects a strong conviction that this is the side that needs telling. Proponents of chemical control have acclaimed its advantages in more pages of the literature than are devoted to any other snail control method. But crucial questions that have challenged expansion in the use of other pesticides, and even caused significant reductions on many fronts, either have never been raised concerning the molluscicides or have been summarily dismissed in the urgency of quickly arresting transmission. With these vital questions still unanswered, there must be no complacency of the type that may be engendered by too much stress on the excellence of molluscicides. Indeed, it is a matter not only of unanswered questions but also of questions that have been answered in the negative. Molluscicides do not provide any complete and perfect control of trematode diseases, and the need for more research into alternative methods is urgent.

III. Biological and Integrated Control Methods

Biological control methods are efforts to encourage the growth and reproduction of natural enemies already present in the region and to introduce foreign species. Such control methods have been applied to insect pests and weeds for many years, and whole volumes on the subject date back to the 1930's (Thompson, 1930; Sweetman, 1936). Foreign introductions (primarily of insects) for biological control have given results ranging from complete failure to spectacular success. Although there have been many more of the former, a new surge of interest in biological control is evident among contemporary entomologists (Sweetman, 1958; 10th Int. Congr. Ent., 1958; XI Int. Kongr. Ent., 1962). This greatly increased interest is not due to the proportion of introductions that prove to have been worthwhile, which will probably always remain less than one-half. Rather, it reflects the superlative advantages gained whenever a biological control experiment "succeeds" and the resulting conviction that each "success" compensates many times over for the costs of several failures.

Two of the major advantages of biological control methods are low overall cost and avoidance of the great disturbances of the ecosystem caused by potent, broad-spectrum pesticides. Once established in the problem area, successful parasites and predators go on working continually without additional expenditures for labor and materials each year. They are self-perpetuating and capable of response to fluctuations in population densities of the pests they attack. All really promising agents of biological control can be introduced without causing any general disturbance of the ecosystem because they are highly specific. Sciomyzid larvae, for example, have greater specificity of action than any molluscicide now in use. Larvae in both major subfamilies apparently feed only on freshwater and terrestrial gastropods, and some are restricted to certain families or even to certain genera.

Proponents of the molluscicides often claim as a major advantage that co-operation of the local residents is not required. However, this is considerably less true of molluscicides than of biological control by natural enemies, whose powers of locomotion and searching capacities distinguish them from all chemical pesticides. Molluscicides can be applied with relatively little local cooperation only if the project is confined to natural watercourses and reservoirs. "In the treatment of irrigation systems where canals and ditches have to be closed, it is necessary to secure public interest and co-operation" (Exp. Comm. Bilharz., 1961). Successful natural enemies, however, would need to be introduced into only one breeding site in each local area. Whether local residents co-operated or not, they would spread from there into all watercourses occupied by their snail prey or host in that area.

A. PROSPECTS FOR BIOLOGICAL CONTROL OF SNAILS

There is no *a priori* reason to doubt that biological control can work on snails if the right predators or parasites can be found, reared and studied. The fact that no spectacular success has yet been made in snail control should not discourage anyone. This is exactly what should be expected at our present primitive level of understanding of the ecological relationships of snails and their natural enemies. It is utterly meaningless as a basis for prediction of future success or failure. As Pepper (1955) has stated, "Without a knowledge of the true relationships of both the host and the enemy to their environment as well as an understanding of the interrelationships between the two populations, there is no basis for postulating any effects, either beneficial or detrimental, of parasites or predators on the host populations." Most biological control efforts on snails have sought to utilize general predators, casually observed to feed on snails in nature or known to do so in the laboratory when no other food was available. Satisfactory results could hardly have been expected from such hastily considered experiments, nor should the possibilities of biological control be dismissed because of them. The involved ecological studies prerequisite to success in biological control are only now getting started in malacology. Success should not be expected until most natural enemies of snails are known and recognized, and much more has been learned about their basic biology and ecological relationships.

Despite the recently renewed interest in biological control among entomologists, the Panel on the Use of Pesticides (U. S. Pres. Sci. Advis. Comm., 1963) reported that biological methods of insect control have received relatively little attention "by comparison with the great emphasis on chemical control." It affirmed its conviction that continued and extensive searches will undoubtedly yield more examples of effective biological control and that this approach should be expanded. It summarized, "The variety of methods that has proved useful for biological control of certain pests, and the indication of potential value for others lead to the conclusion that more active exploration and use of these techniques may yield important benefits... for the protection of health."

This emphasis of the Panel seems appropriate and timely even in entomology. But the biological control of medically important snails has been neglected far more than the biological control of insects. To dismiss these possibilities as impractical on the basis of present knowledge, or lack of it, is to decide against trying untested ideas because they might not work.

B. NATURAL ENEMIES OF SNAILS

Michelson's review (1957) of the literature on predators, parasites, and other organisms associated with freshwater Mollusca includes 148 titles. He pointed out, however, that many of these are notes on casual observations, not supported by experimental data. Some of the organisms discussed inflict no apparent damage. Others are rather general predators which may not feed on snails to any appreciable extent in nature. Although interested in determining which organisms should be investigated as potential control agents for Mollusca of medical importance, Michelson concluded that the knowledge available gave little rational basis for selecting a specific organism or group of organisms for this purpose. He suggested that the mammals, amphibians, reptiles and birds probably show the least promise.

Malek (1958, 1962) added some titles to those of Michelson, but expressed doubts concerning the practical value in biological control of most of the organisms discussed.

More seems to have been published about *Marisa cornuarietis* (Linnaeus), a snail competitor of *Australorbis glabratus* (Say), than about any other agent of biological control of a medically important snail. *Marisa* feeds voraciously, and its limiting effect on *Australorbis* apparently results from the dual processes of depleting the common food supply and of devouring (accidentally?) both egg masses and young of *Australorbis*. A series of papers reporting the results of laboratory and field trials has culminated recently in one concluding that control under field conditions in Puerto Rico has now been demonstrated (Radke *et al.*, 1961).

Biologists at the Harvard School of Public Health differ from the majority of snail control workers in maintaining wide and continuing interests in biological methods. They have experimented with control of *Australorbis* both by *Marisa* (Chernin *et al.*, 1956a) and by leeches of the genus *Helobdella* (Chernin *et al.*, 1956b). More recently (Chernin *et al.*, 1960; Chernin, 1962) they have reported on *Daubaylia potomaca* Chitwood and Chitwood, a nematode parasite of *Helisoma* in Michigan which usually kills its host. Chernin (1962) found that about 85% of the *A. glabratus* experimentally exposed to gravid females of this nematode became infected and that all infected snails died in about 6 weeks. He discussed the possible use of *Daubaylia* in the biological control of this medically important snail.

Michelson's current survey of possible agents of biological control seems to be concentrated in the field of microbiology. He has reported on an acid-fast bacillus that is transmissible to species of *Australorbis*, *Biomphalaria*, and *Bulinus* as well as *Helisoma* (Michelson, 1961). He noted no extensive mortality due to this *Mycobacterium*, which does not seem to be useful for biological control unless its virulence can be increased considerably. Michelson (1963) has since described a microsporidian parasite, *Plistophora husseyi*, which causes extensive tissue destruction and mortality in the three species of Physidae experimentally infected. His efforts to infect snails of other families were unsuccessful, and he concluded that the host range may be restricted to species of Physidae. Michelson recalled other protozoan parasites of snails which cause extensive pathology and mortality and hence may be useful in biological control—the flagellate discussed by Hollande and Chabelard (1953) and possibly the new species of *Haplosporidium* studied by Barrow (1960).

Bequaert (1925) reviewed the literature on arthropod enemies of snails, including several papers on beetles and beetle larvae remarkably modified for snail killing. Keilin (1919, 1921) and Bequaert (1925) presented original observations on snail-killing flies belonging to the Calliphoridae and Sarcophagidae, respectively. Berg (1953) listed thirteen families of Diptera which include species reared from snails.

Knowledge of widespread snail-killing habits in the family Sciomyzidae has developed only during the past decade. Erroneous older reports of larvae feeding on dirt and slime (Dufour, 1849) and on *Lemna* (Gercke, 1876) evidently established thought patterns concerning the food of sciomyzid larvae, and misleading statements based on the resulting generalizations dominate the literature. Most general reference works on the Diptera and on aquatic insects either characterize the sciomyzid larvae as catholic in their feeding habits (Colyer and Hammond, 1951) or present broad, comprehensive lists of alleged "food materials" which indicate the same thing—that the sciomyzid larvae eat a little of everything (Bertrand, 1954). Wanting to be more specific about larval food, Sack (1939) stated that the larvae of both major subfamilies probably are phytophagous. No one even suggested that killing and eating snails is a widespread, if not universal, way of life among sciomyzid larvae until recently (Berg, 1953), and supporting evidence for that hypothesis has been accumulating only since that time. Except for an experiment that resulted as much from accident as by design, these habits probably would still remain unknown. In the light of this situation, one wonders how many more natural enemies of snails, and potential agents of biological control, are yet to be discovered.

C. COMPLEXITIES OF BIOLOGICAL CONTROL

The major disadvantage of biological control is the great amount of research necessary even to identify the natural enemies that may be useful in biological control, to learn their basic biology well enough to maintain them in the laboratory, and to answer the involved ecological questions that must be answered before an experimental introduction can be recommended. Even when a natural enemy has been introduced, has become established in the problem area, and is known to kill and feed on the target species in nature as well as in the laboratory, it may be difficult to determine whether it is really reducing population densities or merely killing off a certain surplus so that fewer will remain to be killed by the natural control factor(s) holding this population in check even before the introduction was made.

The studies of pest species recommended by Mead (1961) and others do not go far enough to answer the questions posed above. Studies "of pest species" may logically be interpreted as including analyses of the effects of temperature, current, light, and other physical factors which impinge on them, but each of the natural enemies considered potentially useful in biological control must be studied intensively as a separate objective. Selection of a species of Sciomyzidae to help control the snail host of *Fasciola gigantica* Cobbold in Hawaii was based on comparative studies of nine tropical sciomyzids with respect to fecundity, generations per year, numbers and kinds of snails killed by each larva, tolerance for the environmental conditions in the problem area, and so forth. Learning to rear and handle sciomyzid flies in the laboratory was prerequisite to all of these studies. This is entomological and ecological, far more than malacological, research. In fact, the fly now colonized on all four of the major islands of Hawaii, and known to kill *Lymnaea ollula* Gould in nature as well as the laboratory, was selected without even seeing the pest species.

In problems of this nature, the investigator whose interests and training are focused on the natural enemies themselves seems more apt to obtain dependable answers to the technical questions being asked than the one whose sphere of interest and competence is centered in the pest species. The organisms that might be of value as natural enemies belong to diverse systematic groups, and the questions concerning each can be answered only by utilization of techniques peculiar to each specialized field. It follows not only that we must study natural enemies as well as pest species but also that the co-operation of experts from the various areas of biology is required for competent pursuit of these involved problems.

The complexity of these problems and the amount of research required to solve them may have figured importantly in the decision of some biologists that natural enemies probably will never constitute effective means of snail control. Malek (1962) seemed to imply this when he refuted "contentions" based on laboratory experiments that several competitors, predators, and parasites might serve in the biological control of medically important snails. He declared, "Under laboratory conditions, however, the biological balance is disturbed to the [1] disadvantage of the snail and it seems too unlikely that these parasites and predators [2] would exterminate the snail hosts in the natural habitat..." The writer has indicated two points in this statement on which comment seems in order.

(1) Malek's point that laboratory conditions never really duplicate those in nature is well taken, but it does not necessarily follow that the unnatural conditions place the snail at a greater disadvantage than the natural enemy. The opposite almost certainly was true in experimental exposures of snail hosts of bilharziasis to dipterous larvae summarized on pp. 294 and 296, and analyses of other experimental exposures in the laboratory probably would indicate that this is not an uncommon situation.

(2) Does the choice of verb imply that control measures have no value un-

less the target snail is exterminated? If so, how many species of snails have been exterminated by molluscicides? Or should the effects of molluscicides be measured by a different standard? One is reminded of a germane and cogent discussion by Wilson (1960): "In the oversimplified view, biological control is 'successful' if other forms of control can be dispensed with, and 'unsuccessful' if they cannot... Such standards are not used for other methods of control. An insecticide which gives a good kill is considered to be successful, even though the agriculturist will be faced with recrudescence of the pest soon afterwards. Chemical control is not expected to provide perfect solutions to pest problems, and it is unrealistic to expect biological control to do so. The double standard is primarily a tribute to the extraordinarily high levels of biological control attained against some pests..."

The goal of snail eradication may be even less realistic than the same goal with respect to insect pests of agriculture. Whereas both may be virtually impossible, the former may also be unnecessary for full attainment of the ultimate objective. Malaria has been eradicated in some areas without destroying all *Anopheles* mosquitoes. Rowan and Rodriquez (to be published) indicate that transmission of bilharziasis may be stopped by improved sanitation well before the egg loads entering streams in Puerto Rico are reduced to zero. If any factor essential for the continued transmission of a disease is reduced below a certain threshold, transmission ceases. No one knows the extent to which snail populations would need to be reduced in various breeding sites to accomplish this end, but eradication probably is not necessary. Recognizing the principle of critical threshold levels, Laird (1959) wrote that the aim of biological control is not eradication but "the economical reduction of susceptible populations... below the level necessary for the continued transmission of human pathogens."

The Panel on the Use of Pesticides (U. S. Pres. Sci. Advis. Comm., 1963) indicated that the goal of eradication is seldom realistic regardless of the pest involved or the control methods being used. It added, "The acceptance of a philosophy of control rather than eradication does not minimize the technical or economic importance of a program, but acknowledges the realities of biology."

D. RISKS OF FOREIGN INTRODUCTIONS

Many invasions of foreign species, both accidentally and purposefully introduced, have had such disastrous results that properly cautious attitudes have developed toward all introductions. Levi (1952) wrote, "Although in the past many introductions of exotic animals have been dismal and expensive failures, and the complications resulting from them are legion, the introduction of such animals is still continued." In summarizing many cases, Elton (1958) could find only a "score or so of very large invasions which had, from a human point of view, a satisfactory ending." These authors have performed a valuable service in warning against irresponsible, hastily considered introductions.

However, neither of the quoted statements had reference to insect species introduced for purposes of biological control. Levi dealt exclusively with "wildlife", primarily mammals, birds, fishes, and plants. Elton specifically excluded "counterpests" from the quoted summary and stated that inclusion of introduced natural enemies would raise the total of "satisfactory invasions" appreciably. As a matter of fact, it would multiple it several fold.

Pest control by carefully planned introductions of thoroughly studied insects is a very different matter from the hastily considered introductions of vertebrate animals and plants that Levi warned against. De Vos *et al.* (1956) wrote, "The introduction of a *mammal* [my italics] for the sole purpose of controlling pests is almost certainly doomed to failure. Food habits are rarely so specialized that a mammal will feed entirely on the pest to be controlled." This appropriate emphasis on the lack of specialized food habits in predatory mammals invites comparison and contrast with the Class Insecta, which has furnished more successful agents of biological control than all other classes put together. Many insect species feed only on a single species of animal or plant, and some entire families (e. g. the Sciomyzidae) apparently confine their feeding to single classes or orders of food organisms.

The quoted statement of Levi (1952) is made in such broad, inclusive terms that its implied appeal to discontinue "introductions of exotic animals" might seem generally applicable to all groups except domestic animals. The danger of such a misunderstanding is suggested primarily because domestic animals are the only ones that the author specifically excluded. It is compounded by the fact that "wildlife," if interpreted literally, would include almost all species of insects. To help everyone keep fundamental distinctions between different cases in mind, a paraphrase of the quoted statement is suggested: "Although in the past many introductions of exotic insects have been moderate to unqualified successes, and the savings effected by them have greatly exceeded their cost, the introduction of such animals might still be opposed if anyone thought no deeper than to put all foreign introductions in the same category." This is just as true as the original statement and, with respect to insect agents of biological control, far more apropos.

People with and without training in biology have asked about the risks of introducing Sciomyzidae into new areas. The supposed dangers mentioned most often are: (1) that the adult flies, like house flies, may carry the organisms that cause other diseases, (2) that the larvae may kill other snails, perhaps desirable species, instead of, or in addition to, the target species, and (3) that the introduced larvae may decimate the snail populations, then make unpredictable changes in their food habits and attack totally unrelated (and perhaps valuable) organisms. Each of these items is discussed separately in the following paragraphs.

Houseflies are notorious carriers of diseases primarily because they combine two unfortunate traits. (1) They breed in filth and excrement, where the chances of contamination with human pathogens are great. (2) They are attracted to man's dwellings and food, where the risk of human ingestion of pathogens they deposit also is high. Sciomyzidae do not have either of these habits. They breed primarily in marshes, swamps, and ponds, and the adult flies characteristically remain in breeding sites of the larvae. They are not attracted to man or to human foods, and they rarely enter buildings of any kind. Since the adult flies feed on snails which the larvae have killed but not completely consumed, the Sciomyzidae apparently have obligate relationships with snails throughout the life cycle.

If it seems at all likely that an introduced sciomyzid may attack snails of a desirable species, careful and thorough studies should be made before that fly is released from quarantine to obtain the most accurate prediction possible concerning the species that would actually be attacked in nature. The species acceptable under starvation conditions are easily determined. But this information is not very meaningful, particularly if tests are conducted in such small containers that larvae and snails are forced into contact with each other. Properly designed tests in larger containers often indicate that some of the edible species really are quite safe from attack in nature. Thus the "host range studies" of *Pherbellia dorsata* (Zetterstedt) prior to its release from quarantine in Hawaii showed that these larvae pose no threat to the Achatinellidae, a family of beautiful tree snails restricted to the Hawaiian Islands. Laboratory test cages just one foot high proved large enough to demonstrate the reality of expected segregation by different habitat selection. While the arboreal snails stayed up on the guava branches set in moist sand and peat moss, the larvae remained in the peat moss just a few inches below them, where they died from starvation (Chock *et al.*, 1960). The same host range studies indicated that carnivorous snails introduced for the biological control of the giant African land snail are not vulnerable to Sciomyzidae, because they turn and devour attacking larvae.

The alleged danger that introduced sciomyzid larvae may make major changes in food habits and attack valuable, non-molluscan species contradicts a basic principle of evolution. All larvae whose food habits are known (105 species) have fed exclusively on snails and slugs. They have died, presumably from starvation, when confined with other invertebrates found in their habitats. Students of evolution recognize such specialization as a one-way street—an evolutionary cul-de-sac. They understand the nature of specialization primarily as the loss of genetic plasticity, hence the loss of all potential for alternative ways of life, including food habits. Genes lost in specialization are not regained. Animals cannot develop a new ability if their gene pools do not even possess the rudimentary basis for change in that direction. As De

Vos *et al.* (1956) wrote, "The more specialized [an introduced] species is,... the less likely it is to become a pest."

The evolutionary pressure placed on an imported natural enemy when it proves successful and depletes its own food supply is no different from pressures that these animals have experienced repeatedly in their native lands. Frequent attrition of food resources, and consequent starvation of some individuals, is commonplace among predators—especially among those restricted to one group of prey animals. Severe stress of this nature is known to recur in predictable cycles in some of the best known predator-prey systems (Keith, 1963) and almost certainly does so also in many systems not so intensively studied. There could be no better indication of the inability of sciomyzid larvae to develop other food habits than the fact that no reared species has done so. The wide spread of this food habit throughout many genera representing different subfamilies indicates that specialization already had occurred in the common ancestor of the family. In the geologic ages that have passed since that time, reversal of these adaptations and exploitation of other food resources has not occurred in any species of known larval habits, despite the extreme pressures brought by recurrent famines. Access to a new biota thus presents opportunities only within the area rigidly fixed by genetic specialization. Foreign introductions cannot cause deviations from long established evolutionary trends unless transportation geographically causes new genes to evolve.

E. POSSIBILITIES OF INTEGRATED CONTROL

Instead of focusing attention on the contrasts between chemical and biological control, and encouraging a debate on their relative merits, Franz (1961) suggested an approach that stresses their identical aims and the possibility of using one to complement the other. He expressed the hope that this trend will help in reducing the present differences of opinion as to the value of chemical and biological methods of pest control and that, "with a basis of sound ecological study, the best features of both methods may be utilized in integrated programmes of control". The evidence that biological and chemical control of insects can be integrated (e.g. Ullyett, 1947; Ripper, 1956; DeBach, 1958; Stern and Van den Bosch, 1959; Smith *et al.*, 1962; Beirne, 1963) has come from many kinds of insect pests in various situations. Beirne (1963) summarized, "Integrated control programmes that utilize the advantages and minimize the disadvantages of chemical and biological controls are the ultimate answer to most pest problems. Rates at which they are developed will be regulated by rates at which the required ecological information is obtained and applied." It does not follow that similar integration of methods is practical in snail control, but this possibility should be investigated vigorously and without pessimistic prejudgement of the results.

The best system of integrated control depends on the use of selective pesticides that are less toxic to natural enemies than to the target species (Ripper, 1956). Systox treatments of alfalfa fields, applied correctly and in suitable concentration, kill the spotted alfalfa aphid while allowing its natural enemies to survive in goodly numbers (Stern and Van den Bosch, 1959). Similarly, ryania is more toxic to the codling moth than to the insect species that tend to hold it in check (Pickett, 1961). If such differential toxicity is possible even within the Class Insecta, conscientious searching for a chemical agent more effective against the Mollusca than against animals in other phyla should have a good chance of success. Indeed, a newly discovered compound claimed to kill snails quickly by attacking their respiratory systems, while being "relatively innocuous" to other forms of animal and plant life, was reported very recently (see *Time* magazine for 5 July 1963, pp. 66-67). Representatives of Shell Chemical Company have kindly identified this compound as *n*-trityl-morpholine and given its code name as Shell WL-8008 Molluscicide (personal correspondence).

With the relatively unselective molluscicides currently in use, integrated control measures would be difficult. We have not determined the tolerance of sciomyzid larvae to the most common molluscicides, but it is likely that they and most other snail predators would be killed by molluscicides at the concentrations used. Both sodium pentachlorophenate and dinitro-cyclohexylphenol are toxic to arthropods (Exp. Comm. Bilharz., 1961), and Gretillat (1962) pointed out that the new molluscicide zinc dimethyldithiocarbamate is a potent larvacide that could be used against both bilharziasis and malaria in a single operation.

In a broad and illuminating presentation of the integrated control concept, Stern *et al.* (1959) discussed three systems of integrated control that do not depend on selective pesticides. One or two of these may be usable in snail control, even with the broad spectrum molluscicides currently in use. DeBach and Landi (1961) obtained good control of the purple scale of orange trees by treating alternate pairs of rows each autumn with an insecticide that killed the major natural enemies as well as the scale. The following year, insecticide treatment was applied to the trees previously left unsprayed. The untreated pairs of rows functioned as reservoirs of natural enemy populations which recolonized the adjacent treated trees in a relatively short time following each spraying operation. One hesitates to compare this situation with the vastly different problems encountered in snail control, yet the two have one important characteristic in common. Like the purple scale, snails also have limited powers of locomotion and dispersal in comparison with the natural enemies (e.g. Sciomyzidae) that might be used against them. Since the success of this system probably depends to an important degree on the fact that the natural enemies can move from the unsprayed to the sprayed areas at least as rapidly as the target species, it may be significant that a similar situation exists with respect to sciomyzid flies and snails.

A system that seems better adapted to snail control is based on the fact that sections of high and of much lower priority with respect to snail control can be recognized in most areas. Whereas bilharziasis snails must be controlled in rice fields, this may not be essential in swamps that are not normally frequented by people. This situation is less like that of citrus groves than that of natural forests in Canada. Sections of forest in which a valuable tree occurs in relatively pure stands are sprayed when threatened by insect pests. Adjacent areas in which this species occurs less abundantly, however, may not be sprayed so that populations of natural enemies can persist and build up (Balch, 1958).

Control thus integrated on an areal basis may have another important advantage that is related to the development of resistance to molluscicides. In insects, the speed of this genetic change varies directly with the proportion of the population exposed to the insecticide. Spot-treated insect populations develop resistance much less rapidly than those subjected to overall coverage. The genetic mechanisms of snails presumably would function essentially like those of insects in this respect.

IV. Natural History of the Sciomyzidae

A. Discovery that Larvae Kill Snails

The discovery that sciomyzid larvae kill snails resulted from placing an unknown aquatic larva into the water in a laboratory bowl with some living *Lymnaea palustris* Müller. An hour later, the larva was found with its anterior end thrust into the aperture of one of the shells and deeply embedded in soft parts of the snail. The snail showed no sign of life. By the next morning, the larva had left the remains, but more than half of the soft parts had been consumed.

Other larvae, differing in specific details but having the same gross appearance, were found in marshes and at the edges of ponds. The warty bodies, fleshy lobes surrounding the posterior spiracular discs, and tapering anterior ends without sclerotized head capsules are very characteristic (Fig. 1). They attacked snails of the genera *Lymnaea*, *Helisoma*, *Physa*, and *Gyraulus*, killed them quickly, and fed on them. Adults representing four genera of Sciomyzidae: Tetanocerinae emerged from the floating puparia formed by these larvae (Berg, 1953).

These rearings indicated that overt predation on pulmonate snails is a widespread way of life in the subfamily Tetanocerinae, which suggested that it may have been reported before. However, a search of the literature indicated that no one had ever seen any sciomyzid larva kill a snail. Although some Sciomyzinae had been associated with snails, no one had even suggested that larvae of the Tetanocerinae feed in this way.

Adults of at least six species of Sciomyzinae had been reared from puparia found in snail shells, and some entomologists seemed persuaded by this circumstantial evidence that these species are parasitic on snails (Mercier, 1921; Lundbeck, 1923). However, Oldham (1912) frankly expressed doubts whether the species he reared "is parasitic on the living mollusc, kills and subsequently devours it, eats the body of the snail which has died from some independent cause, or merely chooses the empty shell as a convenient place in which to pupate." Even Lundbeck (1923), who wrote the most comprehensive of the early papers on biology of the Sciomyzidae, knew relatively little about the life cycles of these flies. Although he reared the adults of twenty-six species from puparia, he admitted that this was "without knowing the larvae and thus not being able to decide anything as regards their feeding habits." The reasonable doubts expressed by Oldham evidently were shared even by those who chose to consider these few species parasitic on snails. Without any compelling evidence that their larvae were feeding on living snails, no one seemed sufficiently convinced even to work through a life cycle in the laboratory and find out.

The occurrence of sciomyzine puparia in snail shells assumed greatly increased significance after the discovery of snail-killing habits in the Tetanocerinae. Positive proof of this way of life in one major subfamily seemed to support the circumstantial evidence on the other, giving a new plausibility to the hypothesis of a parasitic relationship of sciomyzine larvae to snails. This suggested that eating snails may be a way of life that is spread throughout the entire family, not confined to the Tetanocerinae as previously supposed. One implication of this was that complete life cycles of many Sciomyzidae perhaps could be obtained by confining wild-caught adults of various species in laboratory cages for oviposition, holding the eggs for hatching, and giving the larvae their choice of a variety of snails. Another was that immature stages of intimately associated species probably could be found by collecting quantities of snail shells and examining them for concealed larvae and puparia.

With the use of these techniques, it has been possible to observe the complete life cycles of eighty-three species of Sciomyzidae in the laboratory. Twenty-two additional species have been collected as larvae, observed to kill snails, and then reared through for identification. All larvae in these 105 species have fed solely on gastropod mollusks, introduced alive into their rearing containers.

Although all known sciomyzid larvae kill and feed on snails and slugs, they differ greatly in ways of feeding, microhabitats in which they live, and morphological adaptations. In the following discussion, the Sciomyzidae are divided into three groups based on these differences in basic biology.

B. THE AQUATIC, PREDATORY TETANOCERINAE

Most of the Tetanocerinae (but none of the Sciomyzinae) now reared have warty-bodied larvae (Fig. 1) that are found in or near the water. These are the only sciomyzid larvae commonly seen and the sole basis for all descriptions of "sciomyzid larvae" written before this investigation started. Although they do not submerge completely in search or pursuit of snails, these larvae are well adapted for aquatic life, floating at the surface and piercing the surface film with their posterior spiracular plates to obtain air. They kill snails in water as well as on land, maintaining contact with the surface film even while supporting the weight of snails considerably heavier than themselves (Fig. 26). This is accomplished in part by buoyancy from a bubble maintained in the gut by frequently swallowing air. Added support is provided by the hydrofuge properties of oil-covered "float hairs" surrounding the posterior spiracular plates, causing them to resist being pulled down through the surface film. If the surface film is broken, as by wave action, both larva and snail sink. The larva then abandons the snail and is refloated by the air bubble (Berg, 1964).

Observation of encounters on moist substrates is more satisfactory. There is a quick thrust as soon as a larva finds the exposed foot of a snail, evidently setting the sharp mouthhooks in this muscular tissue (Fig. 2). The snail suddenly withdraws into its shell, usually pulling the attached larva along so that only its posterior end extends from the aperture (Fig. 3). Probably in response to other movements of the mouthhooks, the snail retracts more. The larva, either pulled along or following in pursuit, disappears from view. If an opening is made through the body whorl a few minutes later, the larva is found lying on the foot and visceral mass with the anterior third of its body embedded in this material (Fig. 4). Muscular contractions of the snail quickly diminish and stop, and the larva darkens appreciably as it becomes engorged with snail tissues. Whenever larvae attack snails having red pigments in the hemolymph, massive hemorrhaging is quickly evident.

The aquatic, predatory Tetanocerinae have been characterized (Berg, 1961) as not selective of food snails, even attacking amphibious and terrestrial species as well as the aquatic pulmonates. That statement was based on data from many individual exposures of selected snail species to hungry larvae in small rearing containers. It has meaning with respect to the capabilities of confined larvae, but larvae having free access to snails of several species in nature may exercise considerable selectivity.

While developing from hatching to pupation, an individual larva in this group commonly kills and feeds on more than a dozen snails. Since they kill quickly, feed rapidly, leave the snail as soon as their hunger is satisfied, and attack another when they again become hungry, these larvae usually are not found in snail shells in nature. Both they and the floating puparia they form can be collected by pushing the floating and emergent vegetation of marshes, swamps, and pond borders below the surface, thus forcing the larvae and puparia to float free, where they are readily seen.

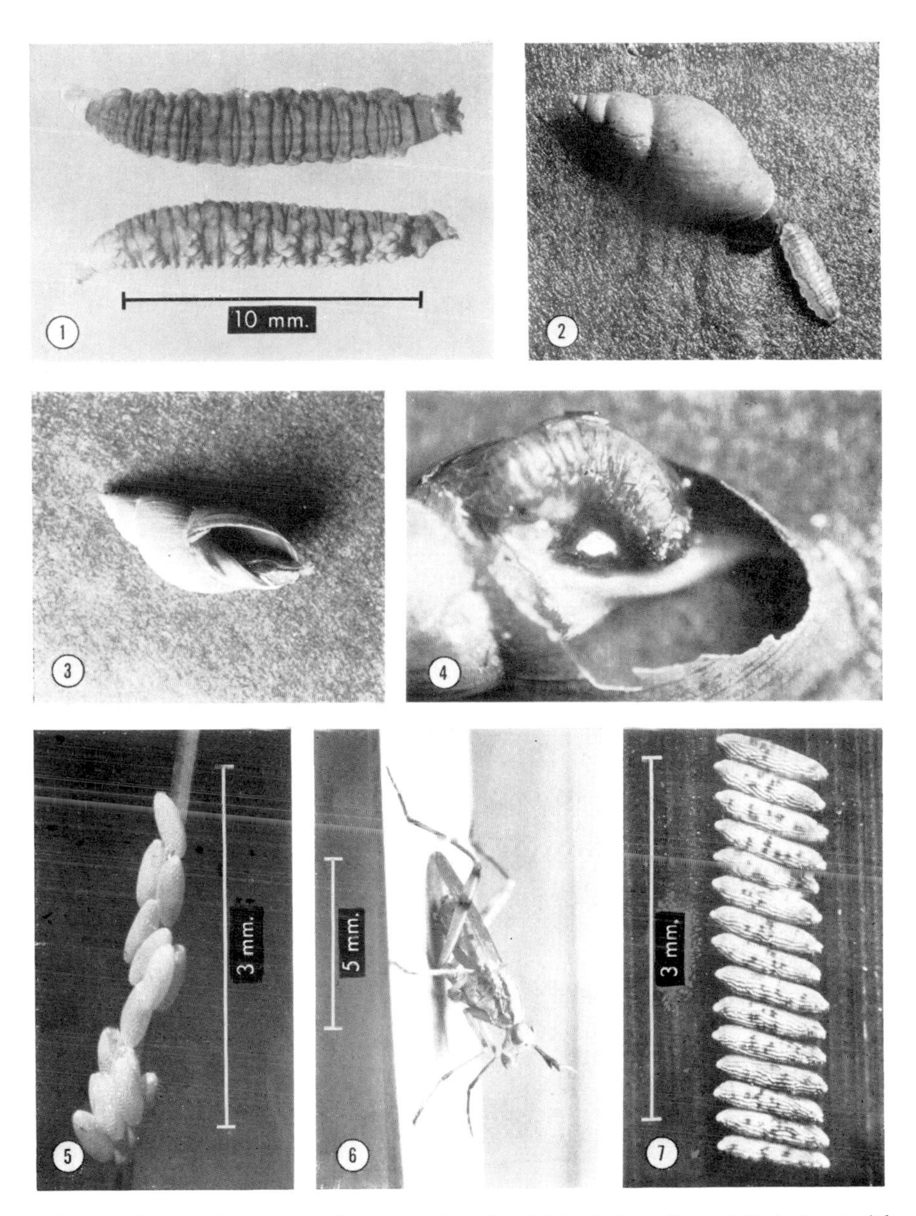

FIG. 1. Larva of *Tetanocera ferruginea*, dorsal and lateral views. FIGS. 2, 3, 4. Sequential stages in attack of *Sepedon macropus* larva on *Lymnaea palustris*. (See text for details.) FIG. 5. Egg mass of *Pherbellia seticoxa*. FIG. 6. Adult of *Sepedon macropus*, in typical head-downward position on emergent vegetation. FIG. 7. Egg mass of *Sepedon praemiosa*.

The puparia formed by most aquatic tetanocerine larvae are clearly adapted for floating. They are curved dorsally at the posterior end, and they float with the posterior larval spiracles projecting into the air (Fig. 8). They are even more buoyant than the larvae, and the hairs surrounding the posterior spiracular plates (the "float hairs") retain their hydrofuge properties.

Adult Sciomyzidae of the species included in this group can be captured by "sweeping" through the grasses and sedges of marshes and swamps, and amidst the emergent and shoreward vegetation of ponds, canals, and streams. Their frog-like resting positions, usually with heads downward, and their long, porrect antennae give them a very characteristic appearance (Fig. 6). We first fed them a mixture of honey and brewers' yeast, which they accepted very readily. Later, we learned that egg production can be increased more than tenfold if crushed snails are added to this diet (Chock *et al.*, 1961).

Eggs of the Tetanocerinae are elongate-oval, with color varying from white to shades of gray, brown, and buff. Ovipositing females of *Dictya* and most species of *Sepedon* usually lay their eggs in neat masses above the water line on emergent vegetation (Fig. 7). Newly hatched larvae drop or crawl down into the water and are capable immediately of killing small snails.

The aquatic Tetanocerinae constitute the largest group in the family and include the species most easily reared. Berg (1961) listed forty-nine species, collected in North America, Central America, and Europe, which were reared by the research team at Cornell University. All species of *Sepedon*, *Dictya*, *Hedroneura* (Auct.), *Elgiva* (Auct.), *Hydromya*, *Pherbina*, *Protodictya*, and *Renocera* reared in this study and most of the species of *Tetanocera* are included in this group.

The lines of demarcation between categories are obscured by larvae with atypical and transitional ways of life. Thus, the European genus *Pherbina* contains some larvae that might better be classified as "shoreline" than as either aquatic or terrestrial (Knutson, 1963). Larvae of *Protodictya hondurana* Steyskal evidently use their adaptations for life in a liquid medium to prevent submersion and asphyxiation in decayed and liquified snail tissues (Neff and Berg, 1961). However, the known larvae and puparia of both *Protodictya* and *Pherbina* are typical of the aquatic Tetanocerinae morphologically.

Detailed studies of the natural history and descriptions and figures of immature stages of aquatic, predatory Tetanocerinae appear in three recent doctoral theses and in the modifications of them that have been and are being published. Two sections of the study of Nearctic, Neotropical, and Palearctic *Sepedon*, *Hoplodictya*, and *Protodictya* (Neff, 1960) have been published (Neff and Berg, 1961, 1962). The final section, on fifteen species of *Sepedon*, is now in preparation. Foote's (1961) study of seventeen Nearctic and one Neotropical species of *Tetanocera* (some not aquatic) is being prepared for publication as a single unit. The investigation of a wide variety of European Sciomyzidae (Knutson, 1963) includes treatment of five genera in which some or all species have aquatic larvae. A section on *Hydromya dorsalis* (Fabricius), an interesting, somewhat atypical species that seems to breed only in flowing water, was published recently (Knutson and Berg, 1963). The exposition and description of three species of *Elgiva* (= *Hedroneura* of authors) has been submitted (Knutson and Berg, 1964).

Since progress on the rearing of this group was summarized as forty-nine reared species, we have added three Australian, four North American, and four European species. Three Ethiopian species reared for us by Dr. W. C. Frohne bring the new total to sixty-three species known to have snail-killing larvae essentially of the form illustrated in Fig. 1. Complete life cycles of fifty-two of these have been observed.

C. THE MOST HIGHLY SPECIALIZED SCIOMYZINAE

Three species which have the most highly specialized and delicately adjusted parasitoid (or parasitic) relationships observed in the Sciomyzidae are discussed here.

Pteromicra inermis Steyskal was described from specimens reared at Ithaca, New York during the present investigation. Puparia were discovered when some shells of *Lymnaea palustris* were candled in a narrow beam of light. All shells containing puparia were dry externally. Their original occupants had either been stranded by receding water levels or had voluntarily left the water to aestivate. A heavy infestation was found in snails aestivating on the vertical face of a concrete bridge abutment, two or three feet above the water. We wondered how larvae could move from one snail to another on that hot, dry surface without dying from desiccation.

Adults emerged and mated. Living snails are always added to breeding cages because the odors they release may be essential stimuli for oviposition (Neff and Berg, 1961). In this species, the flies cemented their eggs in a very characteristic pattern right onto the shells of those living snails (Fig. 9).

Newly hatched larvae entered the shells to which their eggs were attached and remained there continually throughout their larval and pupal stages. For the first 4 days, the snails were not visibly disturbed. The larvae seemed to lie quietly between the shell and the mantle. Then the growing larvae gradually increased the vigor of their feeding motions. They consumed perhaps a third of the soft parts, and usually killed the snail, by the 6th day inside the shell. Larvae fed rapidly for two more days, reducing each snail body to a few drops of black liquid, which they pushed out through the aperture just before forming their puparia. These are relatively long and strangely corkscrew-shaped, practically filling one complete whorl of the shell (Figs. 10, 14).

Each individual that developed directly left the snail shell as an adult fly about 22 days after entering it as a newly hatched larva. Diapausing pupae carry this species through the winter. Individuals in which this pupal diapause intervened remained in their snail shells from several months to a full year.

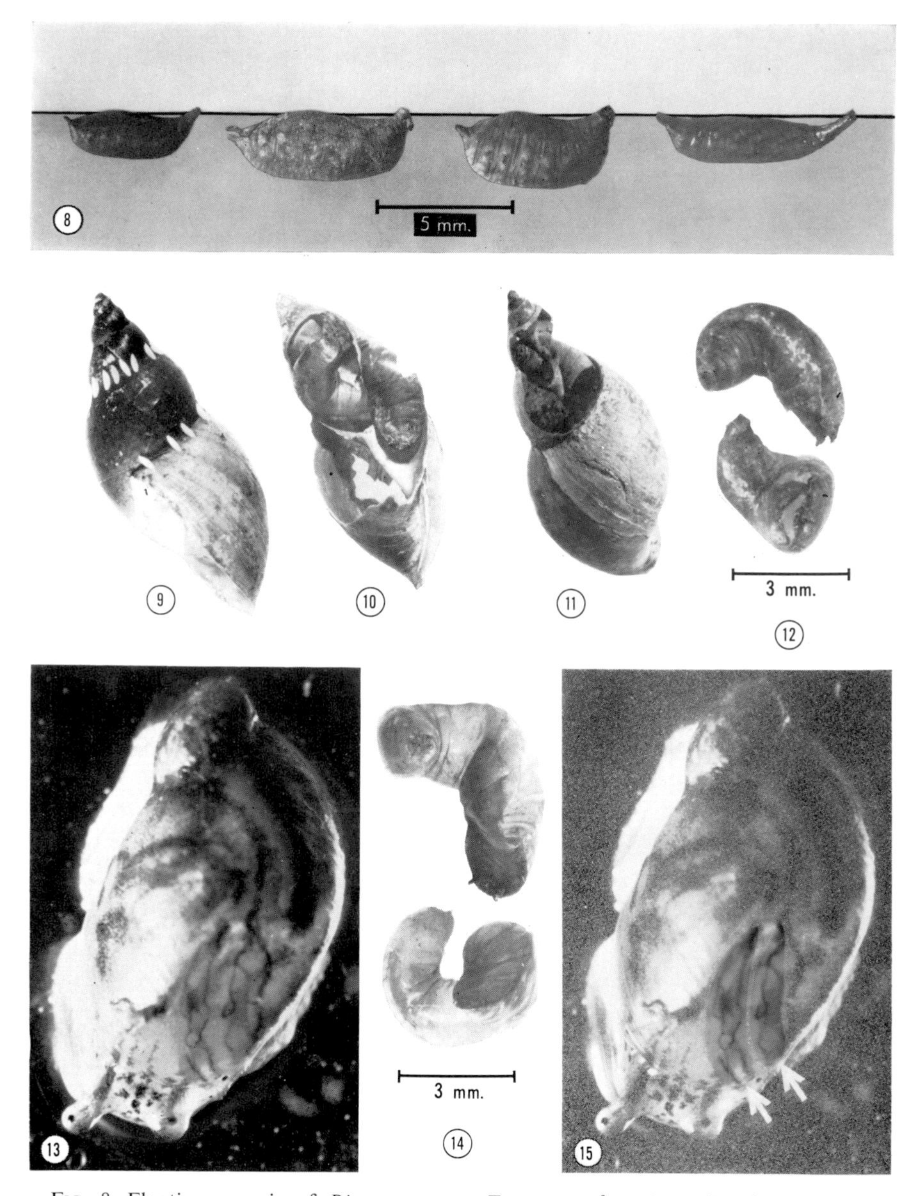

FIG. 8. Floating puparia of *Dictya expansa*, *Tetanocera ferruginea*, *Sepedon fuscipennis*, and *Elgiva connexa* (from left to right), showing orientation to surface film. FIG. 9. Eggs of *Pteromicra inermis* on aestivating *Lymnaea palustris*. FIG. 10. Corkscrew-shaped puparium of *P. inermis* exposed by opening *Lymnaea* shell. FIG. 11. Puparium of *Pherbellia griseola* in shell of *Lymnaea palustris*. FIG. 12. Twisted puparium of *P. griseola*, two views. FIG. 13. *Oxyloma retusa*, harboring two larvae of *Hoplodictya spinicornis* beneath transparent shell. FIG. 14. Puparium of *P. inermis*, two views. FIG. 15. Same photograph as Fig. 13, retouched to show positions of larvae.

No larva has been known to attack a second snail, even when its first victim was only half grown and another *Lymnaea* was lying in contact with it in a small rearing vial. Such undersized larvae pupated and finally emerged as abnormally small flies. It is also remarkable that only one living larva has been found in any snail, even though several eggs have been attached to some shells. This contrasts with other sciomyzid larvae, several of which often feed gregariously in one shell (Fig. 24). It suggests that the first larva to enter may have some way of eliminating its competitors for this limited food supply so that at least one larva can develop to maturity. The finding of dead, first instar larvae on the snail foot near the living larva supports this hypothesis. Thus it appears that the specialization of this species to subsist on aestivating snails has fixed the rules of "one snail per larva" and "one larva per snail." This specialization, avoiding any search for a second snail, combines with oviposition directly onto the shell to eliminate all larval searching for food (Berg, 1961; J. L. Gower, unpublished).

Ovipositing females of *Sciomyza aristalis* Coquillett (Nearctic) and *Pherbellia schoenherri* Fallén (Palearctic) are similar to those of *P. inermis* in finding living snails for their larvae to feed upon and cementing their eggs to the shells. They differ in living on land snails (Succineidae) and in placing their eggs longitudinally in the suture between shell whorls instead of across it (Fig. 27). Foote (1959) published on the natural history of *S. aristalis* and included figures and descriptions of the immature stages. The capture of a mated female on 7 June 1963 provided the opportunity to rear this uncommon species again and to preserve additional material. In the process, Foote's surprising report of the first instar larva wedging itself inward between the mantle and the shell until it loses all contact with the air was checked and corroborated. The tiny larvae crawl a distance equal to several times their own lengths from the aperture and do not return to it until 5–8 days later (about at the time the snail dies). At first, the snail shell and mantle enclose the larva so tightly on all sides that use of an open tracheal system would seem impossible. The young larvae must get their oxygen by direct diffusion from the adjacent flesh and hemolymph of the snail. Later, a small gas pocket develops around each larva. Use of an open tracheal system then becomes physically possible, but since this gas pocket is isolated from the ambient air the larva must still be parasitizing the snail for oxygen as well as food. As reported for *P. inermis*, no larva has been known to leave the empty snail shell and attack another snail. An individual enters a shell as a newly hatched larva and does not leave it again until it emerges as an adult fly from the puparium formed inside that shell.

The young larvae of *Pherbellia schoenherri* are similar to those of the two North American species in inflicting a minimum of injury on their hosts. For the first 2 days the snails remain active and show no visible signs of distress. They live for as much as 9 days after the larvae enter. However, these larvae maintain contact between their posterior spiracles and the air and give other indications of being less highly specialized. Each larva normally destroys only one snail, but if two larvae consume a snail together each seeks and kills another snail (in our rearing cages, at least) to complete its food requirements. Larvae of this species also differ in commonly leaving the snail shell and crawling down into the soil to pupate (C.O. Berg and J. W. Stephenson, unpublished; Verbeke, 1960).

D. SPECIES OF INTERMEDIATE HABITS AND ADAPTATIONS

The most specialized of the parasitoid Sciomyzinae differ from the predatory Tetanocerinae primarily in the following respects. (1) The young larvae inflict only minor injury which increases as they grow larger. Snails live from a few hours to a few days after larval invasion, with larvae tending to remain in the shells. The larvae lack obvious aquatic adaptations and do not venture unsupported out into open water. (2) There is evident host selection, at least on the family level. (3) They pupate in snail shells. (4) The adult flies attach their eggs, in characteristic patterns, on the shells of living snails. (5) Larvae never leave one snail and attack another; each larva consumes only one snail and tolerates no competition for this food.

A part of item 1, the lack of aquatic habits and adaptations, probably characterized the common ancestor of all Sciomyzidae. All other characteristics listed may have been steps which developed successively, and usually but not always in the order given, in the evolution of the most highly specialized parasitoid species. One reason for thinking so is that several living species have taken only step 1; several others, only steps one and two, and so forth. These progressively more parasitoid groups are discussed briefly in the following paragraphs.

More than a dozen species of Sciomyzinae have larval habits essentially as indicated in item 1, but lack all of the other parasitoid traits. The larvae remain in snail shells, intimately associated with the food snail, from one feeding to the next. Yet they freely abandon any shell in which the snail tissues are consumed or badly decomposed and invade the shells of living snails. Many larvae often feed together in one shell (Fig. 24), so snails are consumed rapidly and frequent moves to new food snails are necessary. Although there may be some selection of snails in nature, larvae in rearing jars are able to subsist on a wide variety of aquatic and terrestrial species. Mature larvae usually leave the shells and crawl down into the litter or soil to pupate. Eggs are attached, in no discernible pattern, to mosses at the bottoms of breeding jars (Knutson, 1963).

The thirteen species of Tetanocerinae with known larvae which lack aquatic adaptations illustrate items 1 and 2 but no others. The young larvae are at least as secretive and intimately related to food snails as the unspecialized Sciomyzinae discussed above (Figs. 13 and 15). Some species live inside the shells of snails continually for about 4 days, some for as much as 9 days, before killing the mollusks (Neff and Berg, 1962; Foote, 1961). The larvae are selective of certain families or genera of terrestrial gastropods and unable to survive unless the proper ones are provided. However, older larvae of at least some species act as predators, remaining outside of snail shells much of the time and killing and feeding quite rapidly when they do attack. No known species of Tetanocerinae routinely pupates in, or oviposits on, snail shells. In all known species, at least two snails are consumed during the development of each larva. Species in this group resemble the Sciomyzinae which do not pupate in snail shells in gross appearance of both larvae and puparia and in sites of oviposition and pupation. Like the larvae of all Sciomyzinae, those of the terrestrial Tetanocerinae are more similar superficially to the maggots of muscoid flies than to their aquatic relatives. They have tapered anterior ends and somewhat truncated posterior ends. The integument is nearly white, unmarked, and quite transparent (Fig. 17). The wart-like body tubercles, fleshy lobes surrounding the posterior spiracular disc, and "float hairs" are much reduced or lacking. The puparia are ovoid, compact, and not distinctive when compared whith those of Diptera in other families (Fig. 18). Known larvae subsist entirely on terrestrial snails and slugs, and adult flies occur more commonly in moist woods than over water.

Most of the reared Sciomyzinae pupate within snail shells (Figs. 10, 11, 19, 25). They have been collected largely as puparia, discovered when apparently empty shells have been examined with transmitted light (Figs. 16, 19). Host relationships have thus been learned, and complete life cycles have then been observed in the laboratory. The puparia of some species are remarkably modified into the shapes impressed upon them by the shells (Figs. 12, 14, 20, 22, 23). All except four of the known species oviposit, singly or in irregular masses, on prostrate vegetation at the bottoms of breeding cages (Fig. 5). The larvae of all species have the morphological and behavioral characteristics listed in item 1 (Fig. 21), and most species are somewhat host specific. A few can subsist on a wide variety of snails. They tend to remain in snail shells even when not feeding, but leave shells in which the edible food is exhausted. Larvae which invade sufficiently large snails, without competition from other larvae, may spend the entire larval life in one snail shell (unpublished data, A. D. Bratt and others).

In addition to the three species discussed under Section IV, C, only one other species, *Atrichomelina pubera* (Loew), is known to oviposit routinely onto the shells of living snails. This species is remarkable because of unusual flexibility and adaptability of larval food habits. Parasitoid, predatory, and saprophagous capabilities are combined, and the feeding of a given larva may even change from day to day depending on the food available and the intensity of intraspecific competition. When a larva enters a large snail without competition from other larvae, it kills slowly, finds enough food there

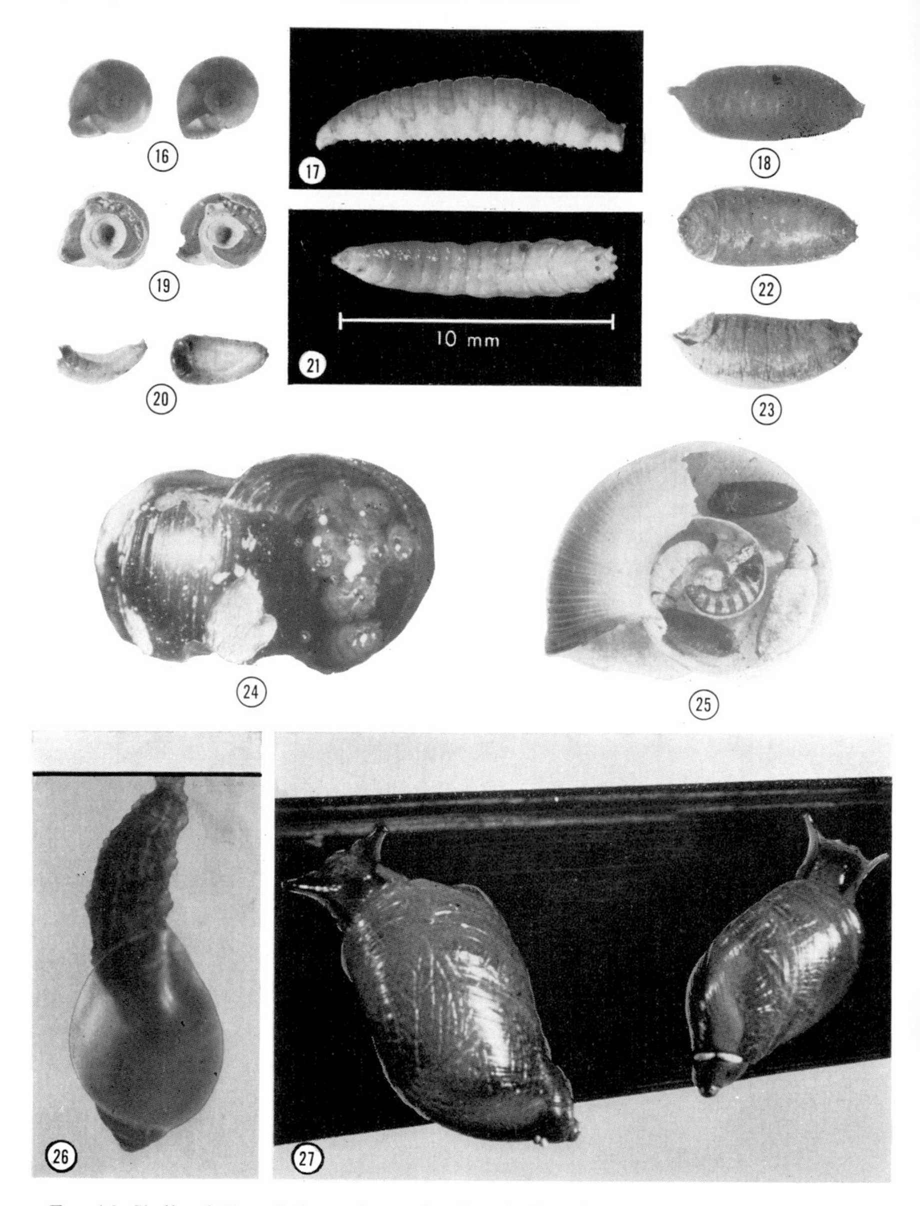

FIG. 16. Shells of *Planorbula armigera*, showing shadows in upper parts when viewed with transmitted light. FIG. 17. Larva of *Tetanocera rotundicornis*, lateral view. FIG. 18. Puparium of *T. rotundicornis*, lateral view. FIG. 19. Same shells as shown in Fig. 16, opened to show puparia of *Pteromicra similis*. FIG. 20. Puparia of *P. similis* removed from shells, lateral and dorsal views. FIG. 21. Larva of *Pherbellia seticoxa*, dorsal view. FIG. 22. Puparium of *P. seticoxa*, dorsal view. FIG. 23. Same in lateral view. FIG. 24. Eight large and four smaller larvae of *Pherbellia humilis*, feeding together on one *Helisoma trivolvis*. FIG. 25. Shell of *H. trivolvis* opened to show five puparia of *P. seticoxa*. FIG. 26. Larva of *Dictya expansa*, hanging from surface film while feeding on *Physa* sp. (retouched). FIG. 27. Living *Succinea pfeifferi*, each harboring two eggs of *Pherbellia schoenherri* (retouched). (Figs. 17, 18, 20, 21, 22, and 23 all enlarged as indicated by scale on Fig. 21.)

to complete its development, and frequently remains in that shell throughout both larval and pupal stages. However, when a larva attacks a small snail, or when several larvae invade the same shell, the snail is killed quickly and consumed before larval development is completed. Nutritional requirements of the larvae can then be completed only by finding and consuming other snails. Larvae of *A. pubera* continue to feed on snails after they die and even invade the shells of dead and decaying snails if living animals are not readily available. They accept a wide variety of snails and show no clear evidence of specific preferences (Foote *et al.*, 1960).

This stepwise progression of intermediate forms seems to present a living record of the course of evolution. However, evolution in the Sciomyzidae almost certainly did not start with the highly adapted species at either end of this series and progress through it in only one direction. The habits and adaptations of dipterous larvae in other families suggest that the common ancestor of the Sciomyzidae probably was a general scavenger in moist situations. Gradually increased feeding on living snails that were stranded and helpless perhaps resulted in the evolution of a snail killer with intermediate, and relatively flexible, habits such as those of *A. pubera*. By divergent evolution, this species may then have given rise to both the aquatic predators and the terrestrial, parasitoid Sciomyzinae (Berg *et al.*, 1959).

E. ON THE USE OF TERMS "PARASITIC" OR "PARASITOID"

Like the early entomologists who reared sciomyzine adults from puparia found in snail shells and suggested that their larvae may be "parasitic" on snails, most present-day entomologists would unhesitatingly apply this term to the species identified here as the most highly specialized Sciomyzinae. To place the intermediate species as either "parasites" or "predators," some would apply the rule that a predator feeds on its prey, "a limited period of time, less than the immature or mature feeding period [being] spent with each victim" (Sweetman, 1958). Conversely, Sweetman defined parasitism as "that form of symbiosis in which one symbiote lives and feeds internally in the host, or lives and feeds externally on the host during the entire immature or the mature feeding stage."

According to these definitions, genetically identical sciomyzine larvae from a common egg mass have to be classified oppositely whenever one larva attacks a large snail, where it finds enough food to satisfy its nutritional requirements, but the other attacks a smaller individual that does not provide enough food. Another larva may enter the shell of a large snail and begin what seems destined to be a "parasitic" life, only to be suddenly transformed into a predator because other larvae also chance to find the same snail!

Most remarkably, this transformation occurs without any discernible change in the essential nutritive relationship of the larva to the snail. Whether a sciomyzid larva attacks two or more snails successively or completes its larval life in one, it needs such great proportions of the snail body for its nourishment that death of the snail(s) is inevitable. To many biologists, this is the essence of predatory relationships. Elton (1927) emphasized that parasites are essentially similar to predators in many respects, "the chief difference being that the latter live on capital and the former on income of food." The fundamental distinction stressed by Elton is that a predator kills because it must kill in order to get enough to eat; a well adjusted parasite does not kill—it does not need such great quantities of host tissues, and its best opportunity for survival (in many cases its only opportunity) depends on the host also remaining alive.

Among entomologists, Reuter (1913), Wheeler (1923), and Bequaert (1925) accepted the principle that animals which consume relatively great proportions of their victims' bodies are predators, not parasites. Reuter recognized "typische Raubinsekten (Raptatoria), parasitenartige Raubinsekten (Parasitoidea), und typische Parasiten (Parasita)." Preferring the term " 'parasitoid' Hymenoptera" for the groups well known as the "parasitic Hymenoptera," Wheeler wrote, "Species that behave in this manner are not true parasites, but extremely economical predators, because they eventually kill their victims, but before doing so spare them as much as possible in order that they may continue to feed and grow and thus yield fresh nutriment just as it is needed. For this reason and also because, as a rule, only the larval insect behaves in the manner described, it is best called a 'parasitoid.' " Bequaert endorsed these interpretations and applied this term to certain snail-killing flies.

The term "parasitoid" supplies a convenient alternative to the indiscriminate "lumping" of insect larvae which routinely consume their prey with parasites which never consume much of their hosts and ideally do not even kill them. It serves a useful purpose in designating specifically the type of symbiote that "...destroys its host so that it functions more like a predator..." (Noble and Noble, 1961). In this relationship, an insect larva lives as intimately associated with its victim as a delicately adjusted parasite, yet typically kills it with a certainty and predictability even greater than that of most overt predators. Interest in the large group of insects characterized by such relationships is growing rapidly because of their increasing use in biological control. A specific term for this important group and their distinctive way of life seems well justified.

It is not suggested that adoption of the term "parasitoid" would remove all uncertainties and disagreements concerning the precise limits of "true parasitism" and "true predatism." Since parasitic ways of life have undoubtedly evolved from predatory ones several times, intergrades between the two are inevitable. No definition should be expected to create a sharp, universally accepted line of demarcation between them. Sweetman's definition is not questioned because of its failure to do so but because of the conviction that it does not focus on the most basic and fundamental difference between parasites and predators.

The most appropriate use of this term may be in the comparative sense of species B having more parasitoid tendencies, habits, and adaptations than species A, but less than species C, and so forth. Its literal meaning (parasite-like) suggests such usage. Furthermore, most biologists are not strongly opinionated concerning the precise relationships that an animal must have with its prey or host in order to qualify as a "true parasitoid." Thus "parasitoid" remains flexible and relative, and is more appropriately applied to sciomyzid larvae for this reason as well as because its stated meaning is far closer to correctly describing their way of life. Whereas some Sciomyzinae clearly are "more parasitoid" than others, it seems very doubtful whether "more parasitic" should be used in this context. Most parasitologists probably would not regard any sciomyzid as parasitic at all.

The use of "parasitoid" primarily as a comparative adjective is especially convenient in interpreting the ethological series presented by the sciomyzid larvae. All biologists probably would agree that the typical aquatic Tetanocerinae are predators. Since there is no natural break in the continuum discussed in the previous section, it avoids hairsplitting definitions if one regards all sciomyzid larvae as predators, some of which simply have more parasitoid tendencies than others.

The alternative of regarding at least some species of Sciomyzinae as either "true parasites" or "true parasitoids", and insisting on a precise line of demarcation between them and the predators in either case, would create many unnecessary problems. Each of the several steps in the progression from the aquatic Tetanocerinae to the most highly specialized and intimately associated Sciomyzinae might be chosen by someone as the logical dividing line if several biologists were asked to define those categories. In fact, transitional species even obscure the lines between designated "steps," which are only man-made divisions in a continuum of morphological and behavioral characteristics. Since any line of separation drawn through such a continuum would be artificial and controversial, it seems pointless to set up and define any categories. The many different mixtures of predatory and parasitoid tendencies exhibited by sciomyzid larvae points up the value of keeping terms as flexible as nature itself.

V. Trials of Sciomyzidae against Medically Important Snails

The possibility that larvae of the Sciomyzidae may have value in the biological control of snails was suggested when direct observation of their snail-killing habits was first reported (Berg, 1953). Since that time, opportunities have been sought to investigate the practicality of that suggestion. Studies have included exposures of snail hosts of bilharziasis to sciomyzid larvae in laboratory tests, and laboratory and field trials of sciomyzid larvae against snail hosts of fascioliasis in Hawaii, Guam, and Australia.

A. LABORATORY TRIALS AGAINST SNAIL HOSTS OF BILHARZIASIS

The preliminary trials against a variety of snail hosts of bilharziasis (Berg and Neff, 1959; Neff, 1964) were conducted at the Laboratory of Tropical Diseases, National Institutes of Health, U. S. Public Health Service, using aquatic, predatory Tetanocerinae of three genera and snails from the colonies maintained by that Laboratory. The snails used in these trials included eight hosts of *Schistosoma mansoni* Sambon: *Australorbis glabratus* (Say), *Biomphalaria boisseyi* (Potiez and Michaud), *B. nigricans* (Spix), *B. paparyensis* (F. Baker), *B. pfeifferi* (Krause), *B. stramineus* (Dunker), *Planorbarius metidjensis* (Forbes), and *Tropicorbis centimentralis* (Lutz). Also included were three hosts of *S. haematobium* (Bilharz): *Bulinus africanus* (Krauss), *B. globosus* (Morelet), and *B. truncatus* (Audonin), and two hosts of *S. japonicum* Katsurada: *Oncomelania formosana* (Pilsbry and Hirase), and *O. quadrasi* (Moellendorff).

It seemed that the most elementary questions concerning the potential value of sciomyzid larvae for control of snail hosts of bilharziasis could be answered without extensive research to determine the most promising species for use against those snails. Therefore, the preliminary trials were made with the larvae of ten species that happened to be available in laboratory culture and in nature at that time. The species utilized were *Sepedon armipes* Loew, *S. caerulea* Melander, *S. fuscipennis* Loew, *S. tenuicornis* Cresson, *Tetanocera ferruginea* Fallén, *T. vicina* Macquart, *Dictya brimleyi* Steyskal, *D. expansa* Steyskal, *D. pictipes* (Loew), and *D. stricta* Steyskal.

"Experimental" jars were made up by placing a single snail and one larva on moist sand in each 1 oz jar. "Control" jars were made up in the same way, each with one snail, but without any larva. To test the ability of larvae to kill snails while unsupported in water deeper than the lengths of their bodies, other "experimentals" and "controls" were made up by pouring water into these small jars to a depth of at least 1 in. Placement of a snail in one of these jars for a 24 h period was called an "exposure day."

The results in experimental jars, in series of trials totalling more than 1,100 exposure days, are summarized in Table I. During those trials, 750 snails and 133 larvae died in experimental jars. There were only two snail deaths in the more than 200 exposure days that snails were held in control jars. It seems reasonable to attribute these two deaths to natural causes and to assume that they probably would also have occurred in the aquaria or finger bowls in which the snail colonies were maintained. At least, natural mortality plus deaths both from transfer to jars of the type used in the trials and from the physical conditions encountered there totalled less than one per 100 exposure days. This mortality rate is so low relative to that found in

Results of Exposing Snail Hosts of Bilharziasis to Sciomyzid Larvae

Snails	Sciomyzidae *Sepedon: armipes caerulea fuscipennis tenuicornis*			*Tetanocera: ferruginea vicina*			*Dictya: brimleyi expansa pictipes stricta*			Totals		
	Exp. days	Snails dead	Larvae dead	Exp. days	Snails dead	Larvae dead	Exp. days	Snails dead	Larvae dead	Exp. days	Snails dead	Larvae dead
Australorbis												
glabratus	154	147	8	92	91	13	52	51	6	298	289	27
Biomphalaria												
boisseyi	51	50	0	75	56	6	56	46	3	182	152	9
nigricans	—	—	—	7	7	0	6	6	0	13	13	0
paparyensis	—	—	—	17	17	2	3	3	0	20	20	2
pfeifferi	1	1	0	4	4	0	22	20	3	27	25	3
stramineus	—	—	—	5	3	0	4	3	0	9	6	0
Planorbarius												
metidjensis	7	7	0	—	—	—	—	—	—	7	7	0
Tropicorbis												
centimentralis	—	—	—	4	2	2	4	3	1	8	5	3
Totals	213	205	8	204	180	23	147	132	13	564	517	44
Bulinus												
africanus	11	0	4	88	72	11	58	17	18	157	89	33
globosus	—	—	—	72	53	10	—	—	—	72	53	10
truncatus	47	29	5	59	42	4	51	19	11	157	90	20
Totals	58	29	9	219	167	25	109	36	29	386	232	63
Oncomelania												
formosana	30	0	10	40	0	2	52	1	4	122	1	16
quadrasi	17	0	4	—	—	—	44	0	6	61	0	10
Totals	47	0	14	40	0	2	96	1	10	183	1	26

experimental jars that it seems unnecessary and unwise to complicate Table I by including data from control jars. Results obtained in them simply indicate that snail deaths in experimental jars were due almost exclusively to attacks by the larvae. Deaths from all other causes presumably were negligible by comparison.

Each major column in Table I is subdivided into three columns: exposure days, snails dead, and larvae dead, for quick appraisal of the number of dead snails per exposure day, the ratio of dead snails to dead larvae, and so forth. Limited availability of both snails and larvae of some species imposed severe restrictions on the numbers that could be used, and certain snails were exposed only to certain larvae. For this reason, comparisons of the effectiveness of larvae in a common genus cannot be made from these tests. When snails of a more plentiful species were exposed to larvae representing more than one species in a genus, no differences in effectiveness of the various larvae were indicated. It is therefore pointless to devote a vertical column to each species of Sciomyzidae. The results appear more meaningful when data for all species in a genus are grouped into a single column.

The most striking contrasts shown in Table I are in the righthand major column, where the totals given for all sciomyzid larvae show great differences in vulnerability of the various genera of snails. The discoidal Planorbidae (Planorbinae) which act as intermediate hosts of *S. mansoni* are incapable of any effective defense against the larvae. There were 517 deaths of these snails in 564 exposure days. Some species in this group should logically be the first target in any field trial of sciomyzid larvae against snail hosts of bilharziasis. The forty-four deaths of larvae during these trials were not due to any counteroffensive by the snails. No antibiotics were used to protect the larvae, and the mortality rate reported is in the same order of magnitude as that experienced when larvae are reared on local pulmonate snails without benefit of antibiotics. Many, and probably all, of the larvae involved in the 47 exposure days on which snails were *not* killed were either moribund or about to molt. Larvae never eat for a few hours preceding a larval molt, nor for at least 24 h before forming the puparia.

Conducting trials in small jars imposes unnatural conditions on both larvae and snails. There are two evident respects in which the snail-killing potential of larvae is abnormally restricted. Both probably prejudice the results on the hosts of *S. mansoni* more than on either of the other genera.

Each larva is limited by the experimental design to killing only one snail per day. Larvae of aquatic Tetanocerinae digest their food rapidly and are easily stimulated by living snails to kill again even before this digestion is completed. A larva left in a jar containing three small snails frequently kills all three in less than 24 h. This may also occur in nature when snails are abundant, but it could not happen in the laboratory trials reported here.

Secondly, there almost certainly was more death of larvae from putrescence and the microorganisms associated with it than there normally would be in nature. In natural snail habitats, products of decay are volatilized, carried off by flowing water, or greatly dispersed and diluted in large quantities of standing water. Larvae are free to crawl away from any situation in which such substances accumulate. In the closed experimental jars, however, decay products from snails killed but not completely consumed accumulated throughout the 24 h period, and dense growths of microorganisms often were evident. Larvae crawled up onto the jar corks, but bacterial films also extended there. The vulnerability of sciomyzid larvae to infection and death from such causes is evident from the improved survival rates in mass rearings after antibiotics were adopted (see p. 302).

Snails of the genus *Bulinus*, hosts of *S. haematobium* principally in Africa and the Near East, showed differential vulnerability depending on relative sizes of snails and larvae. Small larvae sometimes were unable to kill fully grown snails, but large larvae killed snails of all sizes. Whether successful in killing them or not, some of the larvae that attacked large snails died a few hours later. The deaths probably were caused by asphyxiation after the posterior spiracular plates of the larvae were covered by the thick, viscous mucus that is so characteristic of *Bulinus*. Mucus was exuded visibly as soon as a snail was attacked, and the threshing, slime-covered foot frequently contacted various portions of the larva. Larvae, trailing long threads of sticky mucus, were often observed backing away from a dying or still vigorous snail. They were rolling their posterior spiracular discs vigorously in and out as though trying to break the films that covered them. The following day, these larvae were either dead or fully recovered.

Since all larvae attacked the planorbine hosts of *S. mansoni* successfully, and none proved effective against snails of the genus *Oncomelania*, the distinct differences in effectiveness of the various sciomyzid genera in trials against *Bulinus* spp. appear quite remarkable. *Tetanocera* larvae killed seventy-six snails, while experiencing only eleven deaths of larvae, for each period of 100 exposure days. *Sepedon* larvae destroyed fifty snails, with losses of sixteen larvae, and *Dictya* larvae managed to kill only thirty-three snails, while sustaining losses of twenty-seven larvae, in comparable periods of 100 exposure days. However, these differences seem simply to reflect the various maximum sizes attained by *Tetanocera*, *Sepedon*, and *Dictya* (arranged from the largest to the smallest) and to be controlled by them. Species of *Dictya* probably would not be effective in controlling *Bulinus*, but *Tetanocera* and *Sepedon* may be if the right species can be found, thoroughly studied, and properly introduced.

Sciomyzid larvae cannot be used against most operculate snails, including the genus *Oncomelania*. Trials of these hosts of *S. japonicum* for 183 exposure days resulted in twenty-six dead larvae and only one dead snail. The calcareous opercula of *Oncomelania* close their apertures tightly when the snails retract. Larvae whose mouthhooks were already attached to snail tissues were almost severed transversely. Some were dead when the snails relaxed and allowed them to drop out. Others had broken cephalopharyngeal skeletons and were unable to attack and feed again. The snails usually retracted in response to the first experimental probing by a larva and before its mouthhooks were embedded. This resulted in no injury to either, and as long as the larva continued to investigate that snail the operculum was kept so nearly closed that the larva was excluded. The 156 exposure days on which neither snails nor larvae were killed indicate the resulting stalemates, with larvae crawling over the snails and trying vainly to enter closed shells. Snails extended again when larvae withdrew, but each encounter with a larva seemed to result in more rapid retraction when one approached. Almost all instances of a larva being killed or severely injured occurred during the first day of exposure of a snail.

The trials in water deeper than the lengths of the larvae indicate that they kill snails almost as effectively in that medium. Species of *Australorbis*, *Biomphalaria*, and *Bulinus* were exposed, in approximately equal numbers. Data from these trials are compared here with those from trials of the same species, also in approximately equal proportions, on moist sand. Nearly mature larvae of *Sepedon* and *Tetanocera* were used in both media. There were 220 dead snails in 268 exposure days on sand (82%) and ninety dead snails in 118 exposure days in water (76%). Forty-eight dead larvae on the sand (18%) compare with twenty-two dead larvae in water (19%). Such high efficiency of air-breathing larvae when operating in water seems surprising. Despite their open tracheal systems, they evidently have excellent adaptations to live and obtain their food in water (Fig. 26).

Larvae almost certainly can kill larger snails in natural aquatic habitats than in the glass jars used in these trials. The smooth sides of the jars give no support comparable to that furnished by the vegetation and floating debris so abundant in marshes and at the edges of ponds. Larvae congregate on such objects, thus getting the support necessary for attacking snails much heavier than themselves and holding them up to the surface long enough to feed on them. But a larva attacking a snail in a glass jar has no way to keep itself and the snail afloat except by buoyancy from the air bubble in its gut and the tendency of the hydrofuge hairs on its posterior spiracular disc to "cling" to the surface film. These forces are not adequate if the snail weighs several times as much as the larva.

Inasmuch as preliminary tests indicated that the hosts of *S. mansoni* are particularly vulnerable to attack by sciomyzid larvae, a colony of *Australorbis glabratus* was established in the laboratory at Cornell University for convenience in additional testing. Some questions that had not been answered during short visits to the Laboratory of Tropical Medicine perhaps could be if a colony were always at hand. How many of these snails will one larva destroy between hatching and pupation? What sciomyzid larvae (or how many species) will attack and kill this snail? How much larger may a snail be than a larva and still be killed by it?

Several larvae of *Elgiva rufa* (Panzer) (Tetanocerinae) were reared individually from hatching to pupation with living *A. glabratus* of assorted sizes always available. Each larva killed twelve to fourteen snails, ranging from 1 to 10 mm in diameter. This experiment was repeated with a series of larvae of *Elgiva connexa* (Steyskal). Each larva killed from twelve to seventeen snails, ranging up to 13 mm in diameter (Knutson and Berg, 1964). Similar results were reported by Neff (1960), who found that each larva of *Sepedon caerulea* that was reared individually killed twelve to eighteen *Helisoma trivolvis* (Say) ranging from 3 to 15 mm in diameter; individually reared larvae of *S. macropus* Walker each destroyed up to twenty *H. trivolvis*. Foote (1961) reported twelve to sixteen snails killed per larva when he exposed medium to large individuals of *Lymnaea* and *Physa* (up to 13 mm in length) to larvae of *Tetanocera robusta* Loew. However, each of the larvae that had many smaller snails accessible to it killed from twenty to twenty-six individuals of *Gyraulus*, *Lymnaea* or *Helisoma*.

These results are meaningful only as rough approximations. The number of snails killed per larva in nature must be expected to vary greatly depending on average sizes of snails, their population density, intensity of competition among larvae, and other factors. It would be easy to establish a much higher record for snails killed by a single larva than any of those given above simply by keeping a larva surrounded by only newly hatched snails or by eggs almost ready to hatch. Such a record may be quite significant in certain species of Sciomyzidae whose larvae apparently select such unhatched or newly hatched snails in nature (Knutson and Berg, 1963).

The influence of snail density on numbers killed per larva is greater than might be anticipated. A larva that encounters a snail after fasting for two or three days often consumes all soft parts well up into the spire. By contrast, larvae with partly filled food tubes eat less of each victim. Well fed larvae, which evidently attack just because this is their normal behavior pattern, seem to take only a few swallows of the hemolymph. It follows that a larva always surrounded by snails will kill many more than one that manages to find a snail only on alternate days. This is a favorable feature in a species that is being considered as a possible agent of biological control. It means that the predator can subsist on low numbers of prey whenever the prey population ebbs; yet it can destroy relatively large numbers at peak densities.

The list of Sciomyzidae that have attacked and killed *A. glabratus* in the laboratory includes forty-one species, thirty-six aquatic Tetanocerinae and five Sciomyzinae. All of the Tetanocerinae tried against this species have attacked and killed it; some of the highly specialized Sciomyzinae have not. One might conclude from this that finding a species of Sciomyzidae that will destroy a particular host snail usually is not a major problem. Indeed, this seems to be true. For this reason, choice of a candidate species for biological control of a given snail, living under certain sets of ecological conditions, can be based on several less obvious but vitally important considerations.

Larvae operating on moist sand are able to kill individuals of *A. glabratus* weighing several times their own weight. However, no snails of the *glabratus* colony were large enough to determine the maximum size differential between larva and snail at the time that matter was being investigated. Consequently, well grown individuals of *Helisoma trivolvis* were substituted. Mature larvae of *Sepedon macropus* weighing 20·2 mg and 27·1 mg killed and fed upon *H. trivolvis* weighing 385·2 mg and 429·9 mg. The snails, including shells, weighed about sixteen and nineteen times as much as the larvae that killed them (Neff, 1960).

For various reasons, the species of Sciomyzidae used in the preliminary trials reported here are not the most promising ones for bilharziasis control. Of the ten species tried, *Sepedon caerulea* seems to be the only one that may merit further testing. It has larger, more vigorous larvae than most species of *Sepedon*, and it is the only species in the group that is adapted for life in the tropics. It is closely related to *S. macropus*, which was successfully established in Hawaii and Guam for control of the snail host of fascioliasis. No species of *Sepedon* that we have reared has a true diapause, those in temperate zones apparently passing the winter as hibernating adults. In the tropical species that are known biologically, one generation follows directly after the other with no delay at any time of the year.

Before any species is chosen for a field test against bilharziasis host snails, more refined laboratory tests should be run using tropical Sciomyzidae only. This cannot be done until colonies of other tropical species are procured and reared to learn their basic biology. Particular attention should be given to South American species of *Sepedon*, *Thecomyia*, *Tetanoceroides*, and *Protodictya*. These could be introduced into endemic areas of Africa and other tropical regions, freed from the restrictions placed by native enemies on their capacities for population increase.

B. LABORATORY AND FIELD TRIALS FOR FASCIOLIASIS CONTROL

Although sciomyzid larvae had shown considerable promise in laboratory tests against certain snail hosts of bilharziasis, other interests forced an interruption of these studies. Snail hosts of liver fluke presented far greater opportunities for field experiments in snail control because of the excellent cooperation being offered by entomologists, parasitologists, and ecologists in Hawaii and Australia. If we would simply supply living larvae of the species considered most likely to succeed, staffs of biologists with considerable experience in making experimental introductions for purposes of biological control would plan and execute entire projects. This included arrangements with local quarantine officials, preliminary laboratory tests against the target snail, host range studies to determine relative danger to other snails, mass rearing, releases in nature, surveys to check on colonization, and so forth.

At the same time, it seemed important to work out the life cycles and larval habits of biologically unknown marsh flies as quickly as possible. The many Sciomyzidae that remained completely unstudied might well include species that are more effective snail killers or more adaptable to the ecological conditions in problem areas than any species yet reared. It was clear that at least a basic biological knowledge should be acquired of as many of the unreared species as possible, before too much was invested in elaborate and expensive field tests on less effective species. This would demand extensive collecting trips to get breeding stocks of species that do not occur in North America. Thus there was urgency both in greatly expanded studies of basic sciomyzid biology and in applied work on snail hosts of fascioliasis in Pacific regions. The research team at Cornell did not have enough funds and manpower to pursue studies of bilharziasis host snails while investigating these two areas of the problem.

A request for a sciomyzid for biological control of liver fluke in cattle came from the Department of Agriculture of Hawaii in September 1958. It asked for any species that would destroy *Lymnaea ollula* Gould, the intermediate host of *Fasciola gigantica* Cobbold. That snail had never been studied by the Cornell group, but it seemed quite likely that Sciomyzidae which kill and feed on all local species of *Lymnaea* would also be effective against it.

Finding a species that would thrive in the climatic conditions of Hawaii would normally have posed a greater problem. Fortunately, S. E. Neff had recently returned from a collecting trip to Central America, and colonies of nine tropical species were still being studied in the laboratory. On the basis of fecundity, larval vigor, and other considerations, *Sepedon macropus* Walker was selected as probably most likely to succeed. The habitats in which it was collected in Nicaragua and Guatemala seemed quite similar to breeding sites of *L. ollula* in Hawaii.

Five dozen larvae of this species were sent by air mail, with instructions for rearing (Berg, 1959). Entomologists of the Hawaiian Department of Agriculture saw at once that larvae of this species will attack, kill, and feed on *L. ollula*, and soon learned that they thrive on this diet. Studies on the biology and host range of *S. macropus* in the quarantine insectary at Honolulu indicated that it has promise for the biological control of *L. ollula*, without posing any threat to desirable native snails.

After reporting these results and getting the fly released from quarantine, the Hawaiian entomologists found the problem of mass rearing to be relatively easy (Chock *et al.*, 1961). They changed from petri dishes, in which each group of five or six larvae requires daily feeding and care, to rearing in several redwood tanks, each measuring $47 \times 17 \times 15$ in. Adult flies were held in 1 gal glass jars supplied with honey and crushed snails for food, a saturated sponge to provide water, and a bouquet of grass to furnish resting places and sites for oviposition. After 2 days, all adults were transferred to a freshly made up jar. Pieces of grass on which egg masses were cemented were then cut off and floated in the redwood tanks, which were well stocked with *Lymnaea* and *Physa* of different sizes. Larvae which hatched in these tanks required no care; there was no handling from the time that eggs were dropped onto the water until floating puparia were lifted out. The puparia were placed in the gallon jars for emergence, mating, and oviposition. With the change from petri dishes to large tanks, the average production increased from about 400 to 4,500 adult flies per month, with no increase in insectary staff.

Releases in nature consisted of about 400 adult flies liberated simultaneously in each snail breeding site. A check on colonization was made in each swamp or river valley 3 months after the last release there. Adult flies and immature stages were found at seven of the ten release sites on Oahu, six of eight sites on Kauai, and at lesser numbers of breeding places on Maui and Hawaii. At the time of this writing, about 4 years after releases were made, there is no doubt about the establishment of this Central American fly on all four of the major islands of Hawaii and also on Guam, where living material was sent from Hawaii to combat a similar problem. Larvae have been observed repeatedly to kill snails of the liver fluke host species in nature as well as the insectary.

Experimental introductions of other species of Sciomyzidae to Hawaii have been instructive but unsuccessful. The most dense concentration of floating puparia that I have seen was found in two vernal ponds near Latina, Italy (between Rome and Naples) in early April 1960. Almost 300 puparia, not recognized but all of the same species, were collected in less than 5 man-hours. Since there is little chance that associates at Cornell will gain important new information by rearing more than 100 flies of any species, the surplus puparia were offered to entomologists in Hawaii.

Results were similar at Ithaca and Honolulu. More than 80% of the puparia produced adult flies of *Elgiva* (Auct.) *albiseta* Scopoli within 10 days after the puparia arrived. There was hardly any mortality for the first few weeks, and some flies lived as long as 3 months. Yet no eggs were produced, and there was only one observed mating, an evident response to shock treatment. That pair of flies, refrigerated overnight at 45°F, then placed in direct sunlight the following morning, mated 15 min later (Q. C. Chock, *in litt.*). This species finally was reared from eggs laid by flies caught in late August. Like some species of *Tetanocera*, it evidently has a single generation per year, the larvae growing slowly throughout the winter, and the adult flies requiring most of the summer to develop their eggs. It is utterly unsuitable for introduction in Hawaii.

Like many other species of *Lymnaea*, *L. ollula* often crawls out of the water onto banks and mud flats. To extend predatory pressure to these shoreline situations, it seemed advisable to introduce a sciomyzid having terrestrial larvae. That decision was made before the start of our collecting trip to Europe in 1960, but no suitable species was encountered during work south of the Alps in early spring. A species that has most of the desired characteristics, *Pherbellia dorsata* (Zetterstedt), was collected and reared that summer near Copenhagen. It lives on shorelines and floating algal mats, and was observed to feed on *Lymnaea* of three species in Denmark. After finding that it also accepts North American species of *Lymnaea*, and that it develops rapidly and breeds continually, one generation following directly after the other, we sent eighty puparia to the quarantine insectary at Honolulu.

Davis *et al.* (1961) reported that propagation was "very successful," that the larvae appear to feed even more voraciously than those of *Sepedon macropus*, and that the life cycle is 5 or 6 days shorter than that of *S. macropus* under the climatic conditions of Hawaii. They continued, "...it was evident from hundreds of feedings that *P. dorsata* larvae preferred these aquatic snails in the following order: *L. ollula*, *Physa compacta* Pease, *Melania mauiensis* [Lea], and *M. newcombi* [Lea]. The terrestrial snails *Achatina fulica* Bowdich and *Bradybaena similaris* (Ferrusac) were accepted in a lesser degree." Because the larvae would accept terrestrial snails at all, thorough testing of their host range was necessary before this species could be released from quarantine. As indicated above, under "Risks of Foreign Introductions," the terrestrial snails that the Hawaiians wanted to protect proved practically invulnerable to the larvae (Chock *et al.*, 1960). Davis *et al.* (1961) concluded that "many thousands" of adult *dorsata* "have been liberated throughout the major islands of the State."

Despite all of the favorable indications during studies of this species before its release, and the fact that it was liberated in greater numbers and greater concentrations than *Sepedon macropus* had been, *dorsata* has never been recovered in nature on any Hawaiian island. Some may suggest that this species is not adapted for the climate of Hawaii, which certainly is not very similar to that of Denmark. However, *dorsata* also occurs in southern Europe. Although there may be varietal differences at the different latitudes, the strain from Denmark thrived exceptionally well in the insectary at Honolulu. This is a screened building in which both temperatures and humidities remain very close to those outside. Insects inside are shielded from direct rays of the sun, but this is largely true also of the microhabitats that *dorsata* would be expected to select in many of the breeding sites of *L. ollula*.

Terrestrial predators that are unable to operate in the aquatic habitat of *macropus* may take a heavy toll of *dorsata* larvae and puparia. *Pheidole megacephala* Fabr., a predatory ant that has upset the natural biota badly in Hawaii (Zimmerman, 1948), was mentioned by Chock *et al.* (1961) as an important predator even of the aquatic species during propagation of *macro-*

pus in tanks set outside of the insectary. A "production line" of these ants was observed going over the wooden side of a tank, then returning carrying larvae and puparia that had drifted against its inner walls (C. J. Davis, personal communication). In nature, larvae and puparia often drift against emergent vegetation, floating debris, and other objects isolated from the shore, and many aquatic sciomyzids escape for this reason. Larvae and puparia of terrestrial species, by contrast, must be completely vulnerable to these marauders.

If this is the right explanation for failure to colonize *P. dorsata* in Hawaii, most terrestrial sciomyzids introduced there probably would be eliminated in the same way. However, there are at least four South American genera of Sciomyzinae, all completely unknown biologically, that should be investigated in this connection. Since some of them live in areas infested by predatory ants, they evidently have evolved some form of immunity from ant depredations.

In the process of mass rearing, Hawaiian entomologists have made contributions to knowledge of sciomyzid biology of value in future experimental introductions. They pointed out the benefits of adding crushed snails to the diet of the adult Sciomyzidae (Chock *et al.*, 1961), and reported that further supplementing the diet with granular protein hydrolysate results in additional increase in egg production (Kim, 1962). They first reported that sciomyzid larvae feed well on embryonated snail eggs (Chock *et al.*, 1961), a fact which adds significantly to their potential value as predators. Chock *et al.* (1961) listed several predators of both snails and Sciomyzidae observed in tanks used for mass rearing. Finally, Kim (1962) reported that aureomycin, added to the crushed snails that are fed to the immature larvae, has decreased larval mortality appreciably. He added that terrestrial larvae can be held 36–48 h longer before transfer to fresh rearing dishes and that a teaspoon of aureomycin added to each large rearing tank clears the water and effects a considerable reduction in the activity of microorganisms.

The possibility of introducing Sciomyzidae into Australia to help control liver fluke of sheep was first discussed when H. G. Andrewartha, of the University of Adelaide, visited the author in 1959. Correspondence with Professor Andrewartha, with his doctoral student, J. Lynch, and with entomologists of the Commonwealth Scientific and Industrial Research Organization led to an inspection trip to Australia in August 1961. Native Sciomyzidae including undescribed species were then collected, and living flies that initiated the first three rearings of Australian Sciomyzidae through complete life cycles were air-mailed to Cornell University. These three, two species of *Dichaetophora* and one in an undescribed genus, are Tetanocerinae with aquatic larvae. They are predatory on a wide variety of pulmonate snails, including *Lymnaea tomentosa* Pfeiffer, the snail responsible for transmission of *Fasciola hepatica* Linnaeus.

The sciomyzid fauna of Australia is poor in comparison with those of Europe and North America, but the presence of even a few native species complicates the problems of snail population dynamics and population control over those in Hawaii, where there are no native Sciomyzidae. Since the native species clearly are not holding snail populations low enough to keep fascioliasis under satisfactory control in the liver fluke belt in southeastern Australia, one may wonder whether any introduced species can be expected to do so. Yet introduced species which become established at all frequently build up huge populations too familiar to everyone because of notorious introduced pests. Such population explosions are generally attributed to liberation of the introduced species from controls that were exerted by natural enemies in their native lands on their capacities for population increase (Elton, 1958).

The presence of native Sciomyzidae reduces the chances that this phenomenon will occur. The native species probably are being held in check by parasitoid Hymenoptera, which might be expected to attack and destroy introduced Sciomyzidae also. However, no one knows whether parasitoid Hymenoptera exert nearly as important a check on Sciomyzidae in Australia as in North America. Only one puparium, of the more than 100 collected in Australia, was "parasitized." Even if parasitoid Hymenoptera are important enemies of Australian Sciomyzidae, the species concerned may prove to be as host specific as some species of Ichneumonidae in North America, They, or any other natural enemies so decidedly host specific, presumably would be unable to attack any introduced species.

There were too many unknown factors to permit any reliable predictions of fly, snail, and fluke populations that might result from a successful introduction. The decision to introduce a North American sciomyzid was based more on hope than on specific expectations. Although there was no certainty that an introduced species would do any good, there were excellent reasons to believe that it could do no appreciable harm. The species selected would be so thoroughly studied in the laboratory before its release that the only danger that seemed at all likely to occur was loss of the time invested in this project if the introduction proved unsuccessful. In consideration of the possible gains, the Australian biologists decided that this small risk was well worth taking.

The target snail, *Lymnaea tomentosa*, is a small species (6–10 mm in length when fully grown) which many sciomyzid larvae would attack effectively. As was true of the experimental introductions into Hawaii, it was more of a challenge to find a species that would survive in the problem area. *Sepedon praemiosa* Giglio Tos, which occurs in fair abundance in Riverside County, California, was chosen as a likely candidate. The aquatic larvae attack North American species of *Lymnaea* effectively and thrive on that diet. Riverside County has dry, sun-parched soil, sparse herbaceous vegetation, and a eucalyptus-studded landscape that even looks strikingly like much of Australia. Adult flies of all species found there must have evolved remarkable ability to conceal themselves from direct rays of the sun and/or unusual resistance to desiccation.

In the quarantine laboratory of the Commonwealth Scientific and Industrial Research Organization in Sydney, and later in a laboratory at the University of Adelaide, the larvae of *S. praemiosa* proved effective in destroying *L. tomentosa* of all sizes. They grew well on this diet, and a population large enough to permit releases in nature was obtained.

Dr. J. Lynch made this experimental introduction incidentally to working out his doctoral research problem on the ecology and population regulation of *L. tomentosa*. He lacked the assistance of an adequate laboratory staff to handle mass rearings and subsequent checks on colonization. His failure to recover *S. praemiosa* in nature does not necessarily indicate that this species is not suited for conditions in Australia nor even that it is not established there.

Regardless of whether *praemiosa* is or is not colonized in Australia, the trip there evidently was effective in pointing up the value of native Sciomyzidae to Australian biologists concerned with the liver fluke problem. Dr. J. Boray, parasitologist at the McMaster Animal Health Laboratory, C.S.I.R.O., stated concerning *Dichaetophora hendeli* (Kertesz), "...we found a very vigourous sciomyzid fly at Hampton New South Wales .... Large numbers of snails were killed by the larvae at one particular area, where the loss of snails has been estimated." (*in litt.*, 2 August 1962). Dr. Lynch wrote regarding *Dichaetophora biroi* Kertesz, "...at least in some cases, the sciomyzid flies are capable of controlling fluke snails. I am sure this is a major reason for the few outbreaks of fluke disease in the Adelaide Hills and that this fly is controlling the snail population at Meadows South Australia every spring. This means that fluke troubles are likely to occur only in the autumn and then only under very special conditions" (*in litt.*, 29 April 1963).

It would be interesting to know what other seasonal and geographic differences in snail densities are caused by sciomyzid larvae. Numbers of *Australorbis glabratus* fluctuate seasonally in the coastal swamps of Puerto Rico, where both *Sepedon caerulea* and *S. macropus* occur. In Africa, where there are many described and unknown species of Sciomyzidae, there are unexplained regional differences in densities of snail hosts of bilharziasis and well known instances of the total absence of a species from seemingly favorable areas.

VI. Summary and Conclusion

Those who rely on molluscicides to the extent that alternative methods of snail control are virtually excluded from consideration seem to be overlooking important disadvantages of pesticides and the storm of public protest that has developed against their general use. They apparently are disregarding such problems as molluscicide toxicity to other animals, snail resistance to molluscicides, and the accumulation of stable and highly toxic residues in food chains. They seem to forget that little is known about the storage and concentration of molluscicides in mammalian tissues and their longterm effects on man and domestic animals. Perhaps most importantly, they are advocating methods known to be highly destructive of fish life in countries where the economy and nutritional requirements are critically dependent on fish. The situation indicates an urgent need for re-evaluation of snail control practices and concepts, hopefully resulting in the development of new snail control methods.

There is no valid reason to discount the possibilities of biological control of snails. Present knowledge of snails and their natural enemies is too meager to even indicate the most promising agents of biological control. Extensive studies are needed to probe the complexities of ecological relationships and obtain the information needed for success. The integration of chemical and biological control methods may be possible, particularly if a selective molluscicide can be developed that is appreciably less toxic to other animals than to snails.

One group of natural enemies that may have value in biological control, fly larvae of the family Sciomyzidae, was not even known to kill snails until slightly more than a decade ago. Now larvae of so many species have been recognized as snail killers that this food habit appears characteristic of the entire family. Although all sciomyzid larvae probably kill gastropod mollusks, remarkable differences in their habitats, adaptations, modes of attack, and choices of food snails are indicated and described. The known larvae form a behavioral and morphological continuum that may indicate the course of evolution.

Preliminary laboratory trials of sciomyzid larvae against important snail hosts of human blood flukes indicate that the Planorbinae which transmit *Schistosoma mansoni* (e.g. *Biomphalaria* and *Australorbis*) are especially vulnerable. Trials of sciomyzid larvae against snails which transmit fascioliasis include experimental introductions in Hawaii and Australia. A Central American species introduced into Hawaii and Guam unquestionably has established itself in these islands. It is known to kill *Lymnaea ollula* in nature as well as in the laboratory.

Although some Sciomyzidae show promise as agents of biological control, any prediction of the outcome of such a venture based on present knowledge would be premature and meaningless. Up to now, biological research on these interesting Diptera has been predominantly on their basic natural history. Little time has remained for such applied work as experimental introductions and the refined laboratory testing that should precede them. Yet, much more research on the fundamental biology and ecology of these flies is needed. As long as the natural history of about three-fourths of the Sciomyzidae remains completely unknown, laboratory and field testing for biological control cannot be conducted with any confidence that the species being tested are the most promising ones available for any given assignment. Indeed, the vast majority of reared species probably could not be used effectively against snail hosts of human and livestock trematodes. Most of them live in relatively cool climates, and they probably could not be colonized in the tropical and subtropical regions where trematode diseases cause the greatest problems. Extension of this research into the tropics, to study both the species that might be introduced into problem areas and those that already occur there, is of paramount importance and urgency.

At the same time, the results of introductions that have been made should be analyzed as reliably and thoroughly as possible, using modern methods of population ecology. Such analyses would have value in judging not only the situation being studied but also the probable effectiveness of other Sciomyzidae, against other snails, in other regions. It would seem especially desirable to determine the results of the introduction of *Sepedon macropus* in Hawaii. Because many empty *Lymnaea* shells and appreciable reductions in snail populations (judged subjectively) have been noted in areas where this fly was introduced, all Hawaiian biologists involved in this project consider it successful. The importance of getting hard, statistical evidence to establish this point is recognized both here and in the Hawaiian Department of Agriculture. Yet neither group has had the funds and manpower needed for a thorough, dependable survey to determine changes in snail populations since the flies were introduced.

As the best alternative that can be promised positively, a continuing analysis of the incidence of fascioliasis in Hawaiian cattle is underway. Slaughterhouse records of the percentage of infected livers have been maintained since before sciomyzid flies were introduced here. In the long run at least, this will provide an index of transmission rates that is even more accurate than snail populations. It is at least theoretically possible, by reducing the average age of snails to the point that few are old enough to be shedding cercariae, to reduce the transmission rate significantly without reducing snail density at all.

Serious thought also is being given to the possibility that the effects of introduced Sciomyzidae on snail populations can still be determined statistically. The *Lymnaea* breeding sites on some Hawaiian Islands are well isolated from each other, and *S. macropus* may not have invaded all of them. If not, snail populations can be determined in such unoccupied sites, the flies can be introduced, and the censuses can be repeated. The possibilities of such a study will be analyzed carefully this coming summer. Funds for a statistical survey have been obtained recently, and one will certainly be launched if favorable sites are found.

ACKNOWLEDGEMENTS

I acknowledge financial assistance afforded by research grants E-743 and AI 05923 from the National Institute of Allergy and Infectious Diseases, U. S. Public Health Service, and G-7605 and GB-80 from the Division of Biological and Medical Sciences, National Science Foundation.

Much of the routine sciomyzid rearing reported here was done by former graduate students, Drs. S. E. Neff, B. A. Foote, and L. V. Knutson, and by present graduate students, Mr. A. D. Bratt, Miss J. L. Gower, and Mr. J. L. Bath. Before any of these students came to Cornell, I was ably assisted for one year by Dr. J. S. Mackiewicz and the late Dr. E. J. Karlin.

I am indebted to Dr. E. C. Bay, University of California at Riverside, for most of the photographs. Others were taken by the Photo Science Studios, by the Visual Aids Division of the Department of Extension Teaching and Information, and by Mr. H. H. Lyon, all of Cornell University.

REFERENCES

Balch, R. E. (1958). *Annu. Rev. Ent.* **3**, 449–68.
Barrow, J. H. (1960). *J. Protozool.* **7** (Suppl.), 12.
Beirne, B. P. (1963). *Mem. ent. Soc. Canada* **32**, 7–10.
Bequaert, J. (1925). *J. Parasit.* **11**, 201–12.
Berg, C. O. (1953). *J. Parasit.* **39**, 630–6.
Berg, C. O. (1959). *Fm Res.* **25**, (1), 8–9.
Berg, C. O. (1961). *Verh. XI int. Kongr. Ent.* **1**, 197–202.
Berg, C. O. (1964). *Verh. int. Ver. Limnol.* **15**, 926–32.
Berg, C. O. and Neff, S. E. (1959). *Bull. Amer. malacol. Un.* **25**, 12–13.
Berg, C. O., Foote, B. A. and Neff, S. E. (1959). *Bull. Amer. malacol. Un.* **25**, 10–11.
Bertrand, H. (1954). "Encyclopédie Entomologique, Série A., Vol. 31; Les Insectes Aquatiques d'Europe," 547 pp. P. Lechevalier, Paris.
Brown, A. W. A. (1951). "Insect Control by Chemicals," 817 pp. Wiley, New York.
Brown, A. W. A. (1958). *Monogr. Ser. Wld Hlth Org.* **38**, 1–240.
Brown, W. L. (1961). *Pysche* **68**, 75–109.
Carson, R. (1962). "Silent Spring," 368 pp. Houghton Mifflin, Boston.
Chernin, E. (1962). *Parasitology* **52**, 459–81 .
Chernin, E., Michelson, E. H. and Augustine, D. L. (1956a). *Amer. J. trop. Med. Hyg.* **5**, 297–307.
Chernin, E., Michelson, E. H. and Augustine, D. L. (1956b). *Amer. J. trop. Med. Hyg.* **5**, 308–14.
Chernin, E., Michelson, E. H. and Augustine, D. L. (1960). *J. Parasit.* **46**, 599–607.
Chock, Q. C., Davis, C. J. and Chong, M. (1960). (Dittoed Report).
Chock, Q. C., Davis, C. J. and Chong, M. (1961). *J. econ. Ent.* **54**, 1–4.
Colyer, C. N. and Hammond, C. O. (1951). "Flies of the British Isles," 383 pp. Frederick Warne, London and New York.
Davis, C. J., Chock, Q. C. and Chong, M. (1961). *Proc. Hawaii. ent. Soc.* **17**, 395–7.
DeBach, P. (1958). *Proc. 10th int. Congr. Ent.* **3**, 185–97.
DeBach, P. and Landi, J. (1961). *Hilgardia* **31**, 459–97.
De Vos, A., Manville, R. H. and Van Gelder, R. G. (1956). *Zoologica, N. Y.* **41**, 163–94.
Dufour, L. (1849). *Ann. Soc. ent. France, Sér.* **2**, **7**, 67–79.

Elton, C. S. (1927). "Animal Ecology," 209 pp. Sidgwick and Jackson, London.
Elton, C. S. (1958). "The Ecology of Invasions by Animals and Plants," 181 pp. Methuen, London; Wiley, New York.
Expert Committee on Bilharziasis. (1961). *Tech. Rep. Wld Hlth Org.* **214**, 1–50.
Expert Committee on Insecticides. (1957). *Tech. Rep. Wld Hlth Org.* **125**, 1–31.
Expert Committee on Insecticides. (1958). *Tech. Rep. Wld Hlth Org.* **153**, 1–67.
Foote, B. A. (1959). *Ann. ent. Soc. Amer.* **52**, 31–43.
Foote, B. A. (1961). Doctoral Dissertation, 190 pp. Cornell University. Available: University Microfilms, Ann Arbor, Michigan (L. C. Card No. Mic 62–105).
Foote, B. A., Neff, S. E. and Berg, C. O. (1960). *Ann. ent. Soc. Amer.* **53**, 192–9.
Franz, J. M. (1961). *In* "Symposium on the Ecological Effects of Biological and Chemical Control of Undesirable Plants and Animals" (D. J. Kuenen, ed.), pp. 93–105. Brill, Leiden.
Geier, P. W. and Clark, L. R. (1961). *In* "Symposium on the Ecological Effects of Biological and Chemical Control of Undesirable Plants and Animals" (D. J. Kuenen, ed.), pp. 10–8. Brill, Leiden.
Gercke, G. (1876). *Verh. Ver. naturw. Unterh. (Heimatforsh.) Hamb.* **2**, 145–9.
Gretillat, S. (1962). *Bull. Wld Hlth Org.* **26**, 67–74.
Gunther, F. A., ed. (1962–63). "Residue Reviews," Vol. 1, 162 pp., Vol. 2, 156 pp. Academic Press, New York.
Hollande, A. and Chabelard, R. (1953). *Minerva urol.* **5**, 145. (Original not seen.)
Hunt, E. G. and Bischoff, A. I. (1960). *Calif. Fish and Game* **46**, 91–106.
10th International Congress of Entomology. (1958). *Proc. 10th int. Congr. Ent.* **4**, 441–924.
XI Internationalen Kongresses für Entomologie. (1962). *Verh. XI int. Kongr. Ent.* **2**, 669–882.
International Union for Conservation of Nature and Natural Resources. (1961). "Symposium on the Ecological Effects of Biological and Chemical Control of Undesirable Plants and Animals" (D. J. Kuenen, ed.), 118 pp. Brill, Leiden.
Janzen, D. H. (1961). *In* "Symposium on the Ecological Effects of Biological and Chemical Control of Undesirable Plants and Animals" (D. J. Kuenen, ed.), pp. 61–8. Brill, Leiden.
Keilin, D. (1919). *Parasitology* **11**, 430–55.
Keilin, D. (1921). *Parasitology* **13**, 180–3.
Keith, L. B. (1963). "Wildlife's Ten-year Cycle," 201 pp. University of Wisconsin Press, Madison.
Kim, J. (1962). (Dittoed Report).
Knutson, L. V. (1963). Doctoral Dissertation, 390 pp. Cornell University. Available: University Microfilms, Ann Arbor, Michigan (L. C. Card No. Mic 63–4822).
Knutson, L. V. and Berg, C. O. (1963). *Proc. R. ent. Soc. Lond. (A)* **38**, 45–58.
Knutson, L. V. and Berg, C. O. (1964). *Ann. ent. Soc. Amer.* **57**, 173–92.
Laird, M. (1959). *Acta trop., Basel* **16**, 331–55.
Levi, H. W. (1952). *Sci. Mon., N. Y.* **74**, 315–22.
Linsley, E. G., ed. (1956). *Hilgardia* **26**, 1–106.
Lundbeck, W. (1923). *Vidensk. Medd. dansk naturh. Foren. Kbh.* **76**, 101–9.
Malek, E. A. (1958). *Bull. Wld Hlth Org.* **18**, 785–818.
Malek, E. A. (1962). "Laboratory Guide and Notes for Medical Malacology," 154 pp. Burgess, Minneapolis.
Mead, A. R. (1961). "The Giant African Snail: a Problem in Economic Malacology," 257 pp. University of Chicago Press, Chicago.
Meleney, H. E. (1954). *Amer. J. trop. Med. Hyg.* **3**, 209–18.
Mercier, L. (1921). *Ann. Soc. ent. Belg.* **61**, 162–4.
Metcalf, R. L. (1952). *Sci. Amer.* **187**, (4) 21–5.

Michelson, E. H. (1957). *Parasitology* **47**, 413–26.
Michelson, E. H. (1961). *Amer. J. trop. Med. Hyg.* **10**, 423–33.
Michelson, E. H. (1963). *J. Insect Path.* **5**, 28–38.
Neff, S. E. (1960). Doctoral Dissertation, 201 pp. Cornell University. Available: University Microfilms, Ann Arbor, Michigan (L. C. Card No. Mic 60–6491).
Neff, S. E. (1964). *Verh. int. Ver. Limnol.* **15**, 933–9.
Neff, S. E. and Berg, C. O. (1961). *Bull Brooklyn ent. Soc.* **56**, 46–56.
Neff, S. E. and Berg, C. O. (1962). *Trans. Amer. ent. Soc.* **88**, 77–93.
Noble, E. R. and Noble, G. A. (1961). "Parasitology, the Biology of Animal Parasites," 767 pp. Lea and Febiger, Philadelphia.
Oldham, C. (1912). *Trans. Herts. nat. Hist. Soc.* **14**, 288.
Paulini, E. (1958). *Bull. Wld Hlth Org.* **18**, 975–88.
Pepper, J. H. (1955). *J. econ. Ent.* **48**, 451–6.
Pesigan, T. P. (1953). *Santo Thomas J. Med.* **8**, 1–22. (Original not seen.)
Pickett, A. D. (1949). *Canad. Ent.* **81**, 67–76.
Pickett, A. D. (1961). *In* "Symposium on the Ecological Effects of Biological and Chemical Control of Undesirable Plants and Animals" (D. J. Kuenen, ed.), pp. 19–24. Brill, Leiden.
Radke, M. G., Ritchie, L. S. and Ferguson, F. F. (1961). *Amer. J. trop. Med. Hyg.* **10**, 370–3.
Reuter, O. M. (1913). "Lebensgewohnheiten und Instinkte der Insekten," 448 pp. Friedländer, Berlin.
Ripper, W. E. (1956). *Annu. Rev. Ent.* **1**, 403–38.
Rowan, W. B. and Rodriquez, S. (1964). (To be published.)
Sack, P. (1939). *In* "Die Fliegen der palaearktischen Region" (E. Lindner, ed.), Bd. V, Lf. 37, pp. 1–87. Schweizerbart, Stuttgart.
Schwetz, J. (1954). *Ann. Soc. belge Méd. trop.* **34**, 233. (Original not seen.)
Smith, R. F., Beirne, B. P., Reynolds, H. T., Rabb, R. L., Knight, F. B. and Schoof, H. F. (1962). *Bull. ent. Soc. Amer.* **8**, 188–201.
Solomon, M. E. (1953). *Chem. and Ind. (Rev.)* 1953, 1143–7.
Stern, V. M. and Van den Bosch, R. (1959). *Hilgardia* **29**, 103–30.
Stern, V. M., Smith, R. F., Van den Bosch, R. and Hagen, K. S. (1959). *Hilgardia* **29**, 81–101.
Sweetman, H. L. (1936). "The Biological Control of Insects," 461 pp. Comstock, Ithaca.
Sweetman, H. L. (1958). "The Principles of Biological Control," 560 pp. Brown, Dubuque.
Thompson, W. R. (1930). "The Biological Control of Insect and Plant Pests," 124 pp. H. M. S. O., London.
Ullyett, G. C. (1947). *Ent. Mem. S. Afr. Dep. Agric. For.* **2**, 77–202.
U. S. Agricultural Research Service. (1960). *U. S. agric. Res. Serv. ARS* **20–9**, 1–221.
U. S. President's Science Advisory Committee Panel on the Use of Pesticides. (1963). "Report on the Use of Pesticides," 25 pp. U. S. Govt. Prt. Off., Washington.
Van der Schalie, H. (1958). *Bull. Wld Hlth Org.* **19**, 263–83.
Verbeke, J. (1960). *Bull. Inst. Sci. nat. Belg.* **36** (34), 1–15.
Wheeler, W. M. (1923). "Social Life Among the Insects," 375 pp. Harcourt, Brace, New York.
Wigglesworth, V. B. (1945). *Atlant. Mon.* **176** (6), 107–13.
Wilson, F. (1960). *Tech. Commun. Inst. biol. Control, Ottawa* **1**, 1–102.
Zimmerman, E. C. (1948). "Insects of Hawaii" Vol. I, 206 pp. University of Hawaii Press, Honolulu.

Author Index

Numbers in italics indicate the page in the References on which a reference is listed

F

G

H

N

R

S

T

Z

Subject Index

N

O

P

U

V

W

X

Y